AF788667

ERNEST THOMPSON SETON

This is a volume in the Arno Press collection

BIOLOGISTS AND THEIR WORLD

Advisory Editor
Keir B. Sterling

Editorial Board
Kraig Adler
Armin Geus
David M. Knight
E. Gorton Linsley
Hans Querner
Ronald L. Stuckey

See last pages of this volume
for a complete list of titles

ERNEST THOMPSON SETON

Man in Nature and the Progressive Era 1880-1915

John Henry Wadland

ARNO PRESS
A New York Times Company
New York • 1978

Editorial Supervision: MARIE STARECK

Reprint Edition 1978 by Arno Press Inc.

Copyright © 1978 by John Henry Wadland

BIOLOGISTS AND THEIR WORLD
ISBN for complete set: 0-405-10641-6
See last pages of this volume for titles.

Manufactured in the United States of America

Publisher's Note: Plates 3-7 have been
reproduced in black and white in this edition.

Library of Congress Cataloging in Publication Data

Wadland, John Henry.
Ernest Thompson Seton : man in nature and the
Progressive Era, 1880-1915.

(Biologists and their world)
Originally presented as the author's thesis,
York University, Toronto.
Bibliography: p.
1. Seton, Ernest Thompson, 1860-1946.
2. Naturalists--United States--Biography.
3. Naturalists--Canada--Biography. I. Series.
QH31.S48W33 1978 574'.092'4 [B] 77-81136
ISBN 0-405-10736-6

ERNEST THOMPSON SETON: MAN IN NATURE AND THE PROGRESSIVE ERA 1880-1915

ERNEST THOMPSON SETON: MAN IN NATURE AND THE PROGRESSIVE ERA 1880-1915

JOHN HENRY WADLAND

A thesis submitted to the Faculty of Graduate Studies
in partial fulfillment of the degree of

DOCTOR OF PHILOSOPHY

Graduate Programme in History
York University
Toronto, Ontario

November, 1976

For Jacob

INTRODUCTION

Technological progress has, for the best part of the twentieth century, been viewed as a panacea for the material ills that affect mankind. American historians, especially those of a liberal stamp, have fostered this perspective by focussing their attention on fluctuations in the control of political and economic power, on inequalities between peoples and sexes, on the growth and advancement of civilization and on the ideological premises that contribute to or detract from their own definitions of reality. Comparatively little thought has been devoted to the effects of these anthropocentric concerns upon the natural environment which is essential to sustain the conflicts inherent in them. Nature is generally perceived as an object to be used in the service of humanity. Yet recent developments in biology, particularly in the fields of ecology and ethology, suggest that historians have an obligation to question their materialist bias and to analyze the environmental consequences of human thought and action.

The work of Ernest Thompson Seton (1860-1946) affords a unique vehicle for this purpose. Trained in the finest academies of Canada, the United States, Britain and France, Seton abandoned a successful artistic career to learn independently the language and methods of nineteenth century biologists and naturalists. As a young man he was converted from his family's pious Presbyterianism to an equally pious pantheism. The process of conversion was guided not by men but by the landscapes and animal inhabitants of post-Confederation Ontario and Manitoba. Although his paintings and drawings revealed a fascination for the power of wilderness over man, his commitment to the notion of limitation by nature was best articulated in hundreds of realistic animal stories for children which earned for their author international recognition. These were based upon observations made in the field all over North America and are documented in close to forty volumes of manuscript Journals. For his trouble in writing them, Seton was rewarded by the rebukes of many contemporary scientists who felt that his interpretations of animal behavior were anthropomorphic. In the light of modern research in ethology, however, it is clear that Seton belonged to a fringe group of radicals who aspired to demonstrate, through an emphasis on the theories of learned and innate behavior, not only that animals possessed greater intelligence than was believed, but also that man's continued survival required reasoned acceptance of the fact. Although Seton erred in adopting a dogmatic posture to defend his premises, he was in the vanguard of a more sophisticated modern intellectual movement devoted to developing what Aldo Leopold has called an "ecological conscience" and a "land ethic." In attempting to answer his critics on the subjects of ecology and ethology, Seton produced a respectable number of scientific publications in mammalogy and ornithology, among which The Life-Histories of Northern Animals (1909) stands out as a classic. He also initiated a colorful youth movement, the precursor of the Boy Scouts, to which he applied social principles similar to those developed by the Russian anarchist, Peter Kropotkin. The model for this organization was the North American Indian who, in Seton's view, epitomized nobility. Nobility was measured not by patriotism or by good works but by the individual's ability to acknowledge and to live within nature's biological laws. An anti-Spencerian Social Darwinist, Seton despised competition and stressed mutual aid, believing that the example of the animal kingdom, if observed by mankind, would yield a decentralized, diverse and humble international society free of war. His ideas were copied in part by Robert S. S. Baden-Powell, but the latter's additions and revisions completely reversed Seton's original intention. Although Seton himself was the first Chief Scout of the Boy Scouts of America and although he was instrumental in launching the organization in the United States, his disagreements with colleagues over militarism culminated in his unceremonious ouster from the hierarchy in 1915. Through his art, through his science and through his youth work the animal motif remained paramount. Man was to be viewed with suspicion and distrust.

TABLE OF CONTENTS

PLATES

FIGURES

TABLES

Acknowledgements

This study is really the product of many minds and hands. I want especially to thank Ted Yard who first gave me the idea. Dee and Dale Barber of Santa Fe, New Mexico, and Anya Seton Chase of Old Greenwich, Connecticut, welcomed me into their homes for extended periods of time. Without their gracious hospitality I would not have had the opportunity to see much that has proved essential to my work. Kent Bush and Eleanor Pratt of the Philmont Ranch at Cimarron, New Mexico, the site of the Seton Museum, made my stay there a rewarding and memorable experience. Jean Thompson and Elise Thompson gave me many valuable insights into the background and character of their uncle, as did Mrs. Philip Ketchum, Howard Schreiber and Leonard Clark.

Bill Waiser, a graduate student at the University of Saskatchewan, gave me permission to see and to use materials accumulated for his forthcoming thesis on John Macoun. James A. Dolph of Weber State College, Ogden, Utah loaned me microfilmed copies of the William Hornaday letters. Nina Root of the American Museum of Natural History in New York helped me to gain access to the Seton

Journals. I am indebted to Keir Brooks Sterling of Pace University whose guidance and encouragement, especially in the history of natural science, was indispensable to the completion of my work. To the management and staff of the Atwell Fleming Printing Company of Toronto I wish to express my deepest gratitude for their kindness in making the splendid plates which accompany the text.

Had it not been for the reassurance and direction I received from friends and colleagues--Donald Smith, Mary Northway, Flora Morrison, Phyllis Sherrin, Harvey McCue, Bruce Hodgins, Alan Wilson, Lindsay Staples, Paul Gardner, Bryan Graham, Wayland Drew, Jamie Benidickson, Carol Sherman, Lionel Rubinoff, Robert Page and countless others--I might never have managed to finish. Jack Granatstein, Virginia Rock and Robert Cuff made indispensable suggestions about the orientation of the study and kept me on track. To my supervisor, Richard Storr, whose exacting and disciplined scholarly approach has been surpassed only by his warmth and kind understanding, I owe more than I can say. My parents have always helped me through demanding times, but never more willingly than with this. Finally, and most important of all, I want to thank my wife, Lynda, for her continuous help and encouragement. She has made it all worthwhile.

Chapter 1

NATURE KNOWS BEST

The human mind has nothing but human and moral values to cling to if it is to preserve its integrity or even its sanity, yet the vast unconsciousness of nature in front of it seems an unanswerable denial of those values.

Northrop Frye

I

This study is drawn primarily from the books and personal papers of Ernest Thompson Seton. But it is not the biography of a man. The Boy Scouts of America figures largely in its conclusions. But this is not what A. M. Schlesinger, Sr. called an "institutional biography." Rather it is only a chapter in the very fragmentary biography of an issue which has not yet emerged with a standard vocabulary of its own, an issue which is discussed more often than not today under the heading of ecology. As a branch of biology, ecology dates from Darwin's time. Broadly defined, we understand it to mean the interrelationships of organisms and their environments. In a historical context man is perceived merely as an organism relating to an environment composed of other organisms. Ecology also implies such things as cycles, food chains and geographic distribution, all of which are necessary to the preservation of the equilibrium, or harmony, of all organisms. The present study has to do with the contribution made to the discussion of this perspective by Ernest Thompson Seton. To pass an intelligent judgment on his contribution, it is necessary to notice not only what he said and did in a historical setting, but also what the general contour of the issue is today. In the same way, the place of Booker T. Washington in Black History requires not only knowledge of what he said and did, but also some familiarity with the modern "Black perspective." The same must be said of the ideas of John D. Rockefeller or Theodore

Plate 1. Ernest Thompson Seton (ca. 1900)

Roosevelt, to mention others of Seton's contemporaries. Revisionism in all of these instances involves both the facts of the case and the modern point of view of the historian. One can hardly speak of a revisionist look at Seton because very few historians have said anything about him. It becomes quite clear, however, that an ecological perspective on history must resurrect many figures previously dismissed in frivolous anecdotes. For ecology does not easily lend itself to an acceptance of historical interpretations applauding progress and efficiency to the exclusion of their effects on the natural elements which make them possible. To the ecologist, the natural elements themselves have an integral worth which supersedes all anthropocentric concerns. The reasons for this view will become clear in what follows. But for now it is enough to suggest that men like Seton who railed against technology in the name of nature were not banal proselytizers in a wrongheaded cause. Quite frequently they were remarkable prophets.

II

In his monumental study of Social Darwinism in American Thought, Richard Hofstadter claimed to have demonstrated that biological ideas "are utterly useless in attempting to understand society." He maintained that "the life of man in society, while it is incidentally a biological fact, has characteristics that are not reducible to biology and must be explained in the distinctive terms of a cultural analysis." While it is not our purpose here to challenge Hofstadter's view, it is imperative to suggest that recent events could subject the implications of his broad assertion to significant revision. Declining energy resources, environmental pollution, overpopulation and the economic dislocation generated by excessive consumer demand have conspired to give biology a renewed sense of immediacy. Indeed, to imply that social and cultural values have nothing to do with biology is to confess an insufferable ignorance of the foundation which sustains them both.[1]

Where the age described by Hofstadter spoke of conservation and efficiency, implying a large measure of rational, human control over nature,[2] ours speaks of ecology and limitation, implying that man can neither control nor dictate to nature but must see himself as a part of it, inextricably bound by a set of immutable biological rules which render all his ideologies powerless. Where intellectuals at the turn of the century placed their hopes for mankind in technology, our age produces scholars who postulate that technology has created an artificial environment whose

---

[1]Richard Hofstadter, Social Darwinism in American Thought (Boston: Beacon Press, 1955), p. 204.

[2]See, e.g., Samuel P. Hays, Conservation and the Gospel of Efficiency: The Progressive Conservation Movement, 1890-1920 (New York: Atheneum, 1969).

evolution is proceeding at a rate much faster than that of the natural organism, thereby threatening rather than facilitating its existence.[3] Where social Darwinism best served the interests of laissez-faire conservatives like William Graham Sumner, contemporary biological interpretations of society tend to be outspoken critiques of the past. Radical ecologists share with the social Darwinists a zeal for slow, organic change. They differ from the social Darwinists by repudiating an anthropocentric world view. "Contemporary Western man," observes John Livingston, "considers himself fundamentally different and distinct from the living world that gives him both substance and sustenance. This imagined separation between man and 'nature' has provided the conceptual framework for a further doctrine, that of absolute human power and authority over the nonhuman."[4] Because it focusses solely on the works of man, history denies the whole reality. In spite of its own protestations to the contrary, written history is still a whiggish chronicle of man's "conquest" of the wilderness, of his "taming" of the Indian, of his technological "achievements," and of the "development" of his sophisticated political and economic contrivances. In

---

[3]René Dubos, "Man and His Environment: Adaptations and Interactions," in The Quality of Man's Environment: Smithsonian Institution Symposium (Washington D.C.: Smithsonian Institution Press, 1968), pp. 231-250; Barbara Ward and René Dubos, Only One Earth: The Care and Maintenance of a Small Planet (New York: Norton, 1972).

[4]John A. Livingston, One Cosmic Instant: A Natural History of Human Arrogance (Toronto: McClelland and Stewart, 1973), p. 14.

short, the discipline of history is the method by which the affairs of men are measured, mixed and blended. Theoretically it should explain our present. More often than not, however, it is used to justify modernity, to state unequivocally that whatever is, if it is not best, is the inevitable conclusion of what was. This perspective, while it may satisfy the historian, contributes nothing whatever to man's understanding of his relationship with nature. For most of a growing body of radical ecologists, history and man-centered notions of progress are responsible for "undoing the work or organic evolution."

> By creating vast urban agglomerations of concrete, metal and glass, by overriding and undermining the complex, subtly organized ecosystems that constitute local differences in the natural world--in short, by replacing a highly complex, organic environment with a simplified, inorganic one--man is disassembling the biotic pyramid that supported humanity for countless millenia. . . . From an ecological viewpoint, the reversal of organic evolution is the result of appalling contradictions between town and country, state and community, industry and husbandry, mass manufacture and craftsmanship, centralism and regionalism, the bureaucratic scale and the human scale.

Murray Bookchin, the author of these sentiments, maintains that technology has created a penchant for "mechanical standardization" which permeates the fabric of North American society generally. In response to the needs of a biologically determined environment he appeals to us to adopt "a balanced community, a face-to-face democracy, a humanistic technology and a decentralized society." These apparently esoteric ends are analogous to the diversity implicit in nature's laws.

> The process of social development has carried them out of the ethical, subjective dimension into a practical, objective dimension. What was once regarded as impractical and visionary has become eminently practical. And what was once regarded as practical and objective has become eminently impractical and irrelevant. . . .

These are strong words indeed--perhaps stronger than tolerant men can abide. But Bookchin is not alone in sounding the alarm.[5]

The philosopher of conservatism, George Grant, views the problem of technology in a somewhat different light. North Americans, he suggests, have never been able to identify themselves with their land. In giving up the stable security of the old world for the possibilities of the new, we abandoned the known for the unknown. The new continent

> could not be ours in the old way because the making of it ours did not go back before the beginning of conscious memory. The roots of some communities in eastern North America go back far in continuous love for their place, but none of us can be called autocthonous, because in all there is some consciousness of making the land our own. It could not be ours also because the very intractability, immensity and extremes of the new land required that its meeting with mastering Europeans be a battle of subjugation. And after that battle we had no long history of living with the land before the arrival of the new forms of conquest which came with industrialism.
>
> That conquering relation to place has left its mark within us. . . . There can be nothing immemorial for us except the environment as object. Even our cities have been encampments on the road to economic mastery.

---

[5]Murray Bookchin, Post-Scarcity Anarchism (Berkeley: Ramparts Press, 1971), pp. 67-68, 69-70.

The earliest North Americans came armed with Calvinist Protestantism and Baconian scientific rationalism, predisposing them to order the apparently disordered, to make comprehensible the chaotic. Grant calls this meeting of pioneer and alien land the "primal encounter." And even though "calvinism and the pioneering moment have both gone, that primal still shapes us. It shapes us above all as the omnipresence of that practicality which trusts in technology to create the rationalised kingdom of man." What we are faced with in the twentieth century, he argues, is the alliance of technology with liberalism, the purpose of which "becomes the building of the universal and homogenous state--the society in which all men are free and equal and increasingly able to realise their concrete individuality." Our obsession with practical considerations has forced us to deny the European capacity for contemplation so that, in the end, we have narrowed our horizons "to an unmitigated reliance on technique," and to "the belief that human excellence is promoted by the homogenizing and universalising power of technology . . . and that that doctrine must undermine all particularisms."[6]

So far we have touched on the continental perspectives of two philosophers--the American anarchist, Bookchin, and the Canadian conservative, Grant. On the surface, one might

---

[6]George Grant, Technology and Empire: Perspectives on North America (Toronto: Anansi, 1969), pp. 17, 25, 33, 40, 69.

expect their views, like their ideological foundations and their national origins, to be polarized. In fact, however, it should be clear that both men share a distinct distaste for unbridled technology and the homogenized centralism it spawns. Neither denies the need for order. But both are clearly skeptical of the liberal variety. Bookchin, indeed, allows that nature has its own order. In attempting to sever his bond with nature, man has been guilty of a perverse treason. The laws of biology demand diversity. While Grant is not inclined to dwell upon the biological analogy, he employs the land as a vehicle for the same purpose. The thoughts of both men have been seconded of late by countless important thinkers on both sides of the border and on all points of the political spectrum. One cannot help uniting their overriding concerns around ecology, the idea/symbol which has come to dominate much of their analytical inquiry.[7]

Approaches to the study of ecology vary from discipline to discipline. Coined initially by biologists, the term itself has been appropriated (and often obscured) in many of the social sciences, notably by sociologists, anthropologists and psychologists. But nothing has done more to oversimplify and confuse its meaning than the so-called "environmental

---

[7] For an excellent review of the literature, see David L. Sills, "The Environmental Movement and Its Critics," Human Ecology, 3 (January, 1975), 1-41.

movement" of the past decade which has acquired many popular, highly subjective ideological connotations. These in turn have provoked spirited, if frequently extreme, criticism of the movement from opponents like Richard Neuhaus, who argues that "the notion of transcending politics in the name of nature and the natural is strikingly similar to crucial elements of National Socialism."[8] It is not my desire, or my intention, to become mired in that debate. It is imperative, however, to identify it and to acknowledge its significance.

By definition, ecology means that branch of biology which treats the relations between living organisms and their environment. As a living organism, man clearly comes within its purview. For "organism" and "environment" let us read "nature." From the ecologist's perspective, there is no dichotomy between man and nature. In ecological terms, each organism constitutes part of the environment of all other organisms. The complex, cyclical set of interdependencies thus established depends for its complete survival upon the continued existence of its own weakest link. There is, therefore, an immutable set of biological laws--ordered, organized and efficient--demanding the concurrence of all the species that live under them.

Man, by virtue of his unique, reasoning intelligence,

---

[8]Richard Neuhaus, In Defense of People: Ecology and the Seduction of Radicalism (New York: Macmillan, 1971), p. 152.

has reduced a fraction of these laws to a code, which he calls science. And from this code he has taken various laws, juxtaposed them in altered sequence, and arrived at practical applications, which he calls technology. In many cases the altered sequence of the laws has yielded objects which are inorganic and which, therefore, do not fit within the cyclical arrangement. While they may be said to constitute a part of the environment of other organisms, they themselves are not organisms. Nature itself has produced objects which are inorganic. These too are a part of the environment of organisms. In this sense the imposition by man of newly devised inorganic properties is not, alone, a cause for concern. But alarm over the environment emerges from a fear that the altered sequence of natural laws will function, not to increase the complexity of the environment, but ultimately to simplify it. Simplification of the environment results directly in a simplification--that is to say, extermination--of the species which depend upon its complexity for their survival. Societal demands for commodity durability, leading to the manufacture of, say, non-corrosive properties like stainless steel and plastic, has the adverse effect, in the long run, of rendering the whole environment completely inorganic. Ecologists would argue that man must either expand his scientific understanding in order, when altering the sequence of natural laws, to produce an organic technology, or that he must cease altering the sequence of the laws altogether. The first solution slows technological progress, the second eliminates it.

When either of these two conclusions is identified with an ideology--say Grant's conservatism or Bookchin's anarchism--it breeds a predictable hysteria, such as that voiced by Neuhaus. Immediately battle lines are drawn between anthropocentrists and anti-anthropocentrists. The ensuing debate is emotional dynamite, often alienating and counterproductive. At stake is the question of whether ecology has implications for human social, political and economic behavior. Proponents of the ecological view argue that the diversity implicit in the organic whole demands social and ethnic pluralism, regionalism in the national context, nationalism in the world context, in order to guarantee the proper arrangement of mutual interdependencies. Defenders of ecology are defenders of limitation. Their opponents, on the other hand, are apostles of possibilism. For them, technology is the great universal leveller, the homogenizer of conflicts, the sine qua non of world peace, the guarantor of liberty and the enemy of poverty. Man, they argue, is above nature. His achievements, measured in written history, are clear evidence of the superiority of his reason, monuments to the success of his technology. The ecologists maintain that man is not only in nature, but of it. He must, therefore, subscribe to its laws, employing his reason not merely for the purpose of understanding and codifying them, but also for the purpose of recalling that they control him. For ecologists, all species have a right to life, as well as an integral role to play in it. They look upon written history with suspicion, charging that its very anthropocentrism denies its claims to objectivity.

With a few exceptions, most notably Lynn White, Jr. and Fernand Braudel, I have yet to find among my professional colleagues many who have considered the possibility of an ecological interpretation of history. Because ecology resurrects the old spectre of determinism, denying Western man his full measure of free will, this should come as no surprise. One fact, however, has stirred the historian's juices. A surprising number of contemporary philosophers who have espoused the ecological perspective are self-proclaimed anarchists. George Woodcock, Lionel Rubinoff, Wayland Drew and, of course, Murray Bookchin--the first three, Canadians, the last, an American--are four antagonists of what George Grant has identified as the "North American primal." Not only are they anarchists, they are all Kropotkin anarchists. Because of this, and for the reason that Richard Hofstadter all but dismissed Peter Kropotkin from the realm of credibility, it occurred to me, as I think it must occur to all of us, that the debate on Social Darwinism is not over. Biology has come back to haunt us.[9]

III

Unlike Herbert Spencer, Peter Kropotkin refused to accept that competition was the primary mechanism for the survival of most species. His antithetical notion of mutual aid was developed in a series of articles prepared for the Nineteenth Century between 1890 and 1896 and brought together in book form in 1902, shortly after his visit to Canada. Kropotkin's concept grew directly from the work of the eminent Russian zoologist, Karl Fyodorovich Kessler who, in 1880, had suggested that "besides the law of mutual struggle there is in Nature the law of mutual aid, which, for the success of the struggle for life, and especially for the progressive evolution of the species, is far more important than the law of mutual contest." Kropotkin maintained that Kessler's theory represented "a further development of the ideas expressed by Darwin himself in the Descent of Man." But most important, the Kessler theory

---

[9] Fernand Braudel, The Mediterranean and the Mediterranean World in the Age of Philip II, trans. by Sian Reynolds (2 vols.; New York: Harper and Row, 1975); Lynn White, Jr., "The Historical Roots of Our Ecologic Crisis," Science, 155 (March 10, 1967), 1203-1207; George Woodcock, "A Plea for the Anti-Nation," Canadian Forum, 52 (April, 1972), 16-19; Wayland Drew, "Toward an Ecological Conscience: Peter Kropotkin and Mutual Aid," Ontario Naturalist, 13 (June, 1973), 29-35; Lionel Rubinoff, "Nationalism and Celebration: Reflections on the Sources of Canadian Identity," Queen's Quarterly, 82 (Spring, 1975), 1-13.

of mutual aid reinforced Kropotkin's personal experience.[10]

As a young Russian aristocrat, Kropotkin had elected to do his military service in eastern Siberia. His underlying love for science surfaced during his travels there and throughout the wilderness of northern Manchuria. Two aspects of animal life observed during these journeys struck him as being particularly significant. First, he noted that climate, rather than competition, functioned as the primary check on overpopulation. And second, even where population was abundant, he failed to find "that bitter struggle for the means of existence, among animals belonging to the same species, which was considered by most Darwinists . . . as the dominant characteristic of the struggle for life, and the main factor of evolution." On the contrary, there was considerable evidence of mutual aid, without which, Kropotkin concluded, there could be no evolution. He did not view mutual aid as altruism, but as a biologically determined instinct which had itself evolved, and which constituted the basic element of sociability within the animal kingdom. Regardless of how feeble a species might appear to be, its

---

[10] Peter Kropotkin, Mutual Aid: A Factor in Evolution, ed. by Paul Avrich (New York: New York University Press, 1972), p. 19. Kropotkin's life is well documented in his autobiography, and in the work of his primary apologist, George Woodcock; see Peter Kropotkin, Memoirs of a Revolutionist (Boston: Houghton Mifflin, 1899); George Woodcock and Ivan Avakumovic, The Anarchist Prince: A Biographical Study of Peter Kropotkin (London: T.V.Boardman, 1950).

chances for survival were enhanced in direct proportion to the sociability of its members. Sociability reinforced the survival value of factors like protective coloration, speed, cunning and endurance, but it also fostered the growth of intelligence. It followed that even though technology constituted tangible evidence of man's increasing intelligence, it also held the seeds of his destruction if not kept within the bounds of sociability. Sociability was distinguished by the limitations it imposed on physical struggle. Even in the animal world, "food is only one of a hundred of various conditions of existence." Overpopulation of a given area was not greeted by the manufacture of "weapons for snatching the food from the mouth of . . . congeners" but by forming "new habits, [by] moving to new abodes, and [by] taking to new sorts of food." In this way competition was avoided by a process of adaptation. Those animals which had moved, or had adopted a new food supply, would evolve differently from the group they had left. But they would survive.[11]

Kropotkin believed that the lessons learned from his study of animals could be applied to human affairs. When his attention was drawn to existing writings on the relationship between Darwinism and sociology, he observed that "they all endeavoured to prove that man, owing to his higher

---

[11]Kropotkin, Mutual Aid, pp. 17-18, 68-69, 74.

intelligence and knowledge, may mitigate the harshness of the struggle for life between men; but they all recognized at the same time that the struggle for the means of existence, of every animal against all its congeners, and of every man against all other men, was a 'law of Nature.' This view, however, I could not accept. . . ." Having discovered the "importance of mutual aid in various classes of animals, I was bound to discuss the importance of the same factor in the evolution of man."[12]

Kropotkin had regularly pointed to Darwin's own belief that his theory would produce a new philosophy. And, in the Descent of Man, Darwin did elaborate a sophisticated set of ethical principles that continue to haunt all of his interpreters. He began by arguing that duty, or moral conscience, is innate in the lower animals as well as in man, and that it leads them "to take pleasure in the society of [their] fellows." The moral conscience depends for its higher development upon the evolution of the mental faculties. One such faculty, the power of language, leads to the creation, first, of a community opinion which dictates the common morality to the individuals who constitute the society, and second, of a mutual sympathy which elicits the respect of individuals for the common morality. Out of this relationship grows habit, which functions to

---

[12] Ibid., pp. 22-23.

fortify the social instinct, the morality and the community.[13] The only thing that distinguishes the operation of these principles as between animals and man is the question of degree.

> [W]ith those animals which live permanently in a body, the social instincts are ever present and persistent. Such animals are always ready to utter the danger signal, to defend the community and to give aid to their fellows in accordance with their habits; they feel at all times, without the stimulus of any special passion or desire, some degree of love and sympathy for them; they are unhappy if long separated from them, and always happy to be again in their company.[14]

All morality, in Darwin's view, emerged from the social instinct. In sharing this belief, Kropotkin maintained that "Nature has . . . to be recognised as _the first ethical teacher of man_." Spencer, on the other hand, did not consider man's moral sense a developed extension of the social instinct of animals. Rather, "It made its appearance at a much later epoch, originating from those restraints which were imposed upon men by their political, social and religious authorities." The "militant" and "industrial" stages of Spencer's ethics were premised on this conviction, and led him to the conclusion that morality was merely a compromise between egoism and altruism.[15]

---

[13]Charles Darwin, _The Descent of Man, and Selection in Relation to Sex_ (2nd ed.; London: John Murray, 1885), pp. 97-126, _passim_.

[14]Ibid., p.112.

[15]Peter Kropotkin, "The Morality of Nature," _Nineteenth Century_, 57 (March, 1905), 415, 417.

In repudiating Spencer and T. H. Huxley, Kropotkin was forced to extend the implications of mutual aid to the human sphere. Not surprisingly, he began his analysis with a look at "primitive" men who, living "in close intimacy with animals," learned to regard them as "an extension of their own kin--only so much wiser than themselves." Animals of the same species were lumped together as clans or tribes, distinctive, in spite of carnivorous habits, for the fact that they never killed each other. Above all, they were sociable.

> For them all animals--beasts, birds and fishes alike--were in continual communication, warning each other by means of hardly perceptible signs or sounds, informing one another about all sorts of events, and thus constituting one vast community, which had its own habits and rules of propriety and good behaviour. Even today deep traces of that conception of nature survive in the folklore of all nations.

To the primitive mind, the animals were "mysterious, problematic beings, possessed of a wide knowledge of the things of nature." Furthermore, each species understood all others. "They all belong to one brotherhood, into which they may, in some cases, admit man." Man viewed himself with humility before nature. "The fact is that <u>every life</u> is respected by a savage, or rather it was before he came in contact with Europeans." All of life constituted a collectivity demanding the pronoun "We" for its proper identification. The isolated being, or individual, was seen by Kropotkin as a "product of civilisation" and "an abstraction" which countered natural ethics and created competition.

> It is the tribe, with its hard-and-fast rules, superstitions, taboos, habits and interests, which is always present in the mind of the child of nature. And in that constant, ever-present identification of the unit with the whole, lies the substratum of all ethics. . . .

Because competition was injurious to the species and therefore unnatural, it had to be opposed. "That," said Kropotkin, "is the watchword which comes to us from the bush, the forest, the river, the ocean."[16]

Here then is the scientific foundation of Kropotkin's anarchism. The ultimate form of order comes not from man, but from nature, of which man is an integral part. Western civilization has been marked by a process of dis-integration wherein man, by separating his own species from all others, and by internally individualizing each from all, has lost sight of his continuity with the whole. Biology introduced its own irony into this concept--one which has perplexed defenders of the "Global Village" and "Only One Earth" schools. By demanding diversity, mutual respect and mutual aid, Kropotkin's anarchism seemed to deny the fundamental precept of the whole. On the contrary, argues Kropotkin, the whole is, by its very definition, the sum of its parts. Without the parts there can be no whole. The parts must, therefore, not only be retained, but encouraged to pursue

---

[16]Ibid., pp. 419, 421, 424-26; Kropotkin, Mutual Aid, p. 81.

their natural diversity. There is nothing inconsistent with the defense of diversity and the defense of the whole so long as competition is eliminated as the fundamental focus of the struggle for existence. Having separated himself from his tradition through a preoccupation with mere human change, man must now strive consciously to regain it through recourse to mutual aid. That end will be served best, not by centralized governments, but by the balanced community of all living things. In Kropotkin's view, this requires nothing less than the resurrection of Darwin's "mutual sympathy"--respect for the common morality, or simply individual responsibility--through the destruction of the bureaucratic state. Self-government requires the participation of all species.

Peter Kropotkin, then, might well displace Aldo Leopold as the father of the "ecological conscience."[17]

III

In searching for a tangible symbol of ecology in North America, the modern heirs of the Kropotkin tradition have been drawn to the wilderness, the very birthplace of the concept of mutual aid. As a universal term, wilder-

[17] Aldo Leopold, The Sand County Almanac, with Other Essays on Conservation from Round River (New York: Oxford University Press, 1966), pp. 217-69.

ness is an abstraction. Only when it is attached to some geographical location can its diverse parts be pinned down with any degree of accuracy. And, even then, as symbol it suffers from subjective interpretation. Broadly speaking it is considered wasteland, uninhabited (sometimes uninhabitable) by whites, potentially rich in resources or potentially arable. The Western identification of wilderness with potential suggests process. Process in turn implies maturation. In North America the wilderness has been identified, in numerous historical traditions, as a variable affecting the construction of man's past. George Grant's North American primal has guided both the pioneer and his historian to the conclusion that man's success is greater where wilderness, once greatest, is now least. As a geographical fact, the wilderness of Canada has in large measure avoided despoliation, not necessarily because it is loved more, but rather because it permits less. For the wilderness of Canada is identified with North, and "by the 'North' is meant all that territory beyond the line of minimal growth of the known cereal grains."[18] Still, it may be said that wilderness and North remain central symbols in the Canadian imagination.

> We are close to the wilderness, close to the primitive, and despite our pretensions we are proud of the

---

[18] W. L. Morton, "The 'North' in Canadian Historiography," Transactions of the Royal Society of Canada, Series IV, 8 (1970), 32.

> definition they have given us. The North has done much to keep us restrained and circumspect. Its immensity has offset the arrogance of power in us. . . . Have we not always known that we were limited? Beyond the towns and cities, beyond the highways and railroads and all the places where history seethes and flourishes, have we not always been resigned to our containment by the land? . . . Correct and unanswerable, the wilderness which is a part of us must be reaffirmed without fear and without apology.[19]

In America, of course, we must identify wilderness with West. The difference between the United States and Canada, between West and North, is the agricultural possibility or lack of it. In other words, wilderness gains its popular definition by the potential use to which it can be put by man. The Canadian North, of course, is invariably perceived as "a vast storehouse of natural resources," those very staples which have been employed historically not only to explain the existence of a separate and distinctive Canadian nation, but also to demonstrate the foundation of its colonial bondage to the United States. The potential usefulness of the wilderness, as well as its inherent deterministic, limiting qualities, provides us with a paradox: "the ultimate and the comprehensive meaning of Canadian history is to be found where there has been no Canadian history, in the North." The different uses to which wilderness can be put in both Canada and the United States materially affect the landscape and the people who perceive it. Agricultural development leaves behind it a comforting pastoral and the

---

[19] Wayland Drew, "Wilderness and Limitation," <u>Canadian Forum</u>, 52 (February, 1973), 18-19.

economic security associated with renewable staples. Resource development, on the other hand, evokes, with possibility, fear. The scarred perimeter of Sudbury and the last drop of oil from Leduc No. 1 carry inside themselves the reality of limitation.[20]

The glorification of civilization is inevitably paralleled by the demystification of wilderness. Since 1892, when John Muir launched the Sierra Club, the defenders of wilderness have allied around the conviction that wild land can be "saved" by "preserving" it in parks. But the whole notion of preservation is imbued with a sense of defeat, as if the land itself were an artifact. At the turn of the century the preservation movement was probably a legitimate political response to the technological behemoth. But in our own age the park has become just another managerial unit. Parks are not a gauge of love and respect for wilderness. They are recreation facilities, where wildlife is managed and where the annual head count of visiting tourists determines their usefulness. If we are to equate wilderness and ecology, politics and parks will not provide the tools. We must, indeed, learn to live not merely on, but with, our land. This requires nothing less than complete surrender

---

[20] Morton, "The 'North' in Canadian Historiography," p. 40; J. M. S. Careless, "Frontierism, Metropolitanism and Canadian History," Canadian Historical Review, 35 (March, 1954), 1-21; Henry Nash Smith, Virgin Land: The American West as Symbol and Myth (Cambridge, Mass.: Harvard University Press, 1950).

to the limitations and demands which it imposes upon us. In a sense, we must allow the remystification of wilderness.[21]

It has been argued that certain myths have their roots in biology, that they are a means of transmitting, in a memorable way, instinctively known truths. The myth of the Wild Man, which despite material progress continually recurs in Western thought, may be viewed as evidence that the desire of man for unity with his land has survived, in latent form, the apparent conquest of wilderness.

> As one after another of [the] wildernesses was brought under control, the idea of the Wild Man was progressively despatialized. This despatialization was attended by a compensatory process of psychic interiorization. And the result has been that modern cultural anthropology has conceptualized the idea of wildness as the repressed content of both civilized and primitive humanity. So that, instead of the relatively comforting thought that the Wild Man may exist out there and can be contained by some kind of physical action, it is now thought (except by those contemporary ideologues on both sides of the iron curtain who think they can save "civilization" if only they can succeed in destroying enough "wild" human beings) that the Wild Man is lurking within every man, is clamoring for release within us all, and will be denied only at the cost of life itself.

The psychic properties of wilderness may, indeed, lie at the base of a North American schizophrenia where that which

---

[21] J. S. Holliday, "The Politics of John Muir," Sierra Club Bulletin, 57 (October-November, 1972), 10-13; Roderick Nash, Wilderness and the American Mind (New Haven, Conn.: Yale University Press, 1967), pp. 122-140, 158-168; Wayland Drew, "Killing Wilderness," Ontario Naturalist, 12 (September, 1972), 23; Cole Harris, "The Myth of the Land in Canadian Nationalism," in Nationalism in Canada, ed. by Peter Russell (Toronto: McGraw-Hill, 1966), pp. 27-43.

is perceived as reality in fact constitutes a denial of it.[22]

The interiorization of wilderness has animated human history through two quite separate concepts: primitivism and archaism. Primitivism idealizes "any group as yet unbroken to civilizational discipline," while archaism focusses on "real or legendary remote ancestors, either wild or civilized." Archaism may be used either to reinforce the status quo or to disrupt it, while primitivism is fundamentally a radical doctrine holding to the "conviction that men are really the same throughout all time and space but have been made evil in certain times and places by the imposition of social restraints upon them." As a double-edged concept, archaism appears to have the advantage of flexibility that primitivism lacks. But primitivism rejects flexibility as potential compromise. In eulogizing the "savage" and in denigrating the "civilized," primitivists view reform as the repudiation of externally imposed, and therefore artificial, restraint. In primitivism there is no need to reconstruct an original. The original is omnipresent even though its natural qualities have been

---

[22]Hayden White, "The Forms of Wildness: Archaeology of an Idea," in The Wild Man Within: An Image in Western Thought from the Renaissance to Romanticism, ed. by Edward Dudley and Maximillian E. Novak (Pittsburgh, Pa.: University of Pittsburgh Press, 1972), p. 7; R. D. Laing, The Politics of Experience (New York: Ballantine, 1967), pp. 100-130; Wayland Drew, "Wilderness and Culture," Alternatives, 3 (Summer, 1974), 8-11; Claude Lévi-Strauss, Mythologiques (3 vols.; Paris: Plon, 1964-68).

perverted or repressed. Archaism perceives nature as malevolent, primitivism as harmonious. While the Wild Man functions in both interpretations as "a criticism of whatever security and peace of mind one group of men in society [has] purchased at the cost of the suffering of another," only within the framework of primitivism can he assume the mantle of a Rousseauan Noble Savage. Primitivism and ecology are mutually supporting concepts, recognizing freedom as a law of nature, nature as a law of freedom. Only when a society has united with the land which sustains it, has realized and accepted a community of all living things, only then will the Wild Man have been reincarnated, reality restored and repression ended. The links between anarchism, primitivism and ecology should be clear.[23]

The cultures of the United States and Canada have been interpreted, in two distinctly different ways, as products of man's encounter with the wilderness. Defining the Canadian dimension has been the "garrison" thesis which argues that the germ of British and European cultures was brought intact to the new world and imposed upon the landscape. By endeavoring to keep the alien wilderness out and the comforting culture in, Canadians unwittingly created a national schizophrenia, a "violent duality." This "conquest of nature by an intelligence that does not love it," has been reflected

[23] H. White, "The Forms of Wildness," pp. 25-28.

in both the country's literature and its history, marking a singular incapacity on the part of the people to find a unity--either among themselves, or with their land.[24] More recent nationalist writing in Canada, however, has been characterized as an effort to eliminate the garrison, "to abandon defence, to let down the walls and let the wilderness in, even to the wolves," suggesting that in the end a people finds peace only when its wilderness and its culture are reconciled.[25]

In Canada wilderness exists outside of parks, indeed dominates the landscape of the entire country. In the United States, for the most part, wilderness is confined to parks and is proportionately the smallest component of the landscape. This distinction has awesome implications for the future of both countries. Thanks most of all to Frederick Jackson Turner, the American historian has traditionally been incapable of seeing wilderness in a vacuum. It cannot be left free and to itself. And it certainly must not be feared. Indeed, it finds redemption from waste through man's interaction with it. True, man is strongly influenced

---

[24] Margaret Atwood, The Journals of Susanna Moodie (Toronto: Oxford University Press, 1970), p. 62; Northrop Frye, The Bush Garden: Essays on the Canadian Imagination (Anansi: Toronto, 1971), pp. 224-25.

[25] D. G. Jones, Butterfly on Rock: A Study of Themes and Images in Canadian Literature (Toronto: University of Toronto Press, 1970), p. 8. Two recent novels epitomize the new tendency: Margaret Atwood, Surfacing (Toronto: McClelland and Stewart, 1972); Wayland Drew, The Wabeno Feast (Toronto: Anansi, 1973).

by its obvious powers--climate, topography, vegetation, animal constituents, and so on--but he invariably emerges triumphant over all through his reasoned subjugation. By wilderness we are meant to assume land that is untouched by man, or at least by the white man, since the indigenous population of North America was regarded as a savage race, in bondage to the landscape that sustained it. Turner's wilderness is useful because eventually it "interacts" with man. The conceptual line of interaction is located in the West, upon a frontier, that "meeting point between savagery and civilization" which advances ever onward to its culmination in the sea. The frontier is functional by definition. Through the interaction which takes place upon it, man is made better by nature, nature is made better by man. But this functional dualism, while it may create an immunity to the European germ, also dictates the inevitable end of wilderness. The tensions of the interaction will be resolved in the pastoral.[26]

Turner was as well versed in the developments of Darwinian biology as any liberally educated layman of the late nineteenth century. He also shared Herbert Spencer's predilection for Lamarckism and was wholly receptive to the idea of society as an evolving organism with acquired

[26]Frederick Jackson Turner, The Frontier in American History (New York: Holt, Rinehart, Winston, 1962), pp. 1-31; Francis Jennings, "Virgin Land and Savage People," American Quarterly, 23 (October, 1971), 519-541.

characteristics. "On the miraculous frontier, man and society were separated from their past and forced to assume a new physical and spiritual appearance." Unlike their timorous Canadian cousins, attached to mother's apron strings and cowering behind the walled garrison, Americans were "produced" by their environment. The frontier was a form of society rather than an area.[27] Where Kropotkin would have hoped for a society functioning as part of the biological organism, Turner saw the land serving the social organism:

> The wilderness disappears, the 'West' proper passes on to a new frontier, and in the former area, a new society has emerged from its contact with the backwoods. Gradually this society loses its primitive conditions, and assimilates itself to the type of older social conditions of the East; but it bears within it enduring and distinguishing survivals of its frontier experience. Decade after decade, West after West, this rebirth of American society has gone on, has left its traces behind it, and has reacted on the East.

The man of the West was ultimately homogenized by his environment. The arrival of the frontier at the impasse of the sea, given Turner's proposition that the West dispatched her ideals back to Washington in the Congressman's valise, insured the ultimate homogenization of the East as well.

> This, then, is the real situation: a people composed of heterogeneous materials, with diverse and conflicting ideals and social interests, having passed from the task of filling up the vacant spaces of the continent, is now thrown back upon itself, and is seeking an equilibrium. The

---

[27] William Coleman, "Science and Symbol in the Turner Frontier Hypothesis," American Historical Review, 72 (October, 1966), 35-37.

diverse elements are being fused into national unity. Wilderness, therefore, contributes to the compromise. The process of its own elimination ensures the unity of the new inhabitants. The assimilation of diverse peoples, as the assimilation of diverse landscapes, is viewed as an essential good.[28]

Ironically, the primary stimulus for the American preservation movement seems to have been this Turnerian frame of mind. At the end of the nineteenth century the environment which had produced a unique democracy was on the verge of extinction. "Many Americans came to understand that wilderness was essential to pioneering; without wild country the concepts of frontier and pioneer were meaningless. The villain, it appeared, was as vital to the play as the hero. . . ."[29] In other words, wilderness had become an artifact worthy of institutionalization. National and state parks served as monuments to its own destruction. Reflecting on the comparative dearth of such parks in Canada, the American historian Roderick Nash concluded that

> The Canadian public's sensitivity to and enthusiasm for wilderness values lags at least two generations behind opinion in the United States. Canada, that is to say, is currently at a posture regarding wilderness that the United States occupied in the late nineteenth and early twentieth centuries.

---

[28]Turner, Frontier in American History, pp. 205, 221.

[29]Nash, Wilderness and the American Mind, p. 145.

He insisted, quite charitably, that Canadians were in no way to be "blamed" for this state of affairs, and then concluded, reassuringly, that "the lag of which I speak, while a half-century long, can . . . be closed in far less time." The latent whiggery of Nash's assumptions is reinforced by his cause-and-effect analysis.

> I understood that Canada and the United States had for the last half century operated on different wavelengths so far as wilderness was concerned. Americans had passed through its [sic] frontier era to an increasing awareness of the value of wilderness; Canadians were still a pioneering people who retained the old attitudes toward undeveloped places.[30]

This patronizing view supposes that Turner's thesis constitutes a valid explanation of the Canadian past (or present) and presumes that wilderness values are measured out in parks. In fact the existence of parks is alien to the idea of wilderness in Canada. The wilderness park is at once an American creation and a response to an American situation. It represents the ultimate liberal compromise between technology and nature, the absurd extension of the principle of rational, efficient use, otherwise known as conservation.

Historically it has been the vogue to distinguish between conservation and preservation. In fact, however,

---

[30]Roderick Nash, "Wilderness and Man in North America," in The Canadian National Parks: Today and Tomorrow, ed. by J. G. Nelson and R. C. Scace, Studies in Land Use History and Landscape Change, National Park Series, No. 3 (2 vols.; Ottawa: National and Provincial Parks Association and the University of Calgary, 1968), 1: 75, 79-80.

the two concepts are both highly anthropocentric. The maintenance of the pioneering ideal was itself broadly utilitarian. But there were other factors which spurred the creation of wilderness parks, some of them imitated in Canada. For example, the existence of sundry freaks of nature--geysers, hot springs, spectacular waterfalls and the like--at Yellowstone and Banff, dictated to astute politicians the advisability of preventing exploitation by greedy entrepreneurs and of ensuring universal accessibility. There was no implicit desire to protect wilderness for its own sake. The same could be said of parks set aside for endangered species, notably the buffalo, where the rationale employed was invariably the protection of a heritage. Both conservation and preservation reflected the multiple use-sustained-yield mentality of the Progressives; both aspired to service the real and imagined needs of democracy. Because both concepts required technical expertise for their practical implementation, both contributed to the centralist-decentralist debate which animated so much political discussion in the Progressive Era. The political tendency of conservation in America has always been centralist, placing a large measure of control in the hands of the federal government. Since 1909, the year of the North American Conservation Conference, liberal emulators of American policy have aimed at the same kind of control in Canada. But while Washington continues to maintain the upper hand in conservation matters, tensions between the provinces and Ottawa remain particularly acute. The problems associated with conservation are intimately related to the land policies adopted by the two countries. But differences are also explained by the reversed order of residual powers in their respective constitutions. The fate of the American wilderness was determined by the pioneer. What little of it remains depends for its survival upon legislation like the federal Wilderness Act of 1964. The fate of wilderness in Canada, as well as the fate of ecology for which wilderness is the symbol, is inextricably bound up with the resolution of the tension between provincial and federal governments. American experience has demonstrated convincingly that centralized, bureaucratic control is inimical to the existence of wilderness. But in Canada, because there is so much of it and because it carries within itself an implicit decentralization, there is hope that wilderness will dictate to the state before the state dictates to it. Wilderness and park do not necessarily constitute an equation; certainly there is no ground whatever for arguing that by creating parks, one "saves" wilderness.[31]

---

[31] Nash, Wilderness and the American Mind, pp. 109-116, 220-24; Jackson W. Turrentine, "The Creation of Yellowstone National Park," Mississippi Valley Historical Review, 29 (1942), 187-206; Robert Craig Brown, "The Doctrine of Usefulness: Natural Resource and National Park Policy in Canada, 1887-1914," in The Canadian National Parks: Today and Tomorrow, I, 94-110; F. J. Thorpe, "Historical Perspective on the Resources for Tomorrow Conference," in

In Canada the energies of technology have been directed largely at the process of resource extraction. The Laurentian Shield effectively prevents the pastoral from dominating the national mythology.[32] In the United States, however, the pastoral legacy of Turner's westward marching pioneer has replaced the wilderness as the major mythical property and the primary antagonist of technology. In an age when untouched wilderness was the predominant landscape, the pastoral represented progress. But the gradual ascendancy of the machine during the nineteenth century was paralleled by the progressive emasculation of the pastoral. While the garden remained an imaginative touch-

---

Resources for Tomorrow Conference, Background Papers (3 vols.; Ottawa: Queen's Printer, 1961-62), I, 1-13; Hays, Conservation and the Gospel of Efficiency: The Progressive Conservation Movement, 1890-1920 (New York: Atheneum, 1969), pp. iii-vii; J. Leonard Bates, "Fulfilling American Democracy: The Conservation Movement, 1907-1921," Mississippi Valley Historical Review, 44 (June, 1957), 29-57; J. M. S. Careless, "'Limited Identities' in Canada," Canadian Historical Review, 50 (March, 1969), 1-10; Michael McCloskey, "The Wilderness Act of 1964: Its Background and Meaning," Oregon Law Review, 45 (1966), 288-321. On the relationship between provincial resources and political decentralization in Canada, see especially H. V. Nelles, The Politics of Development: Forests, Mines and Hydro-Electric Power in Ontario, 1849-1941 (Toronto: Macmillan, 1974); Robert E. Cail, Land, Man and the Law: The Disposal of Crown Lands in British Columbia, 1871-1913 (Vancouver: University of British Columbia Press, 1974); Morris Zaslow, The Opening of the Canadian North, 1870-1914 (Toronto: McClelland and Stewart, 1971).

[32] See, e.g., W. H. New, Articulating West: Essays on Purpose and Form in Modern Canadian Literature (Toronto: New Press, 1972); Laurence Ricou, Vertical Man, Horizontal World: Man and Landscape in Canadian Prairie Fiction (Vancouver: University of British Columbia Press, 1973).

stone for innocence and tranquility, pastoralism itself became a "curious state of mind" which was "generated by an urge to withdraw from civilization's growing power and complexity." This association of technology with power, complexity and civilization automatically transformed defenders of the pastoral into "radically disaffected . . . dissidents" from "small, ineffectual groups--socialist, transcendentalist, literary, religious--far from the centers of influence and power." Technology was reality, the tangible triumph of reason over the land.[33]

The contradiction of the machine in the garden has permeated much of American historiography since Turner. Increasingly the pastoral is perceived as irrelevant to modernity. It bears discussion only because it recurs, in a metaphorical sense, as one side of the imagined man/nature dichotomy. Dichotomies invariably simplify intellectual discussion. Besides, the principles of debate demand that they exist. Unfortunately, debates must conclude either in the victory of one side over the other or in a compromise between the two. For liberals, the dichotomy provides the thing to be improved and its improver, the place where we have been and the place where we are, the place where we are and the place to which we are going. Underlying

[33]Leo Marx, The Machine in the Garden: Technology and the Pastoral Ideal in America, Galaxy Books (New York: Oxford University Press, 1967), pp. 9-10, 24, 219, 228-29; Richard Hofstadter, The Age of Reform, From Bryan to F.D.R. (New York: Knopf, 1955), pp. 23-59.

the liberal use of dichotomy is a preoccupation with inevitable social change, identified as progress. Progress does not permit tradition or continuity to interfere with its own claims to reality. It allows the memory of the past, but only when it reinforces the present. Wilderness parks are evidence of the barbarous landscape from which Turner's pioneer delivered America. The pastoral, the seat of agriculture and once the sustaining element of civilization, drops in the practical scale of values to "the starting point for infantile wish-fulfillment dreams, a diffuse nostalgia, and a naive, anarchic primitivism." Through technology man is finally "gaining access to the divine plan." The machine will put an end to mystery and render nature more predictable, freeing civilization for the real business of social progress.[34]

In the United States (and in Canada, as well), imaginative literature has done much more than history to keep the debate alive by responding to the mythical properties of wilderness and pastoral.[35] This interest in nature must be viewed not only as the release of interiorized landscapes

---

[34]Marx, Machine in the Garden, pp. 11, 186.

[35]See, e.g., James Dickey, Deliverance (Boston: Houghton, Mifflin, 1970); William Faulkner, "The Bear," in The Portable Faulkner, ed. by Malcolm Cowley (Rev. ed.; New York: Viking Press, 1967); Norman Mailer, Why Are We in Vietnam? (New York: Putnam's, 1967). See also, Thomas A. Krueger, "The Historians and the Edenic Myth: A Critique," Canadian Review of American Studies, 4 (Spring, 1973), 3-18.

and their inhabitants, but also "as a serious criticism, explicit or implied, of the established social order. It calls into question a society dominated by a mechanistic system of value, keyed to perfecting the routine means of existence, yet oblivious to its meaning and purpose."[36] However, while literature helps to articulate what we already know instinctively, it still requires an alliance with history in order to demonstrate how man has employed reason to sever his connection with the land. If the wilderness was located in the West and implied a heterogeneous landscape supporting a diversity of native cultures and animal and plant populations, then, by the end of the nineteenth century, civilization was firmly rooted in an East aspiring to a homogenous culture in a homogenous landscape. Somewhere in between lay the transition zone, the middle landscape, or the pastoral, whose human inhabitants were, at least according to Turner, directly responsible for the decimation of wilderness and the exaltation of civilization. The last fifteen years of the nineteenth century marked the watershed in American history when the tensions that divided East, West and Middle resolved themselves politically in technocratic Progressivism.[37] But before this

[36] Leo Marx, "American Institutions and Ecological Ideals," Science, 170 (November 27, 1970), 949.

[37] John Higham, "The Reorientation of American Culture in the 1890's," in The Origins of Modern Consciousness, ed. by John Weiss (Detroit, Mich.: Wayne State University Press, 1965), pp. 25-48; Henry Steele Commager, The American

acceptance of government by experts, several representatives of the civilized East "returned" to the West with the conscious intention of capturing the values implicit in the wilderness encounter. This meant that they accepted the interiorized existence of the wilderness and sought to find its physical expression. Unfortunately, however, their own history had left word that nature was violent. As a result, while the force that motivated the encounter was primitivist, the encounter itself was archaist. That is to say, their acceptance of the tradition of the legendary, remote ancestor precluded the possibility of their finding unity with nature. The adoption of archaism over primitivism ensured the victory of conservation over the ecological conscience.

Chief among those from the East to embark on the belated encounter was Theodore Roosevelt, soon to become both the President of the United States and the vanguard of the conservation movement. Disillusioned with the prevailing drift of social attitudes and realizing "how ill-prepared or unwilling he was to cope with the style of life evolving in post-Civil War eastern America," Roosevelt chose to "escape to a non-eastern environment." Having always regarded the West "as a stage in the history of American

---

Mind: An Interpretation of American Thought and Character Since the 1880's (New Haven, Conn.: Yale University Press, 1950), pp. 41-54.

civilization," and pioneer values as the source of American manhood and virtue, he believed that he could claim legitimate translation from the past to the present only by participating in the frontier experience. He had already determined that the East would dominate the future, but he remained convinced that the governors of America's destiny would have to understand both tradition and change. The problem, of course, was that when he went to the West the wilderness had already succumbed to the pastoral. As a result, the experience merely reinforced his previous assumptions and gave them added dimensions. The West was not an alternative to society, merely a qualitatively different kind of society.

> Rather than drawing distinctions between the presence and absence of organized society, [he distinguished] between two types of social order--one agrarian, rural, egalitarian and racially homogenous (West), and the other industrial, urban, elitist, ethnically heterogeneous, and racially mixed (East).

The physical contents of the landscape itself were really quite irrelevant to this view. Certainly they played no part in limiting possibility. The West was clearly in the process of becoming the East. The new pioneer was technology, and the new frontier was the pastoral.[38]

Roosevelt's later concern for conservation may have

---

[38] G. Edward White, The Eastern Establishment and the Western Experience: The West of Frederic Remington, Theodore Roosevelt and Owen Wister (New Haven, Conn.: Yale University Press, 1968), pp. 73, 77-78, 184-85.

revitalized symbols of a "nontechnological, nonurban, noncorporate, nonelitist--in short a noneastern society," but the symbols belied the reality. In fact, "the intimate connection between the conservation movement and the antimonopolist tradition" merely reflected the irony of Roosevelt's perceptions. That the conservation movement was focussed on the West obscured the fact that its very _raison d'être_ was to sustain, in the long run, the material wants of the East. For many westerners it represented an imperialistic tyranny which worked to their disadvantage by denying them both the power and the right to enjoy the same prosperity as their brothers. Roosevelt's frontier experience was an attempt to reconcile West with East, an attempt to achieve some sort of balance between wilderness and technology. The failure of his administration to establish the desired equilibrium is partially explained by the incompatibility of the two elements.[39] In reality, the conservation movement was a political compromise between the pastoral and technology. It carried the pastoral trappings of husbandry packaged in the scientific rhetoric of rational, efficient use. But its purpose was the stimulation and preservation of business. The psychic rumblings of the wild man within yielded in Roosevelt an archaist who idealized the pioneer. While this satisfied the needs of conservation,

---

[39] Ibid., pp. 172, 183.

the ecological conscience required a primitivist mind like Kropotkin's, a mind that would learn from nature rather than try to teach or train it. Roosevelt glimpsed the wilderness, but he never understood it.

Ecology is frequently confused with conservation and preservation. It has also been argued that radical ecologists are elitist. I would argue conversely that ecology must not be confused with either conservation or preservation precisely for the reason that it is _not_ elitist. It encourages and supports the existence of all species, including man. Unlike conservation it opposes centralization and encourages diversity. It acknowledges the presence of order in nature, asking of man only that he and his institutions accommodate it. American historians "who dominated later frontier historical writing were in many respects prejudiced against the land, the wilderness and the animal life in it. . . . The tunnelled frontiersman vision of [the] past that has engendered a conquistador attitude toward the land will be revised."[40] This study seeks to play a part in that revision.

IV

It is worth investigating a man whose life and work

[40] Wilbur Jacobs, "Frontiersmen, Fur Traders and Other Varmints, An Ecological Appraisal of the Frontier in American History," _American Historical Association Newsletter_, 8 (November, 1970), 5-11; David Pearce, "Is Ecology Elitist?" _Ecologist_, 3 (February, 1973), 61-63.

reflected the tensions between wilderness and civilization, Canada and America, primitivism and archaism, ecology and conservation, if only to understand the dimensions of the problems we face in our own time. Ernest Thompson Seton is probably best remembered by two generations of adults who recall the stories from Wild Animals I Have Known. For them his name conjures up nostalgic images of Lobo, the King of Currumpaw, or Raggylug, or Krag, the Kootenay Ram. For others of the same generation, Seton is associated with conservation, with aboriginal rights, and with the values of wilderness in an age of "civilized truths." The Progressive Era, traditionally viewed as the great watershed in modern American history, that point in time when an agriculturally oriented society was giving way to the urban, industrial state, has not enjoyed the depth of intellectual historical investigation which seems to me its due. The muckrakers and social realists were not, after all, the only critics who railed against the status quo at the turn of this century. Nor were they alone in capturing large audiences. It is somewhat ironic that a man who had so many diverse talents should be best remembered for the one in which he showed the least ability. But the books Seton wrote then are today being republished in North America, in Eastern Europe and in Japan with a success which should, at the very least, make us ponder our professional neglect. Seton, it may justifiably be argued, was not a good writer. But even if this is so, we must ask why so many people read his stories. We might also ask ourselves why they remember them so well. Is it pure sentimentality, as we have frequently been told, or is there some underlying subtlety which, however poorly expressed, demands a new hearing?

I do not propose to weigh the relative merits of Seton as a competitor in a market apparently cornered by the likes of Frank Norris, Theodore Dreiser and Upton Sinclair. I choose to accept as an article of faith that Seton's popularity with the reading public gives him a claim to our collective remembrance. In assessing the content of his work, I am compelled to avoid generalizations about artistic excellence (something to which muckrakers and social realists alike were frequently immune), looking strictly at the themes which preoccupied his thinking and trying to suggest reasons for their claim on his attention. Most Seton scholarship of any consequence has focussed primarily on his youth work, linking his experience with the Boy Scout movement and its antecedents to the prevailing social concerns of the Progressive Era. In major studies by Peter Schmitt and David Macleod,[41] Seton's work is viewed as a kind of freakish

---

[41]Peter J. Schmitt, Back to Nature: The Arcadian Myth in Urban America (New York: Oxford University Press, 1969); David I. Macleod, "Good Boys Made Better: The Boy Scouts of America, Boys' Brigades and YMCA Boys' Work, 1880-1920" (unpublished Ph.D. dissertation, University of Wisconsin, 1973).

aberration in the larger mosaic. There has yet been no attempt to see in him a powerful challenge to the gathering technological momentum which marked the period. As it is that very challenge which today gains greater and greater credibility, we may perhaps find that it accounts for the renewed popular interest in his work.

Implicit in the interpretations of Schmitt and Macleod is a kind of Whiggish inevitability, a suggestion that Seton was doomed to fail in his stated objectives precisely because he could not assess his own time in a realistic light. Schmitt, in particular, regards Seton with a levity which, while it endears us to Schmitt, colors adversely the man whose work he purports to be investigating. Indeed, in Schmitt's work generally there is an underlying sense of silliness about the so-called "back to nature movement." This perspective is widely shared by all writers who attach "reality" to business and industry, to politics and economics, to a given social order. They seem to suggest that the defenders of nature were slightly off-centre unless they were immersed in the efficiency rhetoric of Gifford Pinchot and Theodore Roosevelt. Anyone who introduced an alternative perception of reality was out of touch with the currents of his time and, therefore, an easy mark. This is particularly annoying to someone who has delved deeply into the life of an individual whose contributions have been so unfairly categorized. If nothing else, it makes one aware of the dangers inherent in consensus history.

As Schmitt and Macleod would have it, because his ideas failed to sustain the allegiance of the Boy Scout movement, Seton therefore failed and should now be laid to rest with those of his misguided cronies who chose to ally themselves with his unorthodox views. By neglecting to look at all sides of the man--and one must demand this of anyone who proposes to make comprehensive generalizations about any individual in any context--they have failed altogether to isolate a certain far-seeing dynamic which motivated all of his actions. This dynamic was rooted in his personal experience of the world which, even by present standards, was extraordinary. While very few men could lay claim to so many personal idiosyncracies, Seton's human foibles must not be allowed to divert us from the ultimate power of his message. For, while this message may have said little to the long-term development of the Boy Scout movement, it certainly speaks in forceful terms to our own time.

This study is not meant to constitute a complete assessment of Seton's life and work. It will, in fact, look solely at a period arbitrarily defined as "the productive years." Broadly speaking this period can be broken down into three component parts, each of them relating to one of Seton's major interests. The first part deals with his childhood and young manhood, a segment of his life devoted in large measure to a fascination with the world of art and the real springboard to the rest of his activities. The second section will treat his career as an author of Children's stories and popular natural history. The third, and last, part will examine his perspectives on youth work and education. Throughout, I wish to convey Seton's essential British North Americanism, his ties to the land, his ecological conscience, and his concomitant "disnationalism." I will outline his thought, insofar as it was an organized body of thought, and relate it to secondary studies of progressive ideas. I suggest that Seton lies outside any of the traditions discussed there and, in fact, that he was representative of a hitherto ignored fraternity of "bio-intellectuals" whose science followed Kropotkin rather than Spencer into the social sphere. Weaving its way through the entire study is the animal motif, or theme. For it was to the wild animal and its natural habitat that Seton owed an almost political allegiance.

Historians tends to devote too little attention to those interpretations which do not bear directly on their own times. In this sense we have had to await the arrival of an "ecological crisis" in order to justify an ecological explanation of our past. Seton, I would like to suggest, had arrived at an ecological interpretation of his present at a time when a crisis did not exist to give it credibility.

Chapter 2

THE VICTORIAN WILDERNESS: FROM ART TO SCIENCE

The Canadian artist serves the spirit of his land and people. He is aware of the spiritual flow from the replenishing North and believes that this should ever shed clarity into the growing race of America and that this, working in creative individuals, will give rise to an art quite different from that of any European people. He believes in the power and the glory, for the North to him is a single, simple vision of high things and can, through its transmuting agency, shape our souls into its own spiritual expressiveness.

Lawren Harris

Plate 2. Self Portrait (Pen and Ink, 1879)

I

Ernest Thompson Seton, the eighth of ten surviving sons born to Alice Snowdon and Joseph Logan Thompson, was born at South Shields, Durham, England, on August 14, 1860. Although his given name was Ernest Evan Thompson, his work bears various signatures--Ernest E. Thompson, E. E. Thompson, Ernest E. Seton Thompson, Ernest E. T. Seton, Ernest Seton Thompson, Ernest Seton-Thompson and Ernest Thompson Seton. The last designation, by which he is best known, was formally adopted in 1901 and remained current until his death in 1946. Seton's preoccupation with his own name was symptomatic of the egotism which marked, and frequently scarred, his career. It also serves to identify an archaist tendency. Not unlike many men of his time, he aspired to greatness by virtue of a claim to titled nobility. Family legend maintained that the Thompsons were descended from George Seton, Fifth Earl of Winton, a highland Scot who had been routed, along with his Stuart retinue, in the Jacobite rising of 1715. What Seton did not know, or what he refused to acknowledge, was that this same Earl of Winton had died unmarried, childless and in exile in Rome. In other words, Seton had no legitimate claim either to the title or to the name.[1]

---

[1]Ernest Thompson Seton, Journal, 1879-1946, American Museum of Natural History, New York, 1:10 (hereafter cited as Seton, Journal, AMNH); Trail of an Artist-Naturalist: The Autobiography of Ernest Thompson Seton (New York: Scribner's, 1940), pp. 391-393 (hereafter cited as Seton, Autobiography); Seton to Austin Seton Thompson, May 23, 1938, Seton Papers, Seton Museum, Cimarron, N.M. (hereafter

Joseph Thompson, Sr. was comfortably situated in the shipping business when a series of misadventures conspired suddenly to reduce both his fleet and the family fortune. In an effort to escape the uncertainties of his precarious occupation, Thompson sold the remaining ships, loaded his family and belongings on an emigrant vessel, and set sail for Canada in 1866. Amply blessed with pious Presbyterian notions of filial duty, he went straight to Lindsay, in the backwoods of Ontario, and prepared, through his rich endowment of sons, to embark on the life of a gentleman farmer. Unfortunately, Seton wrote later, the family "did not fit well into the neighborhood. Father and Mother were too evidently gentlefolk to attract the rude pioneers." Besides, the "work was far too hard" and the older brothers, who apparently balked at the prospect of maintaining their

---

cited as SP); Anya Seton Chase, private interview, Old Greenwich, Conn., Dec. 15, 1971; [Legal Notice of Name Change], New York Evening Journal, Oct. 29, 1901; DNB, s.v. "Cameron, Archibald (1707-1753)," by Thomas Finlayson Henderson; ibid., s.v. "Cameron, Donald (1695-1748)," by John Westby-Gibson; ibid., s.v. "Seton, George, Fifth Earl of Winton (d. 1749)," by Thomas Finlayson Henderson. The issue was confused by Archbishop Robert Seton of New Jersey, an amateur genealogist who was regarded as "the historian of the family." See, Robert Seton, An Old Family; or the Setons in Scotland and America (New York: Brentanos, 1899); Robert Seton to Seton, Nov. 3, 1898; Seton to Robert Seton, Dec. 12, 1898; Dec. 16, 1898, SP. Within the remaining family there is no agreement on the issue. Hereafter Seton's published work will be cited under the name, Seton. The bibliography will identify the name he was using at the time of writing.

father's baronial aspirations, quit in rapid succession. By 1870, when the last of his potential retainers abandoned him, Joseph was forced to sell the new homestead and to move the remnants of his family including ten-year-old Ernest, to Toronto, where he had found employment as an accountant. One by one the prodigal sons returned to the hearthstone, each with a good job, and the Thompsons soon found themselves comfortably placed in an environment more closely approximating the one they had left behind in England.[2]

Young Ernest did not share the adults' enthusiasm for the city. Life on the farm had provided a gradual transition from civilization to wilderness. The landscape around Lindsay reflected something of what he had been and something of what he would become. It offered the comfort and security of the commonplace, the mystery and uncertainty of the unknown. All around him the local settlers were converting the wilderness into the soft and undulating pastoral it has since become. But in the distance,

---

[2]Seton, Autobiography, pp. 3-14, 43, 53, 10; George Seton Thompson to John Enoch Thompson, Feb. 18, 1917; Polly Burfield to Stuart Logan Thompson, Feb. 26, 1930, SP; Violet M. Carr, ed., Ops: Land of Plenty (Lindsay, Ont.: Ops Township Council, 1968), p. 140; Watson Kirckonnel, County of Victoria Centennial History (2nd ed.; Lindsay, Ont.: Victoria County Council, 1967), p. 286.

> the forest was solid and unbroken. It was inconceivable that there should be anything beyond that. My childish fancy made that the end--the rim of things. I knew there was nothing that way, no clearing, nothing but woods and woods and woods.[3]

Bit by bit the young boy had ventured deeper into their darkness. He constructed inconspicuous hiding places where he could observe and sketch the neighboring wildlife, gradually discerning the cyclical pattern of nature.

> As he sat one day in stillness a Minnow leaped from the water and caught a Fly. Almost immediately a Kingfisher that had been shooting past stopped in air, hovered, and darting downward, came up with a Minnow in his beak, flew to a branch to swallow its prey, but no sooner got there when a Chicken-hawk flashed out of a thick tree, struck the Kingfisher with both feet and bore him downward to the bank--in a moment would have killed him, but a long brown creature rushed from a hole in the bank and sprang on the struggling pair, to change the scene in a twinkling. The three strugglers separated, the Hawk to the left, the Kingfisher to the right, the Minnow flopped back into the pool, and the Mink was left on the shore with a mouthful of feathers and looking very foolish.

Sketching began as a ritual affair, performed not for the love of sketching but for its value as a discipline. He learned quickly that in the woods, "<u>the silent watcher sees the most</u>;" drawing from nature forced the eye to see and the spirit to accept.[4]

From 1870 to 1879 Seton lived in Toronto. His elementary education was interrupted by the family's repeated

---

[3]Seton, <u>Autobiography</u>, p. 53.

[4]Ernest Thompson Seton, <u>Two Little Savages: Being the Adventures of Two Boys Who Lived as Indians and What They Learned</u> (New York: Doubleday, Page, 1903), pp. 173-174.

quests for housing adequate to reflect its upwardly mobile social pretensions. His senior years, however, were spent at the Toronto Collegiate Institute, one of the finest secondary schools in the province. Its major competition in Toronto was Upper Canada College, a private boys' school which one local educator characterized as a "hot-house for fastening on our school system an alien and noxious growth of shoddy aristocracy." The Collegiate Institute, on the other hand, represented all that was pristine and true in Ontario society. No right-thinking person who had seen its "gentlemanly lads and refined lady-like girl pupils" could believe that Upper Canada College had "any right whatever to its tacitly urged claim of superiority."[5] Here Seton had a "wonderful sense of being at home among my own kind. Foul language was practically unknown . . . white collars common." Although he hated to admit it, "the 'white collars' seemed of a finer breed than the collarless." Seton tells us that he was an excellent student. He does not, however, shed much light on curriculum, choosing instead to itemize the beatings received at the hands of teachers and classmates alike. While these "taught me self-reliance and self-restraint" and "seemed to clear the air, let off the spleen, puncture the boil and drain the pus," they also contributed to a certain sense of isolation

[5]C. Pelham Mulvany, Toronto: Past and Present (Toronto: W. E. Caiger, 1884), pp. 82-86.

and persecution that followed him through his life.[6]

Seton's love of wild nature, fostered by life in Lindsay, was reinforced by solitary hikes through Toronto's Don Valley, later to become the setting for many of his animal stories. He did not like the city. Even when forced to live in one, or near one, he did everything he could to extricate himself, physically and psychically, from its clutches. When he first reached Toronto, "it seemed that I had left behind all the loved world of the wild things." But this was not really true, he soon discovered, for "the things you love are begotten inside you" and a "man of spiritual interest will surely find things spiritual in the most sordid of surroundings."[7] The beautiful ravine which bordered the Don River wound its way, unseen, from the hills to the north, through the middle of Toronto, and south to Lake Ontario.

> This place became the central thought in his life. He went there at all opportunities, but never dared to tell anyone of his discovery. . . . This was his little kingdom; the Wild Geese had brought him here, as the Seagulls had brought Columbus to a new world--where he could lead, for brief spells, the woodland life that was his ideal. . . . [This] was one place where the great trees should stand and grow as in the bright bygone; where the Coon, the Mink and the Partridge should live and flourish forever.

With this refuge close at hand, Seton could tolerate the

---

[6]Seton, Autobiography, pp. 70-72.

[7]Ibid., p. 62.

isolation, the persecution and the city.[8]

If Joseph Thompson found young Ernest's solitary existence unhealthy, he viewed with outright alarm his son's avowed mission to become a "prophet of outdoor life." Always careful to guarantee his own long-term security, Joseph had managed skillfully to control the family through a judicious manipulation of piety and guilt. However much the experience at Lindsay had proved the shortcomings of this method, the oldest boys had eventually been reunited under his command and were once more contributing materially to the maintenance of his authority. Aside from his heretical pantheistic inclinations, Ernest had begun quite early to cultivate his own independence from the prevailing tyranny. This, it became clear, must be checked if family discipline were to be maintained and another source of income tapped. Having noted the young boy's facility with pen and ink, Joseph determined that he should become a professional artist with proper training at a recognized academy. As far as Seton was concerned, "the prospect was far from unpleasant for I had some talent and was easily the head boy in the subject at each of my schools. . . ." When his son eagerly accepted the challenge of art school, Joseph believed that he had won another victory for God and the church. In fact, however, he had legitimized an avocation that would

---

[8]Seton, Two Little Savages, p. 15.

serve to unite the literary, scientific and pedagogical dimensions of the naturalist he had sought to discourage.[9]

II

In 1872 the Ontario Society of Artists had been founded in Toronto. One of the earliest manifestations of cultural nationalism in Canada, the OSA assumed responsibility for establishing standards of excellence against which the work of local painters and sculptors could be measured. The leading artists in the organization were by birth predominantly British, ensuring that the standards of the new world would bear considerable resemblance to those of the old. When the Royal Canadian Academy was established in 1880, it in turn was dominated by the OSA. Its constitution was based on that of the British Royal Academy of Arts and its honorary senior executive and chief patron was the Marquis of Lorne, Governor-General of Canada, who had urged the creation of the new body in the first place. Among the thirty members of the OSA to be elected, on Lorne's recommendation, to academic or associate status in the RCA were J. A. Fraser, Marmaduke Matthews, T. Mower Martin, Robert F. Gagen and Charlotte M. B. Schreiber. Lucius O'Brien, Vice-President of the OSA was elected President of the Academy. These influential figures were also Seton's

---

[9] Seton, Autobiography, p. 127.

earliest instructors.[10]

In 1876 the OSA had supplemented its ordinary function of sponsoring exhibitions and sitting in judgment on them by opening the Ontario School of Art in Toronto. Supported by a grant from the Ontario government, the Society operated the school under its own auspices until 1884, when it relinquished control to the province. From 1878 to 1879 Seton

---

[10] J. A. Fraser (1838-1898), of Scottish and Danish extraction, had come to Canada in 1856 after studying at the Royal Academy Schools under F. W. Topham and Richard Redgrave. He became a partner in the Montreal firm of the photographer, William Notman and took charge of the Toronto branch in 1868. He had assisted in the founding of the moribund Society of Canadian Artists in 1867 and in 1872 became the first Vice-President of the OSA. He left Canada for the United States in 1883. Marmaduke Matthews (1837-1913) was born in England where he studied under the water colorist, T. M. Richardson at the Oxford Cowley School. He came to Canada in 1860 and served as secretary of the RCA from 1880 to 1890. T. Mower Martin (1838-1934) had studied at the South Kensington School, London, coming to Canada in 1862. Robert F. Gagen (1847-1926) came to Canada from England in 1862. He worked with J. A. Fraser at the Notman Studios in Toronto before becoming a founding member of the OSA. Gagen's Ontario Art Chronicle, unpublished MS, n.d., Art Gallery of Ontario Library, Toronto, Ontario, sheds considerable light on the activities of the OSA of which he was secretary after 1889. See also, Art Gallery of Ontario, The Canadian Collection (Toronto: McGraw-Hill, 1970), pp. 125-127, 299, 301-302; William Colgate, Canadian Art: Its Origin and Development (Toronto: Ryerson, 1943), pp. 20-24; Newton MacTavish, The Fine Arts in Canada (Toronto: Macmillan, 1925), pp. 21-23, 167, 173; Joan Murray, "Victorian Canada: Part 1," Canadian Antiques Collector, 5 (Feb., 1970), 10-14; James Smith, "The Royal Canadian Academy," Canadian Magazine, 9 (1897), 302, 304; Seton, Autobiography, p. 130. Cf. Dennis Reid, A Concise History of Canadian Painting (Toronto: Oxford University Press, 1973), pp. 84-86.

attended classes in this bastion of the Victorian establishment. Although Matthews, Martin and Fraser offered help and encouragement, none of the teachers at the school took as much interest in Seton as Charlotte Schreiber. "A lady of distinguished family and high social position, she was not teaching for a living but merely because, having attained to artistic excellence and success herself, she found pleasure in offering her knowledge . . . to the younger generation about her."[11] She had studied in London under J. R. Herbert and had exhibited at the Royal Academy from 1854 to 1875, the year she immigrated to Canada. She was a close friend of Elizabeth Barrett Browning and had prepared drawings to illustrate her books. Elected to the OSA in 1876, she had immersed herself in the activities of the Toronto art community and had become one of its recognized leaders by the time Seton came under her influence.

Weymouth George Schreiber, Charlotte's husband, was sufficiently wealthy to maintain two homes, one in the city and one at Springfield-on-the-Credit, twenty miles west of Toronto. In each of these Charlotte kept spacious studios to which she could bring promising students for extra guidance. Strongly impressed by Seton's draughtsmanship and convinced that he was "destined to become a great etcher," she welcomed

---

[11]Ernest Thompson Seton, Unpublished Autobiography MS, n d., SP; Autobiography, p. 290; Muriel Miller Miner, G. A. Reid, Canadian Artist (Toronto: Ryerson, 1946), pp. 16, 19; Colgate, Canadian Art, p. 19.

him, first into her studios, and eventually into her household. When it became clear to her that his own family environment stifled his creativity, she encouraged him to spend weeks at a time at the country estate in Springfield.

> In her comfortable studio there my easel stood not far from hers. I came and went as one of the family. I was her nephew. She was always Aunty Schreiber. . . . I learned to consider her my second mother. She had much influence on me and all of that influence was for good--for simple sincerity in life and art.[12]

She was probably the first person ever to treat Seton with kindness and respect. On one hand she could be a confessor, offering sensitive counsel to frustrated adolescence; on the other she reflected--from her social origins to her gracious manner--all of the virtues by which Seton measured superiority. Springfield effervesced with happiness and warmth and afforded a welcome haven from parental tyranny. The wild landscape surrounding the estate was reminiscent of the Lindsay woods. In his spare time, Seton could wander there, collect specimens, observe and make sketches from nature. In other words, his reverence for Charlotte Schreiber was stimulated by a multitude of causes, many of them having nothing to do with her being his teacher. By

---

[12]Seton, Unpublished Autobiography MS, n.d., SP; Howard G. Schreiber, A Brief History of Charlotte Schreiber (1834-1922) and Her Husband, Weymouth George Schreiber (1828-1898), Toronto, Dec. 24, 1971, Unpublished MS in the author's possession, pp. 1-12; MacTavish, Fine Arts in Canada, p. 177; Joan Murray, Ontario Society of Artists: 100 Years (Toronto: Art Gallery of Ontario, 1972), p. 64; J. Russell Harper, Early Painters and Engravers in Canada (Toronto: University of Toronto Press, 1970), pp. 281-82.

filling an emotional vacuum and by fortifying his ego, she won his undivided loyalty. As a result, hers became the standard against which Seton willingly gauged the quality, not only of his art, but of his life.[13]

The history of Victorian art in Canada may be divided into three distinct periods: "an early phase from 1840 to 1875, a high phase from 1875 to about 1890 and a late phase in the 1890's which formed a transition to the modern era."[14] Seton's serious work emerged during the last two phases. However, because his early instructors were products of the first, it could be argued that he was representative of all three. In 1879 he was awarded the gold medal at the Ontario School of Art, largely through the intervention of Charlotte Schreiber, but partly because he had so far stayed within the rigorous, if highly unimaginative, parameters established by the OSA. Furthermore, like the guiding lights of the Society, Seton was English by birth. His family was moderately wealthy and harbored pretensions to nobility. He himself was extremely conscious of his social milieu and craved the

---

[13] Mrs. Phillip Ketchum private interview, Port Hope, Ontario, January 13, 1972. After the death of her husband in 1898, Charlotte Schreiber returned to England. Seton continued to visit her there until her death in 1922. As late as 1900 he was still sending her checks to cover the financial debts contracted during these early years. Seton to Charlotte Schreiber, December 19, 1900, Seton Letterbook No. 2, p. 168, SP.

[14] Joan Murray, "Victorian Canada: Part 2," Canadian Antiques Collector, 5 (March, 1970), 16.

adulation of recognized greatness. Burdened with these Victorian trappings it is remarkable that he was able to accomplish anything unique.

One variable functioned to disrupt the emerging pattern. Unlike most of his teachers, Seton actually knew something about the land he painted. From his experiences in Lindsay, and from his hikes through the Don Valley and at Springfield, he had developed a strong attachment, not to the pastoral or urban landscapes, but to the wilderness. While his art education prompted him to look for the sublime and the picturesque in nature, suggesting that a painter could soften the abrupt contours of the Laurentian Shield, or open lush clearings in a dense forest where none existed, his actual meeting with the land itself denied the legitimacy of such liberties. Of course he took them anyway--but only because admission to the Victorian artists' sanctuary demanded it. This kind of compromise was not particularly satisfying to Seton, whose intention from the beginning was to use his art as a vehicle for other purposes. "He never sketched because he loved sketching; he did not; the motive always was love of the thing he was drawing."[15] Therefore, in an effort to accommodate the dictates of the OSA and the RCA, as well as to give expression to the reality of nature as he perceived it, he turned his attention from the wilderness

---

[15] Seton, Two Little Savages, p. 176.

landscape to the animals inhabiting it. This had the effect of mitigating the compromise while still sanctioning pictures in the Victorian Grand Manner. It also forced Seton to pay close attention to the behavior of the animals he observed.

Born into a humorless family where religious piety replaced simple human affection, Seton was poorly placed to win, or even to sense, the love of others. As a result he spent his emotion on the land, the source of the few joys he did know. For him the wilderness was both a physical entity and a state of mind. And Charlotte Schreiber, who alone was able to penetrate and to comprehend the barriers that walled out human contact, encouraged and guided the expression of the tensions which resulted. On the surface, Seton appeared to be trapped in the garrison culture. But underneath his art he had already let the wilderness in, even to the wolves. This is not to suggest that he ever abandoned his zeal to reach the pinnacle of Victorian society. At the age of twenty-one he set out on paper "The Plan of My Life," an amazing document which predicts, among other fanciful imaginings, that

> In 1890 I shall marry an English woman, or of English extraction. . . .
>
> In 1905, I shall by God's help, have made a comfortable fortune by my pen and pencil, also in part by judicious speculation. I shall then return to England, buy a small estate in Devonshire, and a house in London.

> In 1915 I shall be knighted by the King in recognition of my work as an artist-naturalist.[16]

These aspirations, none of which ever materialized, reflected the collective consciousness of his family, of the OSA and of Toronto itself. If the style of his paintings was a tool for achieving respectability, their changing subject matter gradually exposed a soul contending with serious contradictions. Seton himself came to recognize the duality. In a metaphorical lament from Paris, written several years later, he reflected back on his Victorian training:

> I have spent all the afternoon in the Louvre, and as I went from one great master to another and saw all kinds of peculiarities and extremes--yet all resulting in great pictures--this is the lesson that impressed me more and more: That man who does immortal work develops himself. Here have I, living in Norway, been trying to grow a palm tree, because I saw that African palms were good. And each fresh frost cut down . . . my poor puny sprout. My wretched seedling had to contend with great, strong frost-defying pine that kept springing up. It has only recently dawned on me that I must grow my pine. It is the timber of my soul. What a tree I might have had now had I realized this ten years ago.[17]

But recognition of the duality and the power to resolve it were two quite different things.

The forcefulness of his earliest instructors dominated Seton's style of expression from 1878 to 1896. Many of his colleagues at the OSA had travelled and painted in the United

---

[16] Ernest Thompson Seton, The Plan of My Life, London, March 7, 1881, Unpublished MS, SP.

[17] Ernest Thompson Seton, quoted in Nancy Bell, "The Work of Ernest Thompson Seton," Humane Review, 4 (April, 1903), 12.

States, and most were familiar with the Hudson River School. For example, Homer Watson of Doon, Ontario, who first exhibited at the OSA in 1878, had spent the previous two years in New York. Here he met George Inness, viewed the work of the School and himself painted in the Adirondacks and along the Hudson River. That Seton had absorbed some of this enthusiasm for the descendants of Thomas Cole and Asher Durand is clear from the landscape background for his painting, The Goat Defending Her Kids from the Fox (1881). But the tendency of Canadian painters to look to the United States for subject matter was cut short by the intensified nationalist feeling stimulated by Confederation. The completion of the Canadian Pacific Railway in 1885 lured many of them across the country in search of idyllic settings that could match the Hudson River Valley. Their avowed purpose was to fulfill the promise of Confederation by employing art as a vehicle for "patriotic sentiments." They failed to realize that the mere painting of Canadian subjects had nothing whatever to do with the creation of a "Canadian School." Homer Watson, described by Oscar Wilde as the "Canadian Constable," was one of the few Ontario artists to isolate the problem. Although he had worked with members of the Hudson River School, and although the acceptability of his paintings was attributed by critics to his frequent trips to Britain and the Continent, Watson himself maintained that the sine qua non of "immortal work" was the "feeling its creator had of having roots in his

Plate 3. Goat Defending Her Kids from the Fox
(Oil, 1881)

native land and [of] being a product of its soil."[18] This meant that the style of the work, as well as its subject matter, would suggest the intimate communion of the artist with his land. But Watson had been born a Canadian and his views were not actively solicited by the OSA. Seton had an attachment to the Canadian land far more intense than Watson's. But he was an Englishman, trained by Englishmen, in a tradition which had no bearing on the Canadian experience. Furthermore, he was not a nationalist. Neither the wilderness nor its inhabitants required a specific geographic or national locus to work their inspiration on him. Wilderness was wilderness, loved for its own sake. When he could not inhabit it, it inhabited him. Therefore, while Seton's paintings were always appropriately archaic in style, they were to become increasingly primitivist in content. There was no need consciously to rebel. He was quite simply out of step with contemporary opinion.

If there were discernible British and American tinges to Canadian art, the high phase of Victorian style which accompanied Seton's entrance to the Ontario School was also

---

[18] Homer Watson, Undated Notes MS, Collection of Mrs. H. C. Van Every, quoted in J. Russell Harper, Painting in Canada: A History (Toronto: University of Toronto Press, 1966), p. 216; J. Russell Harper, Homer Watson, RCA, 1855-1936 (Ottawa: National Gallery of Canada, 1963); W. A. Sherwood, "A National Spirit in Art," Canadian Magazine, 3 (Oct., 1894), 500; Harriet Ford, "The Royal Canadian Academy of Arts," Canadian Magazine, 3 (May, 1894), 47; Seton, Autobiography, p. 145.

strongly influenced by the art of the Continent which many Canadians saw for the first time at the Philadelphia Centennial Exhibition in 1876. The relegation of the Canadian collection to a back room convinced several embarrassed artists that their only hope for achieving a measure of excellence lay in sound European instruction. The real accolades at the exhibition were reserved for the French collection--and the general North American consensus of opinion seemed to suggest that Paris would dictate standards of quality for years to come. Almost immediately aspiring Canadian painters began flocking to France. Among them were William Brymner of Ottawa, son of the Dominion Archivist, and Robert Harris of Charlottetown, Prince Edward Island, both of whom were to figure prominently in Seton's later career. What they brought back to Canada was the Grand Manner "which they nearly always applied to a narrative subject matter . . . placed in an earlier, or even purely imaginary, age," and "a descriptive, realistic, unpretentious style which often was accompanied by a fluent brushwork and rather severe lighting."[19] If this narrative genre had been popular among the visitors at Philadelphia, the exodus of North Americans to Paris ensured that it would quickly become an industry. The election of men like Brymner and

---

[19] Joan Murray, "Victorian Canada: Part 3," Canadian Antiques Collector, 5 (April, 1970), 16; "Victorian Canada: Part 2," pp. 18-19; Reid, Concise History of Canadian Painting, p. 91.

Harris to the Royal Canadian Academy effectively meant that "canvases of heart-rending partings, deeds of noble warriors, or any other appealing or romantic incidents" would be institutionalized for the balance of the century. Instead of accepting their own land and its multiplicity of distinctive characteristics, the "willing exiles" sold themselves, body and soul, to the superficial tastes of a pretending public.[20]

While the heirarchy of the Canadian art world was trying to wrestle some sense of direction from this confusing morass of British, American and French influences, home-grown, native-trained painters received relatively little attention. If Seton were to achieve recognition and stature within his own community, it was clear that he, too, would have to observe the ironic ritual of leaving it. Accordingly, while William Brymner was still in Paris, studying under William Adolphe Bouguereau at l'Académie Julian, and while Robert Harris, freshly returned from working in the atelier of Léon Bonnat, was assuming the Vice-Presidency of the OSA, young Ernest was in England, ostensibly as a student at the

---

[20] Harper, Painting in Canada, pp. 217, 227, 231; Colgate, Canadian Art, pp. 31-2. Joan Murray points out that "there were two separate reactions toward the new style: some artists adopted the style as it came from France, using a non-Canadian subject matter; others tried to adjust the style to their own land. . . . Many other Canadian artists also 'grafted something Canadian on to the foreign stock.'" The only truly Canadian art was created by those who stayed at home; see Murray, "Victorian Canada: Part 3," pp. 17-18.

Royal Academy. Charlotte Schreiber, who was skeptical of the new enthusiasm for French genre, suggested this course of action in the sincere belief that her young protégé could reaffirm his British roots and win the benefits of superior instruction at the same time. Where Seton was concerned, emotional considerations would always have priority over the actual work to be done. In fact, it is not at all certain that he spent very much time at the Royal Academy School. Excited by his recent success, the OSA gold medallist left Toronto on June 12, 1879, arriving in England two weeks later. He did not begin to prepare his competition entry piece, the Hermes, until September 1, having apparently visited his South Shields relatives during the summer. When the Hermes was rejected, he began a second piece, the Satyr, on March 5, 1880. This was not formally accepted until December 25. On January 18, 1881, he spent his first day as a student at the school. He did not pass the probation examinations until April 8 of that year and was on his way back to Canada by the following October. Therefore, it is stretching a point to say that he "studied" at the Royal Academy. At the very best he attended classes for less than a year. Even then he spent relatively little time in the studio, much in travelling and visiting. Nevertheless, Seton was still able to make a strong impression on his instructors. Within weeks of his formal acceptance he was one of six students, chosen from hundreds of applicants, to be awarded a coveted seven year RA scholarship.[21]

All told Seton passed twenty-nine months in England. His father, who all the while assumed that his own plans were reaching fruition, sent him four hundred dollars to help cover expenses. For the rest, however, he was forced to call upon his own resources. He found his most lucrative employment with the publishers, Cassell, Petter and Galpin, occasionally illustrating books. In deference to his mother, whom he always perceived as a martyr, Seton attended church regularly. He was reassured, however, to discover that the church in London, like the church in Toronto, "never had any message for me; I found it more profitable to take a long walk in the country." Indeed, Seton the pantheist was more inspired than Seton the artist. One of the privileges attached to the RA scholarship was free admission to the London Zoo, where he "went for weeks, applying to animal drawing the principles learned in the Academy school," and in a rigorous self-taught program of animal dissection, pursued in his own cramped lodgings. But his most compelling distraction was a daily visit to the library of the British Museum.

> Here I saw Audubon and Wilson, made the acquaintance of Richardson, Nuttall, Coues, Baird, Brewer, Ridgway, John Burroughs and Henry Thoreau. I had thought all

---

[21]Seton, Journal, AMNH, 1: 6-11; Autobiography, pp. 136, 144-45.

> naturalists were dead. . . . Now I learned that America was singularly fortunate in its army of younger men, doing new and better work in the glorious world that was the chief of worlds for me. . . .
>
> My days were devoted to my drawing. . . . Art was still to be my calling in life. But . . . [my] heart was with the birds.[22]

Not once does Seton give us an inkling of the names of his instructors at the Royal Academy; nowhere do we learn the things he was taught. The joys of the brief sojourn seemed, with few exceptions, to emerge from pastimes which had no obvious bearing on the education of an artist.

One thing was certain, however: Seton had no fondness for London. It was a wasteland of "shivering cold, gloom and starvation; all the things that I most loved, except books, were absent." His own unique demands upon himself probably did not help. Poverty, combined with a dabbling in Zen had converted him into a vegetarian. Monk-like, he gloried in sexual abstinence, achieving "domination over my thoughts" by "crucifying the flesh." He longed to return to Canada to gratify his "craving for life with the Wild Things." This meant forsaking not only the valuable scholarship, but also the opportunity to work under the guidance of some of Britain's greatest painters. It meant disappointing Charlotte Schreiber and it was guaranteed to provoke his father. Regardless of the negative considerations, Seton

---

[22]Seton, Autobiography, pp. 136-140.

could not control his urge to return.[23]

He arrived in Toronto in November, 1881, and was promptly presented by Joseph with an itemized bill for $537.50 to cover the "disbursements that had been made for me since my birth." Seton made great drama of this event, using it later on numerous occasions to justify the intense hatred he felt for his father. It must be remembered, however, that it was under pressure from Joseph that Seton had first embarked upon his education in art. After subsidizing the London trip to the tune of four hundred dollars, Joseph might logically have expected his son to spend less time roaming the heath or disembowelling animal cadavers and more behind his easel at the Royal Academy. Having enjoyed what could only have been described charitably, in Joseph's terms, as a vacation, Seton was probably asked, not unreasonably, to pay up. In any case, Seton's recollection of the moment demonstrates the consuming power of his own persecution complex. Aside from reinforcing his love for the wilderness--this time by introducing the real possibility that art could be combined with science to legitimize forever a life within it--the London trip was a disaster.[24]

---

[23]Ibid., pp. 146-151. Seton seems always to have been a bit of a physical and mental flagellant, making the most of childhood maltreatment and afflictions; see, e.g., ibid., pp. 140-142; Seton, Journal, AMNH, 1:59.

[24]Seton, Autobiography, p. 152.

III

The year 1881, then, marks the end of Seton's first exposure to a formal training in art. Following the guidelines laid down by Charlotte Schreiber, he had had the best opportunities available to a young man in his position. On the surface, he appeared to have wasted all of them. But it is doubtful, given his feelings for the land, that he could have lasted much longer in London. Even sketching visits to the zoo could not compensate for natural habitats like the Don Valley or Springfield. The one truly redeeming feature of the London trip had been his trips to the library at the British Museum where he had found his own intuitions about animals converted into the prose of the great naturalists. This aspect of the experience led him, shortly after his return, to join the Natural History Society of Toronto, founded in 1880 by Dr. William Brodie for the purpose of bringing together amateur botanists and zoologists interested in discussing the local flora and fauna. The Society met at the Canadian Institute, with which it later amalgamated, and gradually became a significant force for encouraging research in Canadian science. Brodie himself took Seton under his wing, in much the same way Charlotte Schreiber had done at the OSA. He introduced him to his son, Willie Jr., and together the three men spent many hours roaming the swamps and bogs around Ashbridge's Bay in search of new specimens and interesting field notes. It was Dr. Brodie who encouraged Seton to keep a journal of his findings and who first familiarized him with the technicalities of measurement and mounting.[25]

This valued introduction to the practical side of natural history, coupled with the increasingly uncomfortable situation at home, convinced Seton that the time had come to move on once more. He fully appreciated his capabilities as an artist, but he was not prepared to be coerced by his father into using them solely as a vehicle for family support. For his part, Joseph would not entertain the thought of Seton pursuing his interests in natural history while contributing nothing to his own, let alone to the family's, maintenance. For the father, art was an occupation. For the son, it was quickly becoming a supplementary diversion.

Assuming himself still to be in absolute control of the youngest half of his family, and apparently having learned nothing from his experience in Lindsay, Joseph was again itching to try his hand at farming. This time he focussed his dream on Manitoba, sending one of his sons, Arthur, as an advance party to scout the agricultural possibilities of

---

[25]Ibid., pp. 83, 94, 155, 165, 179, 195; Seton, Journal, AMNH, 1:153; Seton to James L. Baillie Jr., Dec. 2, 1930; J. R. Dymond to Seton, March 30, 1927; Seton to J. R. Dymond, April 14, 1927, SP; J. R. Dymond, "Zoology," in The Royal Canadian Institute Centennial Volume, 1849-1949, ed. by W. Stewart Wallace (Toronto: Royal Canadian Institute, 1949), p. 113; J. M. Barnett, "Ashbridge's Bay," Ontario Naturalist, 11 (Dec., 1971), 24-26.

the new Northwest. In 1872 the federal government had won passage of the Dominion Lands Act, the Canadian counterpart of the American Homestead Act. In an effort to divert immigrants from the United States, the Canadian legislation eased regulations, allowing prospective settlers to fulfill residence requirements in three (as opposed to five) years from the date of entry. Males over the age of eighteen were entitled to claim free a 160 acre quarter section. Funds permitting, they might also purchase contiguous quarter sections at market value. Needless to say, the price of land in the new province was considerably lower than in Ontario, a fact which Joseph's calculating mind was quick to grasp. If he and his five youngest sons each patented free homesteads, they could establish a secure foothold which, in time, might be expanded into a family empire. Arthur had actually claimed a quarter section near Carberry in 1879. Having been in residence on the land for three years, he was theoretically eligible for title to it. However, he had made few improvements during his stay and was still living in the original claim shanty.

Invited by Arthur to join in the construction of a frame house on the property, Seton suddenly found himself not only with a legitimate excuse to leave home, but with a wilderness destination to feed his growing appetite for natural history. Armed only with Charlotte Schreiber's etching apparatus, his journal, sixty chickens and twenty dollars in cash, he arrived at the Manitoba farm in April, 1882, prepared to resume the life he had left in Lindsay years earlier. With him was William Brodie, Jr., his new and closest friend.[26]

Within weeks of their arrival young Brodie drowned, leaving Seton beside himself with grief. To make matters worse, Joseph soon showed up with the rest of the family to survey the land and to determine whether the house and the environment would sustain his regal tastes. Hungry for isolation, Seton welcomed the opportunity, provided by his father's own wish to locate several homesteads, to go off on a series of "land hunts" with his brothers. These travels took him over the entire western half of the province, affording unique chances to sketch and to jot down field observations in his journal. On the great plains of the Souris he marvelled at the multiplicity of migrating birds set off against the broad expanse of sky.

> There were thousands and tens of thousands of lap longspurs, flock after flock stretching into dim distance, all headed northward to their far breeding

---

[26] Seton, Journal, AMNH, 1:37; Autobiography, pp. 155-167; Chester Martin, "Dominion Lands" Policy, ed. by Lewis H. Thomas (Toronto: McClelland and Stewart, 1973), pp. 141, 161-62. Arthur Thompson, two years Seton's senior, had left for Manitoba in 1876 after serving a carpenter's apprenticeship with older brothers William and Joseph, Jr., Toronto building contractors. He claimed the Carberry homestead in 1879, remaining there permanently until 1883 when he enrolled at the University of Manitoba Medical School. Before graduating with honors he served as an apothecary with the Winnipeg Field Hospital during the Riel Rebellion of 1885; Jean Thompson, private interview, Nov. 30, 1971.

> place. When one sees the immense havoc that gunners and netmen make among the winter hordes of longspurs . . . in the south, one wonders how their number can stand the drain; but when one notes the myriads in the north, one wonders how any difference can be made.
>
> There were small flocks of brown cranes, and occasionally pairs of white cranes--glorious creatures--trumpeting in the sky, or sometimes seen on a prairie ridge where three or four of them would do a wonderful minuet, flapping their wings, bowing and chassezing in circles around each other, to the accompaniment of a loud bugling.[27]

Certain that the immensity of the surrounding wilderness exacted its toll of humility from every man, Seton was nonetheless surprised by his own gradual mellowing. Each day on the trail brought some long-to-be-remembered human encounter. People were oddly helpful to one another, giving, yet asking nothing in return. While there was "poverty writ on every wagon and team" he passed, "dirt in the sordid cabins built of sods with tarpaper roofing," while there was "every element that in the East would have made a horrid slum, a bed of wretched misery," here "every eye was ablaze with hope."[28]

---

[27] Seton, Autobiography, pp. 168-74, 189, 286; Journals, AMNH, 1:59-87; Julia M. Seton, By a Thousand Fires; Nature Notes and Extracts from the Life and Unpublished Journals of Ernest Thompson Seton (Garden City, N.Y.: Doubleday, 1967), pp. 85-89. For an interesting account of what Manitoba looked like to Seton, see Robert Miller Christy, Manitoba Described (London: Wyman and Sons, 1885). Seton had met Christy in England and this book was the product of a return visit. See also, Canada, Dept. of Agriculture, What Farmers Say of Their Personal Experience in the Canadian North-West (Ottawa: Queen's Printer, 1882), pp. 12-14, 83-87; T. R. Weir, "Pioneer Settlement of Southwest Manitoba, 1879-1901," Canadian Geographer, 8 (1964), 64-71.

[28] Seton, Autobiography, p. 174.

Although the land hunts warmed his own heart, soothed his own soul, Seton could not locate a single homestead that would satisfy his brothers. As the weather grew progressively colder towards Christmas, his mother and father abandoned all thoughts of trying to inhabit this barbaric wasteland where wind and blizzards dictated limitation to the routine affairs of men. One by one the brothers followed their parents back to Toronto, leaving Seton alone with Arthur. Once Joseph had left, Seton had every reason to stay. As soon as the weather cleared he recommenced the land hunts, this time moving into the far northern section of the province, towards the border with present-day Saskatchewan. By the fall of 1883 he had found the piece of land he wanted on a beautiful plain near Duck Mountain, a short trip by wagon from the Cotai Indian Reserve. "The soil was rich, mostly open prairie, but enough timber at hand for immediate use, and plenty in sight on the mountain five miles away." Like the majority of settlers who came to Manitoba from Ontario, Seton fell for all the practical and scenic amenities to which he had grown accustomed in his former province. There was a lake on the property and a brook coursing white through a small valley; the distant hill broke the monotony of the prairie and guaranteed a diversity of animal life. While the agricultural possibility was close at hand, his real comfort came from the mystery of the woods beyond, the source of the brook and the lake. It was Lindsay again, transported West in space and back in time.[29]

Although there was a "sense of finality" about this

---

[29] Ibid., pp. 206, 233. Neither Seton nor his parents would fully trust the recent speculations of the botanist, John Macoun (1831-1920), whose recent book, Manitoba and the Great North-West (Guelph, Ont.: World Publishing Co., 1882), argued that earlier explorers like Palliser and Hind had erred in identifying the Canadian prairies as an extension of the Great American Desert. Beginning as a surveyor with the CPR in 1872, Macoun had been appointed naturalist to the Dominion Geological Survey in 1882. In both capacities he conducted experiments to determine the agricultural potential of the treeless lands, finding them to be more fertile than the rolling, wooded country in the eastern and northern reaches of Manitoba. His discoveries had been instrumental in dictating a southern rather than a northern route for the CPR. Seton, whose idea of wilderness was as yet incomplete without trees, had chosen a homestead well to the north. Nevertheless, he valued Macoun's advice and direction in all matters of natural history. After 1884, when they began to correspond on a regular basis, the two men became close friends, trading information and ideas. Seton quickly learned, through their "common interest in the wild things," that prairies, like forests, yielded a diverse bounty of animal life. The appearance of the landscape, and the contents of the landscape were henceforth segregated in his mind as mutually supporting tools for analytical investigation. See, Ernest Thompson Seton, Introduction to The Autobiography of John Macoun, M.A. by John Macoun (Ottawa: Ottawa Fild Naturalists' Club, 1922), p. x; John Macoun to Seton, Oct. 29, 1888, Macoun Letterbook No. 1, p. 488; John Macoun to Seton, Jan. 31, 1890, Feb. 6, 1890, Feb. 14, 1890, Feb. 21, 1890, March 18, 1890, Macoun Letterbook No. 3, pp. 110, 120, 145, 170, 231-32; John Macoun to Seton, April 21, 1890, Macoun Letterbook No. 4, p. 76; James Macoun to Seton, April 29, 1890, May 5, 1890, ibid., pp. 89-90, 103; John Macoun to Seton, July 4, 1892, Aug. 26, 1892; Sept. 2, 1892, Nov. 8, 1892, Dec. 1, 1892; Macoun Letterbook No. 5, pp. 10, 35-36, 45, 126-7, 146; John Macoun to Seton, April 18, 1894, Macoun Letterbook No. 7, p. 249, all in John Macoun and James Macoun, Papers, Public Archives of Canada, Ottawa, Ontario (hereafter cited as MP PAC); James Macoun to Seton, Sept. 29, 1908, Nov. 11, 1909, Nov. 30, 1909, Dec. 15, 1909, Jan. 13, 1910; John Macoun to Seton, Oct. 29, 1909, Oct. 3, 1910; Seton to James Macoun, Dec. 13, 1909, all in SP. See also John Warkentin, ed., The Western Interior of Canada (Toronto: McClelland and Stewart, 1964), pp. 245-252.

place that led Seton to register his claim at the Birtle land office, doubts lingered in the back of his mind. When he returned to the Carberry farmhouse Charlotte Schreiber's etching press stared back at him across his bedroom, reminding him that he had disappointed his teacher. Several prints and water colors in the pages of the journal attest to the fact that he was trying to maintain at least peripheral contact with the world of art. This time, however, the work reflected the experience of the artist rather than the instructions of his mentors. The subject matter alone indicates that the influence of the Hudson River School and the ritual classicism of the Ontario School of Art and Royal Academy had, to all intents and purposes, been pushed to one side. Human figures are absent. The pages are filled with an overwhelming sense of space--landscapes disappearing in infinity, the fragile frame house and outbuildings portrayed like rudderless boats in a sea of sky and flatland. Birds and mammals appear with increasing frequency, either as illustrations for the accompanying field notes or as individual portraits, unique in themselves. These reflect Seton's growing attachment to the wild things about him and suggest the extent to which he had immersed his life in theirs. But his instincts rebelled against the thought of becoming a pioneer, the only obvious occupation open to him should he choose to remain in Manitoba. The kindness of neighboring farmers, while serving to give him greater faith in his fellow man, was insufficient to offset the compelling urge to be alone. Even when holding the stilts of a plough, trudging along the freshly turned furrows, his eye was following the myriad activities of the birds. His ear caught their chatter and he rejoiced "to realize that a thousand bird voices which, three years ago, were merely tantalizing mysteries, were now the happy understandable expressions of friends whose names and lives I knew as no one else could know them."[30]

In purely economic terms there was no future for a wholehearted naturalist who was also a half-hearted farmer. He either had to overcome his distaste for drudgery or else find other means of supporting his primary interest. Farming, of course, had other disadvantages. In particular it functioned to drive away the very animals whose habits Seton sought to investigate. The irony was that in order to win title to the Duck Mountain homestead he would have to "civilize" its landscape, render the habitat of animals a government-ordained habitat of man. These conflicting purposes re-awakened his enthusiasm for art--not for art's sake, but for the practical reason that his abilities as a draughtsman had already demonstrated their remunerative power. As a student at the Ontario School of Art he had earned his living by coloring photographs and designing Christmas cards. While in London he had found that he could free-lance

---

[30]Seton, Autobiography, pp. 220, 233, 238, 254; Journal, AMNH, 1:153-208.

as an illustrator for publishing houses, working only when he needed the money. He now determined to go to New York where he assumed he would find the largest market and the highest prices for his talents. If the wilderness beckoned, Manitoba would be a train ride away. It was clear, however, that art would merely be a substitute for farming. Seton's ultimate objective was to become a naturalist.[31]

IV

Seton reached New York in late November, 1883, and was pleased to find there several artistic acquaintances from Toronto pursuing their own elusive dreams. The lack of opportunity in Canada forced many aspiring artists and writers to seek advancement in the United States; "indeed it was only by establishing their reputations abroad that they could hope for wide acceptance in Canada."[32] Seton shared an apartment with Charles Broughton from Peel County, Ontario, a promising young water-colorist. Among former students from the Ontario School of Art, he met and mixed with John T. Willing, Frederick Jopling and William Bengough, all of whom found New York a welcome release from Victorian

---

[31]Seton, Autobiography, pp. 128-29; 153, 244.

[32]H. Pearson Gundy, "The Development of Trade Book Publishing in Canada," in Ontario, Royal Commission on Book Publishing, Background Papers (Toronto: Queen's Printer, 1972), p. 20; MacTavish, Fine Arts in Canada, p. 30.

Toronto, but who nonetheless enjoyed the security and comradeship of their countrymen.[33] Having been accepted into the bosom of the Canadian art colony, Seton canvassed likely prospects for work. To W. Lewis Fraser, art director of Century Magazine, he sold a drawing of a scarlet tanager from his Manitoba portfolio. The price of ten dollars paid the rent; but most important, acceptance of the illustration meant that he had established a business relationship on which he could depend in times of need. Later, as it turned out, the Century Company would become one of his major publishers. Within a short time he was hired by the lithography firm of Sacket, Wilhelm and Betzig as a commercial artist, thereby ensuring a steady income. In the evenings

[33] Charles Broughton (1861-1912) had arrived in New York in 1880 and later became a prominent magazine illustrator. He was appointed to the staff of the New York Art Students' League in 1898. John T. Willing (1860-1898), also a water colorist, was elected an associate of the RCA in 1884 after he had moved to New York. An illustrator and ornamental designer he also wrote two volumes of criticism on English portraiture, thereby winning membership in the American Authors' Guild. Frederick W. Jopling (1860-1945) had arrived in New York in 1880. He studied at the Art Students' League under Walter Shirlaw, William Sartain and William Chase, afterwards becoming a commercial artist. He was on the staff of the New York Herald and sold illustrations to Harper's and Collier's until he returned to Toronto in 1912. William Bengough won distinction primarily from being the brother of John Wilson Bengough (1851-1923), the founder of the Toronto political lampoon, Grip. Seton, Autobiography, pp. 240, 246; Harper, Early Painters and Engravers, pp. 46, 334; H. J. Morgan, Men and Women of the Time: A Handbook of Canadian Biography (Toronto: William Briggs, 1898), pp. 118, 1086; Colin S. Macdonald, comp., A Dictionary of Canadian Artists (3 vols.; Ottawa: Canadian Paperbacks, 1971), 1:38-39, 3:581-582.

he attended classes at the Art Students' League "to continue my equipment as an illustrator." Here he met and befriended Daniel Carter Beard, later to figure prominently in the Boy Scout movement, who was himself "a clever draughtsman." Under the direction of William Sartain, Seton enjoyed his "first experience at a life class with models," a method of teaching known to, but highly disapproved of by the OSA. Occasionally he attended the theatre, but he was careful to save as much as possible for his return to Manitoba.

Despite the novelty and exhilaration of these activities, Seton found it almost impossible to adjust. "New York life goes hard against the grain with me," he wrote. "In fact it is not life at all in my estimation. . . ." During the winter months he had "little chance for bird study, being at the office six days a week; and on Sundays all museums were closed." Aside from having learned that "I could go forth alone into the world of commerce and make a steady living with my brush and pencil," he was not at all sure that he had gained anything worthwhile from the experience.[34]

In an effort to clear his head, Seton went back to Carberry to spend the spring and summer with Arthur. When the planting was finished, they went north to the Duck Mountain homestead to build a claim shanty. But this

---

[34]Seton, Autobiography, pp. 246-248; Seton to H. H. Hatten, n.d., SP.

vacillation between the world of the pioneer and the world of the artist merely proved more unsettling. Clearly he had to choose one or the other. If he had enjoyed the luxury of a university training in science, the decision might have come easily. In government service--say with the Geological Survey--he would have been qualified to carry out field investigations for a salary. To abandon farming altogether meant losing both the Duck Mountain homestead and physical proximity to the wilderness. If art were to be a substitute for farming, as he had hoped when he departed for New York the first time, then the wilderness would have to enter his life through some other avenue. It was obviously impossible to run back and forth between the city and the country while trying to hold down a full-time job. Nor could advertisers and publishing houses have been expected to ship their projects to Manitoba when the work could be done in half the time and at less expense by equally competent New York artists. In fact, the problem had really been resolved long before Seton discovered northern Manitoba, indeed even before he went to London. As a youth he had been afflicted with a hydrocele which had failed to respond to repeated treatments and which would have made a solitary farmer's existence precarious at best. He had tried diligently to ignore it, but the black iron truss he was forced to wear around his groin made him feel "as though I had a loathsome reptile on my person, forever boring my flesh." In the end it was his own health that dictated a solution in favor of art.[35]

It was clear that there was plenty of demand for good artists in New York. If he did only free-lance work he could set his own hours and maintain the independence necessary to read at the library or to go off on extended field trips. His intention remained constant. Art would be the vehicle to subsidize research. By adopting this course of action, Seton was forced to sacrifice the beautiful Duck Mountain homestead. He deeply feared that he might also have to sacrifice Manitoba itself. But that was a chance he would have to take. By now the wilderness was firmly imbedded in his soul. At the very least he could call upon his art to give it expression.

Other important factors, many of them tied to his experience in Manitoba, influenced Seton's decision to return to New York and to art. While living in London he had read Darwin's Origin of the Species and The Descent of Man for the first time. Although the work of nineteenth-century naturalists from Audubon to Burroughs had also captured his imagination in the British Museum, his familiarity with the theory of evolution ensured that their work would be subject to his close and well-informed scrutiny. Among the systematic ornithologists whose books best satisfied his high expectations

---

[35]Seton, Autobiography, pp. 140-42, 211, 213, 250-58; Journal, AMNH, 1:59.

was Elliott Coues. He regarded Coues' famous Key to North American Birds (1872) as "the first successful effort to take . . . knowledge from the museum and . . . to place it within reach of all the world of those who loved to hold our birds, not as skins, but as living friends." As one of the founders of the fledgling American Ornithologists' Union, and as editor of its official mouthpiece, The Auk, Coues was the acknowledged technical authority on North American avifauna. Seton had purchased a copy of the Key on his return from England, using it in Manitoba as his major guide to the identification and classification of specimens. In 1883 he mustered the courage to write to Coues, offering his "services as a correspondent in the line of natural history," and hoping that his specialized knowledge of Manitoba might prove to be a unique and valuable asset.[36] Coues forwarded the letter to his colleague and fellow ornithologist, Spencer F. Baird, Secretary of the Smithsonian Institution in Washington, who accepted Seton's

[36] Seton, Autobiography, pp. 140, 166, 222, 226; Seton to George Blake, May 4, 1884, SP. Elliott Coues (1842-1899) was one of a long list of medical doctors in the U.S. who turned his attention from humans to animals. The "key" method of analysis had been current in botany for several years, but Coues was among the first to adapt it to zoology. From 1873 to 1876, he worked on the U. S. Northern Boundary Commission, and from 1876 to 1880 he was attached to the Geological and Geographical Survey of the Territories whose publications he edited. From 1877 to 1887 he taught anatomy at Columbia College in Washington, the same institution from which he had graduated. See, D. G. Elliott, "In Memoriam: Elliott Coues," Auk, 18 (January, 1901), 1-11; J. A. Allen, "Biographical Memoir of Elliott Coues, 1842-1899," National Academy of Sciences, Biographical Memoirs, 6 (1909), 397-446.

offer with great enthusiasm. The specimen collections at the Smithsonian were quite incomplete for Canada and Baird was delighted to find someone who could help to remedy the situation. He went out of his way to encourage the aspiring young naturalist, outlining the proper methods of mounting skins and eggs and promising to repay his efforts "by the transmission of books on the natural history and ethnology of America."[37] Baird, in turn, introduced Seton to C. Hart Merriam, secretary-treasurer of the AOU, who formally accepted Seton's original offer to act as a correspondent and who encouraged him to join the Union. When it emerged that Seton had a gift for animal drawing, Merriam requested a private preview. Seton obliged during his first stay in New York, visiting the naturalist at his home in the Adirondacks. Marvelling at the young artist's craftsmanship, Merriam "gave me first place among animal illustrators in America; and then and there he ordered fifty . . . drawings as fast as I could get the animals to work from. . . ." Although a medical doctor by training, Merriam was already in the process of

---

[37]Spencer F. Baird to Seton, Feb. 15, 1883, March 2, 1883, SP. Baird (1823-1887) had studied medicine before becoming professor of natural history at Dickinson College. Appointed Assistant Secretary of the Smithsonian in 1850, Secretary in 1878, he contributed his vast collection of mammal and bird specimens to help launch the U. S. National Museum. His _Mammals of North America_ (1859) and _History of North American Birds_ (1875-1884) were the classic studies of the period. See Robert Ridgway, "Spencer Fullerton Baird," _Auk_, 5 (Jan., 1888), 1-14; W. H. Dall, _Spencer Fullerton Baird, a Biography_ (Philadelphia: Lippincott, 1915).

trading his profession for a career as first chief of the U. S. Biological Survey. He had seen considerable government service by the time Seton met him, and it was clear that he would be a useful contact. As it turned out, in fact, this meeting "was the beginning of a friendship that lasted through life."[38]

In other words, the chain reaction set off by his first innocent letter to Elliott Coues had brought Seton into direct contact with three of America's greatest living naturalists. Most important, however, was the fact that they were as interested in him as he was in them. After receiving the commissioned animal drawings, Merriam was ecstatic. "They are so beautiful," he wrote, "that without the specimens at hand for comparison, I am unable to offer a single suggestion in the way of criticism." Merriam then insisted that Seton meet personally with Baird and Coues to see whether other scientific outlets might be found for his creative talents. It was in response to this suggestion that Seton returned to New York in the fall of 1885 to attend the annual

---

[38]Seton, Autobiography, pp. 247-250. C. Hart Merriam (1855-1942) had accompanied the Hayden geological survey as naturalist and had served as surgeon aboard the U. S. Fish Commission steamer, Proteus. Henceforth Merriam was the dominant scientific influence on Seton's career, as Chapter 3 will make clear. See also, T. S. Palmer, "In Memoriam: Clinton Hart Merriam," Auk, 71 (April, 1954), 130-36; Hilda W. Grinnell, "Bibliography of Clinton Hart Merriam," Journal of Mammalogy, 24 (Nov., 1943), 436-57; Keir Brooks Sterling, Last of the Naturalists: The Career of C. Hart Merriam (New York: Arno, 1974).

Plate 4. Bobwhite Quail (Watercolor, 1898)

meeting of the AOU at the American Museum of Natural History. There he became acquainted not only with Baird and Coues, but with men like Robert Ridgway and William Brewster who were especially dedicated to the purpose of expanding the scientific horizons of ornithology. On this occasion he was also introduced to J. A. Allen, curator of the department of birds and mammals at the American Museum, who, with Coues and Brewster, had been instrumental in organizing the AOU. Although Baird was prepared to hire him as a staff artist at the Smithsonian, Seton had no desire to be trapped by any full-time occupation. Instead he opted for another proposition initiated by Elliott Coues. W. Lewis Fraser of the Century Company, whom Seton knew from his previous visit to New York, asked Coues to edit the biological material to be included in the forthcoming, multivolume Century Dictionary. Coues agreed to do the work if Seton could be enlisted to prepare the scientific drawings accompanying the text--a condition to which Fraser willingly acceded. Almost immediately after his return to New York, therefore, Seton found himself with a contract for one thousand drawings.[39]

---

[39]Seton, Autobiography, pp. 280-81; C. Hart Merriam to Seton, June 12, 1884, SP. Joel Asaph Allen (1838-1921) had been a student of Louis Agassiz at Harvard. From 1867 to 1884 he was curator of birds in the Harvard Museum of Comparative Zoology, moving to New York in 1885 to assume his responsibilities at the American Museum. He was President of the AOU from 1883 to 1889 and succeeded Coues as editor of The Auk. William Brewster (1851-1919) had

From 1885 to 1890 the excitement generated by these events had important repercussions. Always eager for recognition and acceptance, Seton had become a main attraction overnight. The opportunities furnished by the remarkable men into whose company his art had gained him admission made it possible to abandon all thoughts of homesteading. In an effort to consolidate his position within the AOU, he took his duties as a correspondent seriously, submitting a flood of notes and field observations for publication in The Auk.[40] While most of these were drawn from his Manitoba journal, all of them were based upon an exacting methodology reflecting the influence of Coues. Beginning with the premise

---

no formal university education but assumed direction of the Museum of Comparative Zoology upon Allen's departure. He was deeply revered by his colleagues in the AOU for his encyclopedic mind and prolific output of important ornithological papers. See also, Frank M. Chapman, "In Memoriam: Joel Asaph Allen," Auk, 39 (Jan., 1922), 1-14; H. W. Henshaw, "In Memoriam: William Brewster," Auk, 37 (Jan. 1920), 1-23.

[40] Ernest Thompson Seton, "Nest and Habits of the Connecticut Warbler (Oporornis agilis)," Auk, 1 (April, 1884), 192-193; "Manitoban Notes," Auk, 2 (Jan., 1885), 21-24; "The Swallow-tailed Flycatcher in Manitoba and at York Factory," Auk, 2 (April, 1885), 218; "Nest and Eggs of the Philadelphia Vireo," Auk, 2 (July, 1885), 305-06; "The Western Grebe in Manitoba," Auk, 2 (July, 1885), 314; "Interesting Records from Toronto, Canada," Auk, 2 (Oct., 1885), 334-37; "The Birds of Western Manitoba," Auk, 3 (April, 1886), 145-56; "The Birds of Western Manitoba: Conclusion," Auk, 3 (July, 1886), 320-29; "The Birds of Western Manitoba: Addenda," Auk, 3 (Oct., 1886), 453; "Occurrence of the Evening Grosbeak (Coccothraustes vespertina) at Toronto, Canada," Auk, 4 (July, 1887), 256-7; "Evening and Pine Grosbeaks in Ontario," Auk, 7 (April, 1890), 211.

that "Science is measured survival," he attempted to discern every finite behavioral and physical detail demanding appraisal in that context. Any student of ornithology, he maintained, must

> note the exact time at which a bird arrives, the direction whence it came, the weather at the time, the note it utters, the company it comes in, the exact food it takes, . . . the time of singing and of pairing, the time occupied in building the nest, . . . the composition and position of the nest, the date of the laying of each egg . . ., the growth of the young, the number of times they are fed daily and the kind of food, how often the parents breed, . . . their sleeping places, and finally, their time of departure.[41]

However, as his confidence grew with each published article in The Auk, Seton began to fancy himself an expert ornithologist. His rapidly inflating ego was pumped up by Willie Brodie's father, who frequently called upon him to submit reports to the Canadian Institute in Toronto, and by the government of Manitoba, which published some of his most immature scientific speculations.[42] Continual contact

---

[41] Ernest Thompson Seton, To the Student, MS Note, n.d., in Seton Letterbook No. 2, SP, p.28.

[42] Ernest Thompson Seton, "The Prairie Chicken or Sharptailed Grouse, Pedioecetes phasianellus (Baird)," Proceedings of the Canadian Institute, 1 (1879-1883), 405-12; "Outlines of Ornithology," Proc. Can. Inst., 3 (June, 1886), 180; "Strawberry Finch," Transactions of the Canadian Institute, 1 (1889-90), 41-42; "Sturnella magna Wintering near Toronto," Trans. Can. Inst., 1 (1889-90), 42; "Disappearance of Forest Birds," Trans. Can. Inst., 1 (1889-90), 47; "Linota cannabrua at Toronto," Trans. Can. Inst., 1 (1889-90), 54-55; "Rare Birds in Toronto University Museum," Trans. Can. Inst., 1 (1889-90), 55-56; "Spring Notes," Trans. Can. Inst., 1 (1889-90), 41-42; "First Sialia sialis at Lorne Park," Trans. Can. Inst., 1 (1889-90), 56; "Report of the Occurrence of the Evening Grosbeak (Cocco-

with Baird, Coues, Allen and Merriam went to his head, grafting untenable notions of distinction by association. Seton remained a victim of the Victorian pox. He assumed that by mixing with the great, he himself would somehow absorb their greatness. But while his methods were often sound and his discoveries significant, he had certainly not been conferred with a spontaneous comprehension of the subtleties and intricacies of zoology.[43]

In his desire to cultivate the American scientists, Seton wholly abandoned the Canadian colony in New York. Contact with other painters was restricted now to the American Museum where, with the assistance of J. A. Allen, who outfitted a special studio for him, he obtained the bird and mammal specimens from which to make the illustrations for Coues. Next door to his studio was the office of Frank M. Chapman, another ornithologist, a member of the AOU, and Allen's newly appointed assistant. The two men warmed to each other from the beginning, Chapman going out of his way to encourage Seton's scientific interests by familiarizing him with the current literature in ornithology. One of Chapman's closest friends and later a regular visitor at the Museum,

---

[43] See, e.g., Ernest Thompson Seton, "Critical Note on Mr. J. B. Tyrrell's Paper Entitled 'Catalogue of the Mammalia of Canada, Exclusive of the Cetacea'," Proc. Can. Inst., 7 (1888-89), 178-80; J. B. Tyrrell, "Reply to Mr. Thompson [Seton's] Critical Note," Proc. Can. Inst., 7 (1888-89), 281-84; Ernest Thompson Seton, "Reply to Mr. Tyrrell's Note," Proc. Can. Inst., 7 (1888-89), 285-86; Wallace, Royal Canadian Institute, p. 153.

was the famous bird painter, Louis Agassiz Fuertes, from whom Seton picked up many important pointers on anatomy and coloration. Together the three men sometimes had occasion to dine at a neighboring restaurant where they shared their personal experiences and discussed recent discoveries in the field. Bird song fascinated Seton. Convinced that it was a form of animal communication, he had worked hard at mastering several specific calls which he proceeded to teach to his friends over dinner. The effectiveness of these lessons varied in proportion to the amount of alcohol consumed, but on at least one occasion the three men became wholly immersed, lustily screeching in the "languages" of the Pied-billed Grebe, the Macaw and the Barred Owl until "neighboring guests changed their seats to move to distant, and what they probably considered safer, parts of the room." While the trio always remained on the best of terms, Chapman maintained consistently that Seton's bird drawings possessed "a greater artistic and spiritual quality" than those of Fuertes. He later lamented the fact that Seton allowed his success as an author to interfere with his art and repeatedly tried to lure him back to it by offering generous contracts to illustrate his own books. Despite their long and intense friendship, however, Seton obliged him on only two occasions--and then primarily because he needed the money.[44]

---

[44]Seton to Frank M. Chapman, June 25, 1898, SP; Seton, Autobiography, pp. 140, 218, 281, 344, 367; Frank

Paid in regular installments by Century for the illustrations he was making for Coues, Seton had sufficient independence to make sporadic forays into the wilderness of Manitoba and Ontario. Indeed, he still considered Canada his home and continued to visit Charlotte Schreiber and his parents on a regular basis. He supplemented his income by writing popular magazine articles about these excursions, and for a time was employed by his brother, Joseph, as resident manager of a real estate development outside Toronto. Nothing, however, was permitted to obscure his ultimate objective. Despite an outward show of arrogance in scientific matters, he was still quite unsure of himself. Most men who had made peace with their land might have been satisfied simply to live with it, on it. But Seton, who was to become an outspoken polemicist in defence of it, required nothing less of himself than a complete grasp of its biological complexities. His personal abrasiveness in

---

M. Chapman, Autobiography of a Bird-Lover (New York: Appleton-Century, 1933), pp. 63, 77, 83-84; Bird Life (New York: Appleton, 1900); Handbook of Birds of Eastern North America (New York: Appleton, 1895); "Naturalist, Artist, Author, Educator," Bird-Lore, 37 (July, 1935), 245-47; "In Memoriam: Louis Agassiz Fuertes, 1874-1927," Auk, 47 (Jan., 1928), 1-26. Frank M. Chapman (1864-1945) remained at the American Museum until he retired in 1942. He was Curator of Birds from 1908. Founding editor of the journal, Bird-Lore, member of the New York Zoological Society, and President of the AOU (1911), he received an honorary degree, D.Sc., from Brown University (1913). Like Seton he was awarded both the Daniel Giraud Elliot Medal and the John Burroughs Medal. See W. K. Gregory, "Biographical Memoir of Frank M. Chapman," National Academy of Sciences, Biographical Memoirs, 25 (1948), 111-45.

Plate 5. Northern Shrike (Watercolor, 1898)

the company of amateur naturalists constituted nothing more than a kind of notice-me image. Beneath this image lay a spirit and an intellect determined to give it substance. Yet even as he was doing the drawings for Coues, writing his articles, managing one brother's business or another brother's farm, and trying to become an informed zoologist, he still required regular moments of renewal in the wilderness. These he took away with him in the journal, his most loved and only constant companion; and these he attempted to translate into art and into science.[45]

Ever since Spencer Baird had first enlisted his help as a collector for the Smithsonian, Seton had also been hard at work preparing what he hoped would be a definitive treatise on the birds of Manitoba. By 1890, when it was finally submitted for publication, the project had consumed

---

[45]Seton, Autobiography, pp. 282-83; Ernest Thompson Seton, "A Carberry Deer Hunt," Forest and Stream, 26 (June 3, 1886), 366-68; "Moose and Bear Queries," Forest and Stream, 26 (June 24, 1886), 427; "Domestication of the Buffalo," Forest and Stream, 26 (July 8, 1886), 467; "Canadian Game and Fish Resorts," Forest and Stream, 28 (Feb. 10, 1887), 42-43; "Do Squirrels Hibernate?," Forest and Stream, 28 (Feb. 17, 1887), 65; "A Record of Failures: Part 1," Forest and Stream, 28 (March 24, 1887), 178-79; "Part 2," Forest and Stream, 28 (March 31, 1887), 198; "Hibernation of the Hare," Forest and Stream, 28 (April 7, 1887), 226; "An Exhibition of Snakes," Forest and Stream, 28 (June 16, 1887), 451; "Ontario Game Laws," Forest and Stream, 30 (March 29, 1888), 185; "Nights With the Coons," Forest and Stream, 30 (July 19, 1888), 518-20; "The Big Buck We Didn't Shoot," Forest and Stream, 37 (Sept. 10, 1891), 143. The owner and editor of Forest and Stream was Dr. George Bird Grinnell, a fellow member of the AOU. See, A. K. Fisher, "In Memoriam: George Bird Grinnell," Auk, 56 (Jan., 1939), 1-12.

much of his spare time throughout the intervening eight years. Always a believer in learning by doing, he presumed that the discipline required by this monumental task would expand, and then unite, the small fragments which initially constituted his knowledge of ornithology. Despite emphatic opposition from the AOU's Canadian migration committee, whose chairman, Thomas McIlwraith, believed the study to be premature, Seton trusted his own judgment and barged ahead anyway. The resulting monograph, almost two hundred pages in length, was a curious blend of science and anecdote, with emphasis on the latter. Seton's limited self-training simply had not armed him with adequate tools for analysis. Although the observations themselves are quite sound--and still serviceable for ecological historians--they are buried in obscure digressions which have no scientific bearing on the subject. Indeed, the major failing of the paper lies in its loving treatment of the environment in which the scientific investigation is set. While the reader is treated to Seton's most extravagant descriptive detail, especially as it touches the questions of behavior and habit, he comes away from the study not so much with a profound understanding of the birds of Manitoba as with an insight into the character of the author. Seton emerges more as storyteller than as scientist, leaving on every page compelling evidence of a personal attachment to the wilderness. The birds become a part of the landscape itself. They drop in on the author's life just long enough for him to isolate their distinguishing characteristics and to establish an intimate relationship (often captured in a lively anecdote) with them. Aspiring to make them live for his readers, Seton overextends his analysis, ultimately identifying himself with the birds. Science demanded an objectivity which his perception of nature could not admit.[46]

The unhappy task of editing this unwieldy bulk of manuscript fell to Robert Ridgway, Curator of Birds at the Smithsonian Institution. A student of Baird and one of the fathers of the AOU, Ridgway had begun his own career as a painter of birds, learning as a young man how to manufacture tints from materials in his father's drug store. For any scientific ornithologist, accuracy in coloration was essential to a faithful representation of birds. But the wide diversity of North American species guaranteed that a comprehensive grasp of the problem would be almost impossible. Nevertheless, Ridgway came closer than any other artist to

---

[46]Ernest Thompson Seton, "The Birds of Manitoba," Proceedings of the United States National Museum, 13 (May 29, 1891), 457-643; Seton to Thomas McIlwraith, May 4 1885, Seton Letterbook No. 2, SP, p. 25. Thomas McIlwraith was a Hamilton, Ont., businessman whom the AOU recognized as Canada's senior ornithologist. He had attended the founding meeting of the AOU in 1883 and remained a prominent member of its migration committee thenceforth. He revised his opinion of Seton after the publication of "The Birds of Manitoba," employing him to illustrate his own Birds of Ontario (2nd ed.; Toronto: William Briggs, 1894). See also, Morgan, Men and Women, p. 741. At least one other author found Seton's monograph extremely useful. See John Macoun, Catalogue of Canadian Birds (Ottawa: Government Printing Bureau, 1909), passim.

conquering these obstacles, eventually publishing his Nomenclature of Colors for Naturalists (1886), a definitive answer to "the need for a uniform series of named colors and [for] the standardization of color names so that there might be no uncertainty as to the colors indicated in written terminology." On several occasions Seton had requested Ridgway's advice on matters of coloration and over the years they had maintained a cordial correspondence. Ridgway obviously respected Seton's superior gifts as a draughtsman, regularly soliciting his drawings for scientific use. But when "The Birds of Manitoba" landed on his desk in 1890, he was forced to separate the artist from the scientist and assume the role of severe critic.[47]

A quiet, kind and humble man who despised controversy, Ridgway elected to make very few alterations in the manuscript, allowing it to ride on the author's introductory disclaimers. He found the work amateurish but probably felt that the Smithsonian should publish it out of gratitude

---

[47] Robert Ridgway to Seton, Oct. 12, 1887, Dec. 7, 1888, May 20, 1889, SP; Alexander Wetmore, "Robert Ridgway, 1850-1929," National Academy of Sciences, Biographical Memoirs, 15 (1931), 61. Robert Ridgway (1850-1929) replaced Coues as America's leading ornithologist with the publication, in 8 vols., of his Birds of North and Middle America (1901-1919). A member of the AOU's committee on classification and nomenclature, he played a central part in building the first edition of the Check-List of North American Birds (1886). See also, Harry C. Oberholser, "Robert Ridgway: A Memorial Appreciation," Auk, 50 (April, 1933), 159-69; Harry Harris, "Robert Ridgway, with a Bibliography of His Published Writings," Condor, 30 (Jan., 1928), 1-118.

for Seton's efforts in building its Manitoba specimen collection. Seton, who considered the monograph a launching pad for his scientific career, had tried, and failed, to sell it to a number of Canadian publishers before taking it to the Smithsonian. And even Ridgway declined to pay for it. Humiliated by this abuse, Seton had worked himself into a foul mood well before the offprints arrived for his scrutiny. In looking them over carefully he unearthed a few typographical errors which he then used as evidence of further mistreatment. He condemned his editor for failing to let him proof the copy and accused him of slashing valuable extracts from the text. "I am so utterly disheartened by [the job]," he wrote melodramatically, "that I regret even having wasted my time on it." Ridgway was astonished by Seton's petulance, replying with uncharacteristic candor that "the manuscript was, without exception, the worst I ever saw." The "thankless task" of editing "was unsought by me [and] was particularly distasteful." It had been deemed necessary to cut "superfluous" material because "the repetition of trivial and often wholly impertinent details was so excessive" that their inclusion would have rendered the costs of publication prohibitive.[48]

Ridgway's abrasive rebuke was obviously necessary to

---

[48]Seton, Autobiography, pp. 279, 284; Seton to G. Brown Goode, June 17, 1891, July 6, 1891; Robert Ridgway to Seton, June 30, 1891; Seton to Robert Ridgway, July 15, 1891, SP.

bring Seton to his senses. After all, he had won favor among the professional naturalists, not for his scientific work, but for his artistic gifts. He had deceived fellow amateurs with bravado and posturing long enough and his come-uppance was well overdue. A love for wildlife and for wild land did not constitute science, and no list of typographical errors could obscure his limitations as a systematist. The cost of this particular lesson was the friendship of Robert Ridgway, a man from whom Seton might well have learned more valuable things. But the basic problem, which was nothing less than the conflict between objectivity and subjectivity, remained fundamentally unresolved. There was a certain irony about being accepted as an artist by a coterie of scientists. But science itself demanded qualities that Seton was either unprepared or innately unqualified to cultivate--the abandonment of feeling on the one hand, and a rigid adherence to empiricism on the other. "The Birds of Manitoba" had demonstrated his inability to separate self from the landscape that self loved. Seton was overwhelmed by his own feelings; and his scientific data were committed to paper with an artist's pen. He had embraced, and had been embraced by, the wilderness. Unless there was some way of finding an outlet for subjectivity in science, art afforded the only possible medium of expression that also guaranteed recognition and approval for Seton.

V

The Victorian side of Seton's character required the commendation of society, but his love affair with the wilderness, which now seemed to defy reasoned explanation, pressed out from within his soul. The clues to this kind of schizophrenic tension were deeply rooted in a friendless, loveless childhood dominated by Calvinist piety. Yet the Biblical wilderness, with its ominous satanic overtones, had left no stain whatever. Indeed, for Seton, the wilderness had become the savior. It was a sanctuary from parental despotism. It was innocence and purity, beauty and joy. It was naturally--and therefore tolerably--limiting, demanding the observation of its laws, imposing humility. He felt no compelling urge to conquer or to humanize it, to convert it into the pastoral. The wilderness was imbued with an innate morality which man might wisely emulate, not subjugate. This morality was legitimized by its own survival value, yet was more charitable than anything the Christian church could substitute for it. Above all, it inspired and sustained Seton's sense of justice by tacitly acknowledging and fostering the equality of every living thing.

By 1890 Seton had completed the drawings for Coues.[49] He had fulfilled a childhood dream by writing a serious ornithological study, and he had lined his pockets from

---

[49]Century Company, The Century Dictionary; an Encyclopedic Lexicon of the English Language, comp. by William Dwight Whitney (6 vols.; New York: Century, 1889-91).

the sale of several magazine articles and privately commissioned animal illustrations. On the surface he had achieved an enviable independence. But from his own point of view, he was a failure. Seton's inability to place "The Birds of Manitoba" with a Canadian publisher was extremely disheartening, giving him cause to ponder the adequacy of his scientific work well before the ugly confrontation with Ridgway. He was powerless, however, to isolate the problem which eroded the quality of his analysis. Desperately wanting to become a respected naturalist, his Victorian pretensions would not permit the luxury of failure in other men's eyes. Seton reserved all his humility for the land itself. Instead of wrestling with his shortcomings, he merely disguised them under a mask of hostile arrogance, pushed them to one side, then sought refuge in his art. Allowing both the freedom of expression and the adulation of society, it alone guaranteed absolute security. Unfortunately, however, the polemical current latent in "The Birds of Manitoba" now surfaced with a vengeance in the subject matter of his painting.

Perhaps at the suggestion of Charlotte Schreiber, with whom he had managed to maintain close contact during his frequent visits to Canada, Seton chose this precise moment to join the OSA exodus to France. He went to Paris late in the summer of 1890 but found all the studios closed for the summer. To put in time while awaiting the end of the vacation period, he returned to London "where at least I could study animals in the zoo." Shortly after his arrival there he was introduced to the German painter, Joseph Wolf, who, following the death of Sir Edwin Landseer in 1873, was considered the finest animalier in England. This propitious meeting resulted in a warm friendship between the two men which lasted until Wolf's death in 1899. His home at the Primrose Hill Studios was a short walk from the Zoo. Seton, who had long admired his work, made a point of visiting the "genial old Santa Claus, of noble countenance and gracious manner" as often as he possibly could, always coming away from their discussions "helped and encouraged." Wolf had grown up on a farm near Coblenz, in Prussia. Like Seton, however, he was more interested in birds than in the manual rigors of farming and had spent much of his childhood and youth sketching wildlife in the comforting forest while others contended with the menial chores. Like Seton also, Wolf had worked for a number of lithography firms before and after taking formal instruction, and continued to sustain himself as a painter by illustrating scientific texts. It has been said, in fact, that there was hardly an English book on natural history published during his creative lifetime to which he did not make some contribution. A quiet, gentle man who lived almost like a hermit, Wolf must have enjoyed talking to Seton most about their similar background and ornithological interests. Certainly no two personalities could have been more polarized. But regardless of the bond that united them, they thrived on each other's company for the entire autumn of 1890, sketching at the zoo or simply reminiscing about their respective lives. Wolf analyzed Seton's portfolio, instructing him in the techniques of bird feathering and fortifying his interest in animal anatomy. By the time he returned to Paris in December, therefore, Seton had found not only a male counterpart for Charlotte Schreiber but also the reinforcing example of a figure who had been able to bridge the gap between art and science.[50]

Once back in Paris, Seton mixed as much as possible with the artistic establishment. Among his earliest acquaintances was Gustave Wertheimer, an Austrian animalier

---

[50]Seton, Autobiography, pp. 284-85, 291-92; Sir Walter Gilbey, Animal Painters of England (3 vols.; London: Vinton, 1900-1911), 3:274-79; Bryan's Dictionary of Painters and Engravers, 1964 ed., s.v. "Wolf, Joseph," by George C. Williamson. Joseph Wolf (1820-1899) illustrated such important works as G. R. Gray's The Genera of Birds (1844-1849) and Daniel Giraud Elliot's The Life and Habits of Wild Animals (1873), A Monograph of the Birds of Paradise (1873) and A Monograph of the Felidae (1883). Seton had had some instruction from Elliot (1835-1915), who was later (1894) to become Curator of Zoology at the Field Museum of Natural History in Chicago, and who was himself an accomplished painter. Because Elliot had learned from Wolf the art of reproducing plumage through variant imbrication, he may well have influenced Seton's decision to stay with the master. Wolf was a close friend of Sir Edwin Landseer. See Frederick George Stephens, Sir Edwin Landseer (London: S. Low, Marston, Searle and Rivington, 1880; A. H. Palmer, Life of Joseph Wolf (London: Longmans, 1895).

whose work was the toast of the Salon. He also solicited the help of Henry Mosler, the American genre painter, whose Le Retour (1879) had won him an international reputation. The members of his own "herd," however, included a peculiar blend of Americans and Canadians who had abandoned the art schools of the New World for those of the Old. Most of them had tried and failed to win acceptance at the Ecole des Beaux-Arts, settling eventually in one or other of the several ateliers which comprised l'Académie Julian. By 1890 Paris had received many of Canada's dispirited sons. William Brymner, Blair Bruce and Robert Harris had come and gone. Paul Peel of London, Ontario, who had also attended the Royal Academy School with Seton, managed to gain entry both to the Beaux-Arts and to Julian's. His painting, After the Bath (1890) was awarded the Salon gold medal. Its purchase by a member of the Hungarian royal family represented a significant landmark for fellow members of the OSA and RCA, functioning, at least for them, as a clear demonstration of the superiority of a European training. George Reid, another of Charlotte Schreiber's students, was leaving Julian's just as Seton was entering. Maurice Cullen from Montreal and Curtis Williamson from Toronto had arrived in 1889. Another Montreal painter, James Wilson Morrice, would join the expatriate group in 1892. Seton did not even bother to try the rigid entrance examinations for the Beaux-Arts. If he could not be on the land he loved, he wanted at least to be in close contact with those whose roots were there. Furthermore, the informality of Julian's left him the freedom to come and go as he pleased, enabling him to make a ritual daily sketching expedition to the zoo--something Joseph Wolf had insisted upon.[51]

---

[51]Seton, Journal, AMNH, 2:238(g); Autobiography, pp. 285-86; Bryan's Dictionary of Painters and Engravers, 1964 ed., s.v. "Wertheimer, Gustave;" Dictionnaire critique et documentaire des peintres, sculpteurs, dessinateurs et gravures, Nouvelle édition, 1966, s.v. "Mosler (Henry)." Blair Bruce (1859-1906) of Hamilton, Ont., studied under Robert Fleury and Bouguereau at Julian's. William Brymner (1855-1925) had led the Canadian exodus to Paris. At Julian's he had studied under Robert-Fleury and Bouguereau from 1876 to 1885, returning to Canada to assume direction of the Art Association of Montreal in 1886. Elected OSA and RCA in 1886, he was President of the latter body from 1909 to 1918. Maurice Cullen (1866-1934) studied at the Beaux-Arts under Gérôme. A close friend of J. W. Morrice, he travelled and painted with him in Italy and France, returning to Canada in 1895. He also taught with William Brymner at the Art Association of Montreal before being elected ARCA in 1899, RCA in 1907. Robert Harris (1849-1919), the first Canadian to follow Brymner to Paris, was a charter member of the RCA and its President from 1893 to 1906. Elected to the OSA in 1879, he served as its Vice-President from 1880 to 1881. J. W. Morrice (1865-1924) of Montreal travelled with Brymner and Cullen throughout Europe. He studied only briefly at Julian's, but mixed with the greatest of the internationally known painters and writers in Paris. More than any other artist he was responsible for liberating Canadian painting from the strictures of the academic tradition. Paul Peel (1860-1892) of London, Ont., studied at Julian's under Constant, Gérôme, Lefevre and Boulanger for five years. A member of the OSA from 1880 to 1891, he was elected ARCA in 1882 and RCA in 1890. George Reid (1860-1947) of Wingham, Ont., had known Seton at the Ontario School of Art. He attended Julian's from 1888 to 1889, studying Bouguereau and Constant. Elected to the OSA in 1885, he was its President from 1897 to 1901. Elected to the RCA in 1890, he was its President from 1906 to 1907. Curtis Williamson (1867-1944) of Brampton, Ont., studied under Constant and Lefevre at Julian's. He was elected OSA in 1892, ARCA in 1894 and RCA in 1907. See Harper, Early Painters and Engravers, pp. 49-50, 80, 148, 229-30, 247,

L'Académie Julian was not really an academy but a loosely organized art school which engaged distinguished professors from the Ecole des Beaux-Arts to come weekly to one or other of its cramped ateliers to "look over" the work of the students. Founded in 1873 by Rodolphe Julian, a retired prize-fighter with artistic inclinations, the academy had achieved considerable stature by 1890. Although it catered primarily to foreign students, many of whom had been rejected by the Beaux-Arts, it all but guaranteed exposure at the prestigious Salon exhibition each year. The internal politics of the Paris art world, built upon an ethic of reciprocal back-scratching, required that the Salon jury members, elected directly by the students themselves, should foster good feelings by pointedly admitting the work of the pupils of fellow teachers. This practice carried over from the Beaux-Arts into Julian's where many of the jury members also taught. There was very little alteration from year to year in the composition of the jury with the result that academic genre painting continued to monopolize exhibited fare at the Salon just as it dominated instruction in the schools. Initially Seton was probably

---

262, 334; Miner, G. A. Reid, pp. 48-53; Moncrieff Williamson, Robert Harris, 1849-1919: An Unconventional Biography (Toronto: McClelland and Stewart, 1970); Kathleen Daly Pepper, James Wilson Morrice (Toronto: Clarke, Irwin, 1966); H. Mortimer-Lamb, "The Art of Curtis Williamson, RCA," Canadian Magazine, 32 (Nov., 1908), 65-71; Newton MacTavish, "Maurice Cullen: A Painter of the Snow," Canadian Magazine, 38 (April, 1912), 537-44.

quite unaware of this smug Machiavellianism. As far as he was concerned, Paris was the home of great art and great artists. He would learn what he could, as quickly as he could, and then return to his cherished North America.[52]

When Seton first ventured into Julian's studio on rue du Dragon he was "disconcerted" by the "loud talking, filthy pictures on the wall, chorus singing, even guitar playing," which conspired to divorce the room from its purpose. A nude model posed seductively on the dais, but the moment she stood up for a rest, pandemonium erupted. Another pupil described the "hot, airless" atelier as "stifling, the noise at times deafening. . . . There was merciless chaff among the students, and frequently practical jokes, some of them very cruel." After suffering the humiliation of having a three foot loaf of bread broken over his head, James Wilson Morrice unceremoniously gathered up his paints and brushes and left forever. For the staid and sombre Canadians, the atmosphere of Julian's proved less than congenial.[53]

Seton himself took a long time to adapt to this "howling wilderness," often lamenting his departure from

---

[52]John Rewald, Post Impressionism from Van Gogh to Gauguin (New York: Museum of Modern Art, 1962), pp. 272-73; William Rothenstein, Men and Memories: Recollections, 1872-1900 (London: Faber and Faber, 1931), p. 39.

[53]Seton, Autobiography, p. 286; Rothenstein, Men and Memories, pp. 36-40; Pepper, James Wilson Morrice, p. 21.

home.

> I do not know why, but I know it is long since I felt so utterly downcast. I try in vain to fall back on my philosophy. I sit down and deliberately map out my environment. I have health, strength, a measure of youth, good prospects, fair encouragement, enough money for the present, nothing on my conscience, friends. . . . And yet for days I have been glad when night came, and sorry at the morning light. . . . It isn't merely homesickness either, for with it comes also an overpowering craving to be once more on the prairies and amid the Sandhills of Manitoba. You may find it hard to believe that in so short a stay I should beget so deep an affection for that country. But often as I wandered alone in the endless solitudes of the Carberry Sandhills, I used to say to myself: "These are the best days of my life."

Surprisingly, most of Seton's letters to family and friends are filled not so much with observations about Paris as with reminiscences of Manitoba. This is not to suggest that the city had no effect at all. Paris, he had to admit, was more relaxed than either London or New York. "Here," he wrote, "one feels a sense of freedom from the high strung civilization that pervades the air of America and England." He was especially pleased to discover that there was "less piety to the square inch" than anywhere else he had been. "The contents of a single print seller's window would be enough in New York or London to endanger the personal liberty and soul's salvation of fifty . . . citizens," but in Paris, where "churches, brothels, priests, whores and infidels meet one at every turn," nothing so mild was cause for special attention. In a sense Paris became a tool for hacking away at his own inbred puritanism. Seton relished silly moments that made a mockery of his Presbyterian upbringing. He found it "charming" that a man accompanying a lady on a Sunday afternoon stroll might stop at a pissoir and, while "standing straddling, and hosing the barrier before him, . . . [lingering] still for the fullest enjoyment of his relief," would continue their conversation over his shoulder. Nevertheless, Seton's loneliness for Manitoba preoccupied his thoughts. He was "sick of this wretched much belauded village. . . . All I [can] bluntly affirm," he confessed, "is that I take far more pleasure in watching the habits of our commonest birds than I do in most works of art."[54]

Despite the boisterous ribaldry of Julian's, Seton admitted learning there "many things never hinted at in my art life in Canada, or even in the Royal Academy of London." Joseph Wolf had been skeptical of the French schools, warning

---

[54]Seton to "Rufus," Dec. 13, 1890, Dec. 17, 1890, Dec. 31, 1890, Jan. 7, 1891, SP. "Rufus" was Henry Milford Steele of Oswego, N.Y., Art Director for Scribner's Magazine, whom Seton had met while studying at the Art Students' League. In the Seton Papers, the "Rufus Letters" appear in one file and constitute a weekly account of Seton's first two years in Paris. Many have been recopied by Steele and by Julia Seton and several of the originals are missing. The tone of the letters suggests that Seton intended them for publication. See also Seton, Autobiography, pp. 280, 351; A Visit of Satan, unpublished MS, SP. Seton maintained his relationship with George Bird Grinnell during this period, submitting articles on activities of interest to American readers. See, e.g., Ernest Thompson Seton, "'Sport' in France, 1," Forest and Stream, 38 (June 16, 1892), 560; "'Sport' in France, 2," Forest and Stream, 38 (June 23, 1892), 586; "'Sport in France, 3," Forest and Stream, 38 (June 30, 1892), 611.

Seton that he would have difficulty finding in them anyone who shared his fascination for animals. At the outset the prospects were bleak indeed. Among the instructors at Julian's only Jean Léon Gérôme, a distinguished academic genre painter from the Beaux-Arts, showed any particular interest in his portfolio. And even Gérôme advised him to "lay aside animals for the present, and work at the [human] figure to learn painting." As a member of the Salon jury, Gérôme had consistently voted against Impressionist entries. Yet despite his reputation for conservatism, he was not kindly disposed to the pedantic style of the OSA. At least part of Seton's education would be directed at un-learning what he had been taught in Canada. It was not at all clear to Seton what the human figure had to say about animals. Reluctantly, however, he took the advice. "Gérôme is Gérôme, I know," he wrote to a friend. "But Ernest Thompson [Seton] is Ernest Thompson [Seton], and I am not sure that Gérôme is right."[55]

---

[55] Seton, Autobiography, p. 286; Seton to "Rufus," Dec. 27, 1890, Jan. 14, 1891. Jean Léon Gérôme (1824-1904) had won the medal of honor at the Salon of 1847 for his painting of The Cock Fight. A student of Paul Delaroche and Charles Gleyre, he was known chiefly for his large ancient historical scenes and for his studies of the modern Near East. His paintings were notable for their eroticism, even after the mid-1860's when he turned to objective realism and became one of the most famous artists in the world. Although often accused of discouraging young avant-garde painters through an abuse of behind-the-scenes politics, he was probably the most influential teacher of the radical American artist, Thomas Eakins. Seton may have gone to Gérôme first on the advice of his friend,

Seton's arrival at Julian's coincided precisely with the beginning of the final phase of Victorian art in Canada. The instructors at the Ontario School of Art, who by now had transferred their aesthetic allegiance from England to France, would have joined Charlotte Schreiber in urging Seton to follow Gérôme's lead--regardless of the master's estimation of them. Such was the self-effacing nature of the colonial mentality. Other academic painters like Bouguereau, Cabanel and Meissonier, all of whom mirrored Gérôme's conservatism, had mammoth Canadian followings by 1890. The four men were distinctive for their huge narrative canvases, constructed around a comprehensive knowledge of animal and human anatomy and capturing, almost over-zealously, the subtlest variations in color and light. Although historical in inspiration, their work was often didactic in intent and unpoetic in result. Their attachment to figure drawing as the proper vehicle for instruction had permeated American art schools through the efforts of former students like Thomas Eakins. Although Seton himself had encountered the phenomenon in the 1880's while working

---

George Reid, a student of Eakins, or simply because he knew of Gérôme's skill in painting animals, particularly dogs. See Clara Erskine Clement and Laurence Hutton, Artists of the Nineteenth Century and Their Works (2 vols; St. Louis, Mo.: North Point, 1969), 1:290-91; Albert Boime, The Academy and French Painting in the Nineteenth Century (London: Phaidon, 1971), pp. 16, 50, 56, 99-100, 109, 120; Gerald M. Ackerman, "Gérôme: Reassessing a Scorned Painter," Art News, 72 (Jan., 1973), 32-35.

under Sartain at the Art Students' League in New York, he was still saddled with the backward prudery which masqueraded as morality in middle-class Toronto. The life classes at Julian's eroded this built-in fear of the human body--and, of course, Seton's diligence in attempting to master the problems they posed helped to satisfy Gérôme, whose blessing he craved. Nevertheless, they did little to further his interest in art. The highlight of each day still came at noon when he rushed off to the Jardin des Plantes menagerie. This is not to suggest that Gérôme's influence was entirely lost on him. As we shall see, Seton was intrigued by the possibilities of narrative genre and anatomy--as long as the focal subject matter was animal and not man.[56]

Although he quickly mastered their language, Seton never managed to cultivate many French friends. By 1892 he was moving almost exclusively with American painters like Robert Henri, Edward Redfield and A. A. Anderson. With the exception of Curtis Williamson, who himself abandoned Paris for the countryside around Barbizon, Seton saw less and less of the Canadians--not because he did not like them, but because they, finding it difficult to adjust to the mayhem at Julian's, frequently wandered off in search of more

---

[56]Reid, Concise History of Canadian Painting, p. 92; Lloyd Goodrich, Thomas Eakins: His Life and Work (New York: Whitney Museum of American Art, 1933), pp. 14-17.

peaceful surroundings. A. A. Anderson, remembered today for his searching portraits of public figures like Thomas Edison and Elihu Root, had arrived in Paris in the 1880's to study at the atelier of Léon Bonnât. Independently wealthy, Anderson wandered in and out of the studios of many French painters, including Cabanel, Cormon, Collin and Godin, before finally setting up his own studio in the boulevard Montparnasse. Here he called together students from the United States living in the Latin Quarter to form the American Art Association. Himself a Salon gold medallist, Anderson wielded considerable influence in Paris by the time Seton arrived.[57] Robert Henri, who attended classes at Julian's in the morning and at the Beaux-Arts in the afternoon, was rapidly losing interest in academic genre and had begun independently to explore Impressionism. Unlike Anderson, who never showed much imagination as an artist or as a thinker, Henri was a self-professed anarchist who "rejected laws and rules whenever they called for blind obedience" and who celebrated the freedom of the individual "to act without having to be shackled by established institutions and formalized codes of behavior." His enthusiasm for the writing of Bakunin and for the activities of Emma Goldman made him an unlikely

---

[57]Reid, Concise History of Canadian Painting, p. 120; A. A. Anderson, Experiences and Impressions: The Autobiography of Colonel A. A. Anderson (Freeport, N.Y.: Books for Libraries Press, 1970), pp. 1-23.

candidate for the friendship of Ernest Thompson Seton. At least on the surface. In fact, the two men enjoyed each other's company immensely, passing the hours in animated debate over the relative merits of anarchism and socialism. Seton, Henri, Williamson and Redfield met at Madame Déligant's inn at Brolles, in the Forest of Fontainebleau, for several weeks in the spring of 1891. Here Seton introduced Henri to the utopianism of Edward Bellamy, whose controversial book, Looking Backward, became the subject of many of the group's discussions.[58] Seton's friendships in Paris thus reflected the tensions gnawing at his own soul. Anderson, the academic portrait painter, appealed to his Victorian conservatism and to his instinct for scientific exactitude. Upon his return to New York in the late nineties, Seton would

---

[58]William Innes Homer, Robert Henri and His Circle (Ithaca, N.Y.: Cornell University Press, 1969), pp. 5, 59, 61, 78, 173, 180-81, 193, 197. Robert Henri (1865-1929) had studied under Eakins' successor, Thomas Anshutz, at the Pennsylvania Academy of Fine Arts in Philadelphia from 1886 to 1888. At Julian's Henri received criticism chiefly from Bouguereau and Robert-Fleury, never having a high regard for Gérôme. After 1891 he moved back and forth between Europe and the U.S., although his primary base was New York from 1900 onward. He abandoned Impressionism when it became commercially fashionable in America and was represented in the Salons of 1896 and 1897 by portrait studies. Eventually he became famous for his realism and for his advocacy of freedom of expression. Henri is probably best remembered for his revolutionary teaching methods at the New York School of Art, at the Henri School and, from 1915 to 1928, at the Art Students League. See, e.g., Robert Henri, The Art Spirit (Philadelphia: Lippincott, 1930). Although Seton never mentions the fact, it is probably true that the two men remained in contact after returning to New York through their mutual friends, Will and Mary Roberts, and through their membership in the Players' Club.

establish his studio in Anderson's famous Beaux-Arts Building on Fortieth Street, suggesting that his need for security within acceptable society always remained strong. But Henri's ideas and the iconoclastic atmosphere of the Latin Quarter combined to reinforce his attachment to the wilderness. While the Victorian aspect of his make-up remained constant, life in Paris introduced the necessity, and the possibility, of spiritual and intellectual growth. The resulting contradictions became increasingly apparent in his art.

Henri contributed something to Seton's innate disenchantment with Gérôme, but his personal forcefulness was inadequate to lead the young Canadian to an intellectual appreciation of the artistic problems associated with the academic tradition. Besides, and unlike Gérôme, Henri had no particular interest in animal painting as such. For his part, Seton could not transpose the analytical intensity of their conversations about political and philosophical radicalism into Impressionism. He continued to regard style and technique as the tools of expression, believing that it was the subject matter of a work which conveyed the painter's vision. If the OSA and the RCA wanted a demonstration of Gérôme's influence, they would find it in Seton's use of color and light, in the texture of his brush strokes and in the quality of his draftsmanship. They would not, however, find it in the content of his narrative. Although he recognized Seton's ability as an animal painter, Gérôme continually tried to discourage him from exercising it, directing his attention instead at the human figure. Seton's skill with birds and mammals owed more to Joseph Wolf, to personal experience in the wilderness, and to his own intuition than to anything he learned at Julian's. His radical narrative, on the other hand, would reveal the increasingly polemical substance of his thought about man's place in nature--suggesting that his split with Robert Ridgway may have done more to arouse his sense of injustice and to provoke his anger than to make him admit the limitations of his science. Henri had stimulated, or at least fortified, some ambiguous quest for independence. The circumstances were laden with ironies which probably escaped Seton's notice. But there was no denying that the influence of Gérôme, the animal painter who forbade the painting of animals, was leavened by that of Henri, the radical anthropocentrist whose interest in animals was peripheral. Seton absorbed the painterly qualities of Gérôme and the intellectual radicalism of Henri, blended them judiciously with his own scientific attitudes to anatomy and with his love for the wilderness, and arrived finally at a Victorian art which was at once true and false, beautiful and grotesque.

Seton chose one particular mammal--the Wolf--as his symbol for nature. En route to Carberry in 1882, he had watched from the window of his snowbound train near St. Boniface as a pack of dogs held a Wolf at bay. Each time the pack dashed in, "the great gray form leaped here and there, and _Chop_, _chop_, _chop_ went those fearful jaws, no other sound from the lonely warrior." The Wolf had already killed several of its attackers. Even as those that were able sprang back to renew the assault, he remained "statuesque as before, untamed, unmaimed and contemptuous of them all."[59] Slowly the train approached the scene, slowly it passed. And young Ernest, his nose pressed hard against the glass, grasped every detail of the encounter, printing it indelibly on his mind. This was "_le Garou_," or the Winnipeg Wolf, whose legend he dutifully researched and later published. The contrast between the wild and the domesticated descendents of the same genus, one perceived as man's enemy, the other as his friend, locked together in mortal combat, demonstrated the absurdity of the man-nature dichotomy. The dogs had been trained by man to track down and to annihilate a species to which they themselves were closely related by blood. The skillful defensive posture of the Wolf was noble, not because it resulted in the death of dogs, but because it was clear evidence that the strength of the instinct for survival was proportionate to the distance between animal and man. As far as Seton was concerned, the dogs were stupid extensions of man, humanized beasts obeying Pavlovian reflexes, unable even to muster the wit

---

[59]Seton, _Autobiography_, pp. 161-62.

to withdraw before the power of the nature that controlled them. In essence, the dogs had been pitifully victimized by man, forced to service his vision of himself as the conqueror of wilderness. In Seton's story of the Winnipeg Wolf, only a child is permitted the luxury of love for the wild animal. But innocence was indigenous to youth as it was to wilderness. The crime of man was the crime of taming. Translated into landscape terms, the dog became the equivalent of the pastoral. For the same reasons that Seton admired the Wolf and loved youth, he could never have become a farmer.[60]

In spite of Gérôme, Seton submitted nothing but Wolf pictures for adjudication at the Salon during his entire stay in France. Now the Salon, which opened every year on May Day, was still one of the most important attractions on the Paris social calendar. It was attended by painters, sculptors, critics, prominent public figures, "including the President of the Republic, foreign diplomats . . . and by everybody residing in Paris who made any social pretensions and was able to obtain a card of admission." Most young artists, including Seton, assumed that any work accepted by the Salon jury was the equivalent of a ticket to future success. Consequently, they would spend the year in perfecting one or two paintings to be offered for consideration.

---

[60] Ernest Thompson Seton, "The Winnipeg Wolf," Ladies' Home Journal, 20 (Dec., 1902), 11-12.

As many as ten thousand pieces might be forced under the scrutiny of the jurors, of which no more than two thousand would be hung.[61] Seton's first offering, The Sleeping Wolf, was delivered to the jury in March, 1891, shortly after his trip to Brolles with Henri. The subject was a North American species which Seton had located at the Jardin des Plantes and which obliged him by taking an afternoon nap in the same place each day. Seton merely set up his easel on the pathway facing the cage and painted the model from life. The Sleeping Wolf was partly an attempt to avoid what Seton felt was "painting of a debased sort," built upon "scenes of brutality and terror" so characteristic of academic genre. It seemed also to reflect a conscious desire to eliminate stereotyped views of the animal itself. The Wolf was carefully drawn, each bristle on its bulging mane singled out for special attention. The peaceful pose, however, fails to mask the obvious power of the body. Even in sleep the animal is alert, its pricked, pointed ears absorbing sights apparently forsaken by the closed eyes. The composition and skillful draftsmanship demonstrate the force of Joseph Wolf's influence, but the dark colors and almost excessive, photographic finish suggest the intervention of Gérôme. It is the poetic flow of the lines which

---

[61] Anderson, Experiences and Impressions, pp. 23-24; Harrison C. White and Cynthia A. White, Canvases and Careers: Institutional Change in the French Painting World (New York: John Wiley, 1965), pp. 27-33.

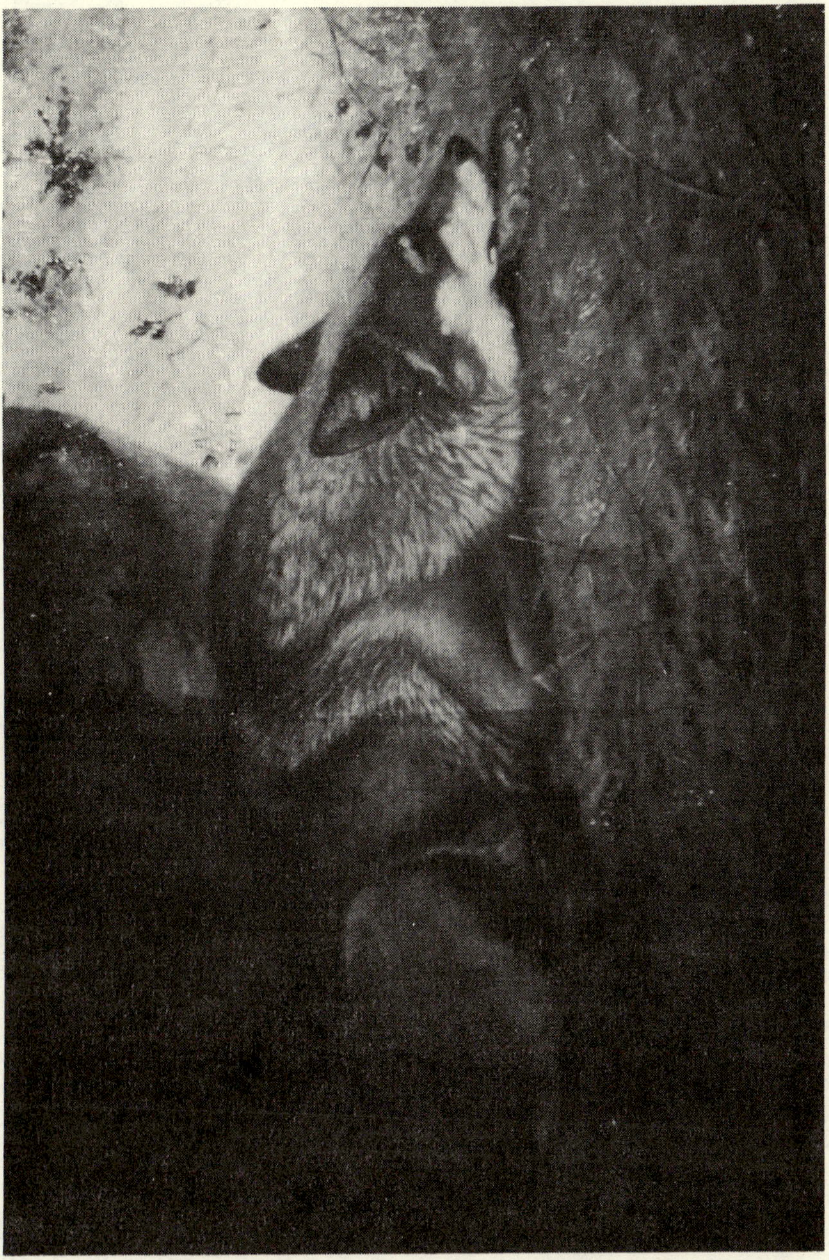

Plate 6. Sleeping Wolf (Oil, 1891)

reveals Seton's attachment to the animal, arguing convincingly that in the power of nature there is also peace. All nature is unity. The fur of the Wolf blends splendidly with the surrounding landscape, seeming almost to dissolve into it. The light is dim. The viewer must adjust his eyes and apply them directly to the subject--just as he would do if accidentally happening upon the animal in the wild. That The Sleeping Wolf was not only accepted by the Salon jury "but hung 'on the line'--waist high, not skyed . . . was an astounding event" for Seton, especially since both of Henri's offerings, Les Rochers and Paysage were rejected out of hand. "This," he observed later, "seems to prove something, but I don't know what it is." In fact, it probably proved nothing but the tight-fisted, iron grip of the academicians over the Salon. Henri was certainly no less a painter, as Seton might have liked to believe. By selecting an animal subject matter Seton had disobeyed Gérôme in only one particular. Henri, on the other hand, had committed the cardinal sin by dabbling in Impressionism. The politics of the Salon ensured that such effrontery would be punished, not sanctioned or encouraged. Seton's little mistake was simply forgotten in the popular enthusiasm for animal paintings, regardless of their inspiration, and because the style and technique of the artist were so clearly derivative.[62]

---

[62] Seton, Autobiography, p. 287; Homer, Robert Henri,

Although encouraged by this first victory at the Salon, Seton felt he had sacrificed too much of his own thought to the jury's preoccupation with form and color. On the other hand, he must have realized that by satisfying the jury with his draftsmanship he was already assured of certain reverence in Canada. Nevertheless, he believed that in order to prove himself fully, he must come away from Paris with one of the Salon's coveted medals. Almost immediately, therefore, he set about painting his second Wolf picture, this time with the avowed intention of employing the genre tradition to convey his own radical vision of man in nature. The power of the sentiment, coupled with his observation of the academic rules, would yield a principled work pleasing to both the artist and his critics. The compromise implicit in this dual purpose had several revealing consequences. These require detailed examination because they provide so clear an understanding of Seton's complex character.

Seton did not have to look far afield to find the subject matter he needed for his grand design. The Paris newspapers were alive with a story about a woodsman in the Pyrennees who "had distinguished himself and profited by killing" a number of Wolves responsible for the decimation of local sheep herds. "One evening," it was recorded, "he did not turn up at his cottage as usual. Next day a rescue party learned from the tracks in the snow that the

---

pp. 59-60; White and White, Canvases and Careers, pp. 91-92.

Wolves had had their revenge." The hunter's body "had furnished the banquet for their feast of celebration." This bit of poetic justice was ideal for true genre. It told a dramatic story and it carried a moral. The subject matter left the artist ample room to exercise his skills as a draughtsman while at the same time allowing the representation of nature in all her devastating superiority over man. Without even bothering to verify the tale, Seton launched his project with the greatest enthusiasm. Urged by Henry Mosler to make the painting large, he purchased a four and a half by seven foot stretched canvas. By January, 1892, he had begun to make the preliminary sketches at "a suitable landscape setting in the snowclad park of St. Cloud, five miles out of Paris." Armed with "some Wolf skins, a skull and some human bones, some peasants' clothes, a pair of sabots" and a "bucket of hot blood bought at the near-by slaughterhouse," Seton attempted to recreate the scene of the tragedy from his imagination. When the local gendarmerie happened by accident upon this ghoulish panorama they arrested its author, apparently bereft of his senses, and forced him to abandon his tasteless sport, eventually "surmising that it was some sort of an American joke." Quite unperturbed, Seton returned with his "colors and sketch easel on the river boat each day in late afternoon and made tone, value and color studies" at the same location. "Then, in the zoo, and from dead dogs in my studio, I made sketches of each Wolf in black-and-white, separate studies of legs, heads, bodies, all elaborately detailed." These preliminaries took two months to complete, the canvas itself only one. "I began," said Seton, "like a gladiator who, after long training, goes into a big fight."[63]

Although finely executed, the painting was as sensationally reprehensible as the scene it depicted was improbable. The remains of the hunter lie crushed and broken on the blood-spattered snow. Torn clothing and trampled ground, a discarded shoe, suggest the horrors of the earlier struggle. Stripped bare of every particle of flesh, the skull is gnawed by the smallest of the ravenous Wolves. (Symbolically it is the source of reason which furnishes the least appetizing morsel.) The remaining members of the pack, presumably gorged, keep a careful vigil. Off in the distance the hunter's pathetic cottage stands out like a blemish on the landscape. A thin plume of smoke drifting upward from the chimney intimates the presence inside the cabin of a wife, perhaps even children, and a warm dinner, waiting. Initially Seton had planned to focus the jury's attention on the nemesis which had inspired the painting by calling it The Triumph of the Wolves. On the advice of friends, however, he altered the title to read Awaited in Vain, believing it would hold a greater appeal

[63]Seton, Autobiography, pp. 287-88.

Plate 7. Awaited in Vain (Oil, 1892)

for the conservative Salon. But he soon regretted the change, noting that "the form of it switched the interest to the cottage. My title made the Wolf the hero," while "the new one transferred the heroism to the man." In other words, "It was no longer a proper retribution by the tortured and persecuted animal, but a cruel, tragic death for the brave man in sight of his home and loved ones."[64]

When news came in April that the painting had been rejected, Seton was crushed. Characteristically, however, he refused to admit the possibility that the work had failed, instead fostering the illusion that the Salon jury had merely demonstrated its own ignorance of natural realism. Immediately he arranged to have Awaited in Vain shipped to Toronto, hoping to win its acceptance to the Canadian exhibit at the forthcoming World's Columbian Exposition in Chicago. Probably on the strength of The Sleeping Wolf, Seton had been elected to the OSA in 1892.[65] Because the selection committee was to be drawn from the Royal Canadian Academy, still in large

---

[64] Ibid., p. 289.

[65] Art Gallery of Ontario, The Canadian Collection, p. 417; Joan Murray, Ontario Society of Artists: 100 Years (Toronto: Art Gallery of Ontario, 1972), pp. 10, 54; Seton to Robert F. Gagen, Dec. 5, 1898, Art Gallery of Ontario Library, Toronto, Ontario. In this letter Seton formally resigned from the OSA, sending Gagen for permanent inclusion in the OSA collection, a Wolf sketch done for his painting, La Poursuite (1895). This remains the only piece of his work in the Art Gallery of Ontario. In the letter to Gagen, Seton notes that because the sketch was "hung on line in the Salon of 1895, the unrighteous cannot afford to scoff at it."

measure dominated by the OSA, he probably assumed that his recent honor would all but guarantee a favorable reception for the gothic monstrosity. In an effort to win local support, and perhaps an added political advantage, he himself returned to Toronto, organized a public preview and invited comment from the press. The reaction, it seems, was divided. "Crowds came every day, and every one who saw it had strong views--either it was a stupendous masterpiece of animal understanding or a disgusting and degrading display of coarse brutality." In the end, public ambivalence worked to his disadvantage for the jury of selection dismissed the painting "by a majority of one--not because it was bad art, but because it gave a false impression of life in Canada."[66] Members of the RCA would have been ill-advised to drop the work of a student of Gérôme on stylistic or technical grounds, especially when that student had exhibited at the Salon. But the demands of public relations and the assumption that the painting would offend Ottawa by discouraging American tourists from heading north were hardly aesthetic criteria worthy of such a body. Although he was promptly elected to Associate status within the Academy,[67]

---

[66] Seton, Autobiography, p. 290; Toronto Week, Nov. 20, 1892; Toronto Truth, Nov. 20, 1892; Toronto Globe, Nov. 26, 1892. Cf. Toronto Globe, Feb. 16, 1893, March 30, 1893.

[67] Art Gallery of Ontario, The Canadian Collection, p. 417; Murray, Ontario Society of Artists, p. 54; Seton to Gagen, Dec. 5, 1898.

presumably in an effort to soften the blow and to defuse any protest, Seton refused to be bought. Fortunately, too, there was a significant clamor against the Academy's stated reasons for rejecting the painting. Young Hector Charlesworth, a journalist for (later editor of) the prestigious weekly, Saturday Night, believed Seton was the victim of political intrigue.

> Artists are as a class poor, and for that reason art politics sometimes get rather near to the gutter. The painter of real brilliance and originality who suddenly appears on the scene is apt to fare much as does the boy with a new suit of clothes who gets into a gang of urchins.

This view, of course, fortified Seton's own assessment of his entire life. Once Charlesworth had "made a fight" for him on these grounds, others who knew of his circumstances also "took up the cry."[68] The cry was loud and long. But the chief crier was Seton himself.

Within days of the initial rejection he wrote to William Brymner, Franklin Brownell and Homer Watson, all members of the jury, demanding a statement explaining "whether it was the subject . . . or the size that prevented its acceptance."[69] When it became clear that the OSA had voted in

---

[68]Hector Charlesworth, Candid Chronicles: Leaves from the Notebook of a Canadian Journalist (Toronto: Macmillan, 1925), p. 335; Jesse Edgar Middleton, The Municipality of Toronto: A History (3 vols; Toronto: Dominion Publishing Co., 1923), 2:58-59.

[69]Ernest Thompson Seton, Copy of a Circular Letter Sent to Messrs. Brymner, Brownell, Harris and Watson, MS Note, n.d., SP.

favor of Awaited in Vain and that the dissenting ballots had been cast by Montrealers, probably led by Brymner, Seton appealed directly to Robert Harris who had been charged by the federal government with the unenviable task of mounting the Canadian collection for the Exposition. He asked Harris to

> remember that this though a trifle to you is a serious matter to me; it means total exclusion from the Fair--Remember also that the Ontario Group of Artists [sic] is unanimous in my support [,] that the whole Toronto press has come out in my favor in spite of myself [,] for I endeavoured to keep the matter out of print . . . Remember also that while no one liked the picture on barely seeing it as you did--all came to admire it on the second or third sight. So that while all Toronto approves it--Montreal disapproves. . . .[70]

Harris had warm personal ties with both cities. He had succeeded Lucius O'Brien as Vice-President of the OSA in 1880 and later, as Director of the Montreal Art Association, had worked closely with Brymner. In 1893 he was about to assume the presidency of the RCA and had no particular desire to be distracted by an unpleasant political controversy which might make him seem to favor one side over the other. Not surprisingly, therefore, he determined to abide by the democratic decision of the jury, angrily deriding Seton for attempting either "to influence the judgement of the jury or [to] ask for reconsiderations which do not apply to other authors of unaccepted pictures."[71]

---

[70] Seton to Robert Harris, n.d., quoted in Williamson, Robert Harris, p. 151.

[71] Robert Harris to Seton, March 14, 1893, April 4, 1893, SP.

Despite their respect for Harris, the Ontario Victorians continued to defend Seton. Lucius O'Brien, for example, believed that Awaited in Vain would "supply an element that without it is entirely wanting. The picture," he hazarded, "is essentially a strong one--intense in feeling and pathos, boldly drawn and powerfully painted." He told the jury frankly that "there will be enough milk and water in our collection" and that, although Seton's was not a "pleasant" picture, "one can see at a glance that it is painted by a man . . . who knows, a man who feels intensely, by a man who can paint." Indeed, he went on, "If we had a few more of this calibre there would be a better future for Canadian Art."[72] Bolstered by such outspoken praise from a former President of the RCA, Seton held his ground. Because Brymner had emerged as the most vocal opponent of the work, he knew that the Toronto-based artists would rally around O'Brien

---

[72]Ernest Thompson Seton, Copy of L. R. O'Brien's Opinion, MS Note, n.d., SP. Lucius O'Brien (1832-99) was born in Shanty Bay, Ontario, the son of a retired British army officer. A graduate of Upper Canada College, he did not become a professional artist until 1872. He succeeded J. A. Fraser as Vice-President of the OSA in 1874 and in 1880 went on to become the first President of the RCA, a position he held until 1890. Influenced at times by Americans like Thomas Eakins and Albert Bierstadt, O'Brien nonetheless shared Seton's zeal for wild nature, frequently introducing Indians into his landscapes. He had travelled widely in Canada and was a pronounced nationalist. He served as art editor and supervisor of publication for G. M. Grant's two volume Picturesque Canada (1882). A man of aristocratic demeanor and aspiration, O'Brien was innately attracted to Seton. Reid, Concise History of Canadian Painting, pp. 85-88; Art Gallery of Ontario, Canadian Collection, p. 339; Harper, Early Painters and Engravers, p. 239.

to make the issue a cause célèbre as long as he himself did not back off. Brymner wrote to Seton from Montreal and confessed that his entry "took up more of [the jury's] time than any other and it was rejected only after having been considered with the greatest care and attention." As the self-appointed defender of the genre tradition, Brymner personally found Seton's work blasphemous. The "struggle of the Laocoon with the Serpent is a very different thing from parts of what had been a man lying on the snow." As far as he was concerned there was nothing remotely noble or heroic about the painting. The thought behind it was "ghastly" and "loathsome." To view man as the soulless victim of an insensitive and retributive nature simply would not wash with Christian doctrine. Even the Crucifixion, noted Brymner, "has the resurection [sic] of the body 3 days later."[73]

As the debate between Montreal and Toronto gained momentum, Seton realized with some satisfaction that traditional artistic considerations had ceased to function as touchstones for the acceptance or rejection of Awaited in Vain. The central topics of discussion had become the subject matter of the painting and the religious persuasion (if any) of the painter. Brymner's suggestion that the work was anti-Christian positively warmed his heart and reinforced the

[73]William Brymner to Seton, April 6, 1893, SP. See also Franklin Brownell to Seton, April 3, 1893, SP.

impress of Henri's iconoclasm. Regardless of the unlikelihood of the event recorded by his painting, Seton was prepared to go to any lengths to defend its underlying theme. In a sense he must have believed that a small falsehood serviced a greater truth. In strictly scientific terms, Wolves did not attack men. But man's legendary fear of the Wolf was a useful tool for demonstrating the power of limitation implicit in nature. The violent deed is allegorical. In the light of present knowledge the substance of the thought informing the picture is as valid as the event pictured is apocryphal. The killing of sheep by Wolves did not automatically justify the killing of Wolves by men. If anything, the Wolves in the painting had learned their violence _against_ man from the violence inflicted upon them _by_ man. Man's inability to recognize apocalypse as the grand prize for anthropocentrism constituted Seton's notion of evil. Viewed from his perspective, William Brymner was a fool.

The Toronto group looked equally silly for adopting a strong nationalist posture during the controversy. But Seton was not about to challenge his friends. As long as the subject matter of _Awaited in Vain_ figured strongly in the debate, he was prepared to become aggressively Machiavellian, employing whatever vehicle worked to win acceptance of the painting. Through his biological research and writing, Seton had acquired a good many acquaintances in the federal civil service--among them Alfred Selwyn, Director of the Geological Survey, and John Macoun, his Assistant.[74] When it was clear that Harris would not intervene with the jury, Seton decided to exploit the nationalist issue by using these men as intermediaries between the OSA and the government. This plan to reroute the decision-making process through purely political channels, and the apparent willingness of members of the OSA to acquiesce in it, reflected the concern of Ontario artists generally that Montreal was becoming too powerful within the RCA. Besides having backers in the OSA and in Ottawa, Seton was able to bring to the cause his own share of official clout. After returning from Paris he had gone to Manitoba for a short draught of wilderness tonic before unveiling his painting in Toronto. While there he travelled to Winnipeg to visit James Wickes Taylor, the American Consul, and Taylor's daughter, Elizabeth, who had also been studying with him in Paris. In Winnipeg he was introduced to a certain Mr. Scott, "the government advertising and immigration agent," who at the time "was preparing the Manitoba Exhibit for the great Fair to be opened at Chicago." From Scott, who was duly impressed by the company he kept, Seton was able to extract a letter of recommendation to Thomas Greenway, Premier of the Province.

---

[74]Macoun counselled Seton to "force the exhibition of your picture at Chicago . . . as nearly all objects of merit have to be forced to the front." John Macoun to Seton, March 14, 1893, Macoun Letterbook No. 6, p. 316, PAC MP.

Greenway, in turn, acceded to Seton's request that he be appointed

> Official Naturalist to the Government of Manitoba, without salary, except when on specially appointed missions, when I am to have $5 a day and my expenses. I would undertake to assemble and display in Chicago a natural history collection that would be new in its methods and tremendous in its advertising value.

In the end he played only a minor role in assembling the exhibit--but that took nothing away from his new status as a "government official."[75] Thus armed with a provincial sinecure, friends in (or at least close to) federal power, and the loyalty of the OSA, Seton seemed well placed to remove his destiny from the hands of the Montreal reactionaries.

Alfred Selwyn launched Seton's appeal by writing personally to John S. Larke, Canada's Executive Commissioner at the Fair. Referring to the jury's stated reason for

---

[75]Seton, Journal, 4:5-15, 27-41, 173-366, AMNH; Manitoba, Department of the Provincial Secretary, Order-in-Council 4214, 1893; Julia Seton, By a Thousand Fires, pp. 103-04. See also Ernest Thompson Seton, "Notes of a Trip to Manitoba, Part 1," Forest and Stream, 41 (Nov. 18, 1893), 424-25; "Part 2," Forest and Stream, 41 (Nov. 25, 1893), 446-48; Autobiography, pp. 297-300. Seton had met Taylor through Spencer F. Baird in the early 1880's and considered his daughter, Elizabeth, a close friend. In the spring of 1892 she outfitted an expedition to explore the Mackenzie River--a feat which Seton himself did not duplicate until 1907. For a good review of Elizabeth Taylor's extraordinary achievements and Seton's part in stimulating them, see Grace Lee Nute, "Paris to Peel's River in 1892," Beaver, 278 (March, 1948), 19-23; "Down North in 1892," Beaver, 279 (June, 1948), 42-46; "To Edmonton in 1892," Beaver, 281 (June, 1950), 3-5. See also Hartwell Bowsfield, ed., The James Wickes Taylor Correspondence, 1859-1870 (Winnipeg: Manitoba Record Society, 1968).

rejecting the painting, Selwyn noted that "The statement as regards the impression of Canada seems absurd in as much as the picture is . . . a scene in the Pyrenees, and has nothing whatever to do with Canada, except that it was painted by a Canadian artist."[76] Larke, whose other good works included superintending a Methodist Sunday School, refused to abuse his office. "I have written to Mr. [James] Smith, the Secretary of the Royal Canadian Academy," he replied to Selwyn, "but as the whole work of selection is in the hands of the Art Associations, unless very strong reasons can be given, it would not be possible for me to interfere in the matter."[77] Senator J. N. Kirchoffer of Souris, Manitoba, an acquaintance of Seton's from the homesteading days, also pressed Larke to take action. But again Larke refused to compromise himself.[78] Meanwhile, Seton continued to work vigorously behind the scenes. He visited Homer Watson at his home in Doon to guarantee his "sympathies in favour of my picture 'Awaited in Vain' which the Jury of Montreal wish to exclude from our Ontario exhibit at Chicago."[79] He also convinced Watson, George Reid

---

[76]Alfred R. C. Selwyn to J. S. Larke, March 14, 1893; Alfred R. C. Selwyn to Seton, March 20, 1893, SP.

[77]J. S. Larke to Alfred R. C. Selwyn, March 15, 1893, SP; Morgan, Men and Women, p. 563.

[78]J. S. Larke to J. N. Kirchoffer, March 27, 1893, SP; Morgan, Men and Women, pp. 541-42. Eventually Seton himself wrote to Larke, but to no avail. J. S. Larke to Seton, March 30, 1893, April 3, 1893, SP.

[79]Seton, Journal, 2:274, AMNH.

and F. M. Bell-Smith to sign a testimonial letter describing the painting as being "of a high order of merit." The letter went on to "beg" Mr. Larke to "decide the present deadlock . . . by ordering the reception of the work in question."[80] Watson and Reid then moved their colleagues in Toronto to pass a motion that "the Ontario Society of Artists express its entire approval of the action of the three Ontario members of the Committee of Selection in their endeavour . . . to secure the reception of Mr. Ernest Thompson [Seton's] picture, for the World's Fair."[81] Still, the scrupulous Mr. Larke refused to budge.

Awaited in Vain, which Robert Gagen had humorously styled "The Bones of Contention,"[82] was becoming a headache for everyone concerned with assembling the Canadian collection. But Seton had no intention of giving up until the painting had been accepted. When Macoun, Selwyn and Kirchoffer

---

[80]F. M. Bell-Smith, G. A. Reid, Homer Watson to J. S. Larke, April 3, 1893, Copy, in Seton's handwriting, SP. F. M. Bell-Smith (1846-1923) was another English-born artist, trained at the South Kensington School, who came to Canada with his father in 1866. Elected ARCA in 1880 and RCA in 1886, he went to Paris in 1891, studying under Courtois and Dupain and at l'Académie Colarossi. From 1905 to 1908 he was President of the OSA. Harper, Early Painters and Engravers, p. 26; MacTavish, Fine Arts in Canada, p. 162; Reid, Concise History of Canadian Painting, p. 83.

[81]Motion by J. H. Wilkinson, Seconded by W. E. Atkinson, April 3, 1893, Copy, in Seton's handwriting, initialled by "M[armaduke] M[atthews]," SP.

[82]John C. to Seton, March 31, 1893, SP.

failed to find a political solution, he used his own rank as a Manitoba government official to gain an audience with the distinguished Toronto Senator, Frank Smith. Smith, who had served successively in the Cabinets of five Prime Ministers, was known to have the ear of many of his former colleagues, including Auguste Angers, the Minister of Agriculture and Mr. Larke's immediate superior. By this time Seton had let it be known that he had altered the painting to make it more acceptable to the jury of selection. Whether changes really were effected is not known, but Smith, who eagerly championed Seton's cause, was clearly of the impression when writing to Angers that "it is a picture of a supposed reality of one of the cruel misfortunes of the early days of the North-West." In his zeal to give the painting greater credibility in the eyes of the Ontario nationalists, it is not unlikely that Seton had stretched his recollection of the Winnipeg Wolf legend to encompass the subject matter of Awaited in Vain. For all that Smith knew about art, any ruse was possible. "I called this morning and saw it," he wrote to Angers, "and I think it would be a great hardship to shut Mr. Thompson [Seton] out from the Exhibition. . . . Will you kindly see to this and do not shut down on this young and enterprising artist."[83] Enclosing this recommendation from Smith, Seton himself

[83] Frank Smith to Auguste Réal Angers, April 7, 1893, SP; Morgan, Men and Women, pp. 22-23, 946-47.

wrote to Angers on the letterhead of the Government of Manitoba, noting that initially the "sole objection to the picture was the ghastliness of the foreground." This, he assured the Minister, had been "changed to meet the views of the committee to such an extent that they now admit that had this been done before, no difficulty might have arisen."[84] He made no effort to destroy the illusion that Awaited in Vain depicted a Canadian scene.

In the meantime Robert Harris was beginning to soften under the pressure exerted by the OSA and by the influential political figures whose support Seton had enlisted. "Mr. Brymner," wrote Harris, "has just shown me your letter and one from H. Watson stating that you have made some important changes in the picture." Although he requested a photograph of the painting and "a paper signed by those of the committee who will agree to the admission of the picture," he left no doubt about the outcome. The federal government was obviously quite unconcerned about the grotesque subject matter. Since that had been identified as the sole criterion for rejection by the jury, it was impossible to find another pretext to justify excluding the work. Harris, who must have understood the artistic sacrilege entailed in altering a painting to satisfy the views of others, decided that it was most important for the jury to appear to be in control

[84] Seton to Auguste Réal Angers, April 10, 1893, Copy, in Seton's handwriting, SP.

of the selection procedure. Without even mentioning the names of Smith, Angers or Larke, he wrote to Seton that "If we join the others in agreeing [,] which is likely if we understand the letter aright as to changes--the entry can be hurried through here and the picture taken from Toronto." Awaited in Vain was then duly accepted on the reconsidered vote of the jury, at least half of whom never saw the revised work. William Brymner, upholding his integrity to the end, cast a lonely negative vote.[85] The art of politics had triumphed over the art of the artists. But neither the artists nor the politicians appreciated that Seton's painting, true or false, altered or not, was really the radical philosophical statement of an aspiring scientis who had used both his métier and his friends as tools for its expression.

Some critics have viewed Awaited in Vain as "evidence of the endurance of the Victorian mode."[86] Others, like Hector Charlesworth, disagreed, arguing that the painting showed Seton to be among "the earliest of Canadian painters to treat the problems of snow in the analytic manner of the French impressionists." While he agreed that "the subject was ghastly, and candidly handled, the beauty and veracity of the colour effect, and the perfect painting of the animals,

---

[85]Robert Harris to Seton, April 10, 1893; William Brymner to Seton, May 15, 1893, SP.

[86]Murray, "Victorian Canada: Part 3," p. 19.

gave the picture a haunting, sinister beauty I have never forgotten."[87] One hundred thirteen Canadian works were represented in Chicago. John George Bourinot, FRSC, editor of the Halifax Herald, clerk of the House of Commons, and sometime critic of arts and letters in the Dominion, singled out the work of Harris, Reid and J. A. Fraser for honorable mention. Ignoring Seton's painting altogether, Bourinot felt there was "a want of individuality of expression and of perfection of finish, in the work of Canadian artists. . . . The tendency to be imitative rather than creative is too obvious."[88] Robert Harris, on the other hand, reported that the national exhibit showed "so marked an advance on what would have been possible in Canada some years since that it was felt to be gratifying to those taking an interest in the state of our Art." Large format pictures dominated the Exposition and Seton's monstrosity had cost a small fortune to transport. As federal funding for the shipping and mounting of the Canadian collection had been minimal, Harris was doubly angry that Seton's work had been included while other, more important pieces had to be left behind.[89]

---

[87] Charlesworth, Candid Chronicles, pp. 334-35.

[88] John G. Bourinot, Our Intellectual Strength and Weakness (Montreal: Foster Brown, 1893), pp. 56, 89. Bourinot, who thought Seton had "made some good efforts at prairie subjects," probably believed that Awaited in Vain was a "Canadian" painting; ibid., p. 87.

[89] Robert Harris, quoted in Williamson, Robert Harris, p. 152; Harper, Painting in Canada, p. 231.

In his autobiography Seton tells us that Awaited in Vain was "given a central place in the space assigned to the Canadian Exhibit."[90] In fact this was not true, as his own defender, Charlesworth, reveals. Once the jury of selection had been "forced by the government to reconsider its decision," Brymner, Harris and other antagonists ensured that the painting was "hopelessly 'skyed.'"[91] Regardless of the critical reviews and regardless of his role in sustaining or disrupting a tradition in Canadian art, Seton's chef d'oeuvre went all but unnoticed by the public at large. There was a certain irony in the fact that such an uncompromisingly philosophical subject matter should have required so many compromises to accomplish its own exposure. The nature of the victory carried its own nemesis.

On the surface Seton responded to the rebuke with characteristic indignant petulance, resigning his association with the RCA and leaving Canada "with bitterness in his heart, . . . disgusted with the unkindness of his fellows."[92] With the exception of Charlotte Schreiber (herself soon to return to England), he banished the artistic élite of Canada from his mind. By 1893, perhaps in part because of the

---

[90] Seton, Autobiography, p. 290.

[91] Charlesworth, Candid Chronicles, p. 335.

[92] Ibid., pp. 334-336; Seton to Robert F. Gagen, December 5, 1898, Art Gallery of Ontario Library, Toronto, Ont.

dishonorable strategy he had employed to manipulate the acceptance of his painting but more likely because of the content of the painting itself, Seton and his father had broken completely, their relationship dissolving in open hostility.[93] Although he would return frequently to the Canadian wilderness, he would never really return to Canada. Beneath the surface, however, Seton seems to have learned something about the compelling integrity of art and science. The emptiness of his political victory all but destroyed any sense of fulfillment he might have achieved from the intuition he had at last been able directly to express. Furthermore, he now knew that sensational extravagance in the manner of expression would not necessarily imply, much less duplicate, the truth. The truth told by a lie was very much like the tale told by an idiot. A society preoccupied with Protestant morality, which viewed the new technology--at Chicago and everywhere else--as evidence of man's mastery over nature, proof of the end of human want, had no desire to be reminded of limitation. Not that people would have believed Wolves incapable of attacking, or even of devouring a man. Indeed, here was another justification for technology, another reason for mastery. To speak the truth of man's limitations within nature required that one also speak the truth of Wolves.

---

[93] Seton to William Snowdon Thompson, October 31, 1897, Seton Letterbook No. 1, p. 237, SP.

There was a certain irony in the fact that a work intended to evoke public sensitivity for the natural world was so contaminated by falseness. Yet Seton clearly remembered the experience, never again allowing his integrity to dissolve in personal ambition. Here was a new beginning.

VI

By the summer of 1894 Seton was back in Paris, again with the intention of proving himself. Having failed in the eyes of artists and scientists alike, he was now determined to satisfy both. To this end he embarked on two separate projects, taking up where he had left off at Julian's during the mornings, and devoting his afternoons to the preparation of his first book, Studies in the Art Anatomy of Animals.[94] While working with Elliott Coues on the illustrations for The Century Dictionary, Seton was continually faced with problems of animal and bird anatomy. "While there were plenty of works on animal dissection . . . these had little bearing on my needs and the needs of all artists who depict animals."[95] At the time he was working in London with Joseph Wolf, Seton had met other animaliers like Briton Riviere and John Swan who had been forced, through lack of available guides, to construct

---

[94] Ernest Thompson Seton, Studies in the Art Anatomy of Animals (London: Macmillan, 1896); Journal, 2:311, AMNH; Seton to Macmillan and Co., September 4, 1895, Seton Letterbook No. 1, p. 21, SP.

[95] Seton, Autobiography, p. 291.

wax models and to prepare whole portfolios of sketches from personal dissection. Swan had frequently lamented the fact that "he had not had some reliable book to shorten the labor." Seton's first term at Julian's had been devoted exclusively at the study of the human figure, yet his most important paintings during the period focussed on animals. There was, he reasoned, ample room for a kind of combined emphasis which would bring the work of animal dissection to the service of the painter. It seemed rather ironic to him that the most prominent animaliers of the nineteenth century--Landseer, Barye, Mène, Géricault--"with one voice send the student to the study of Anatomy," by which, presumably, they meant human anatomy. In Seton's view, the formal study of animal anatomy should "advance a step [beyond surgical and zoological considerations] by treating primarily the visible form of the living animal." Dissection alone was "unnecessarily thorough. It is of much greater importance that the student should model from the living animal, using as guides the best anatomical diagrams obtainable. . . ." He obviously realized that for any artist "Anatomy is like a virulent poison--when judiciously administered it is a powerful stimulant to art, but in an overdose it is death." Echoing the advice of Joseph Wolf, he argued that one should see "but little of Anatomy in the living animal except in silhouette; . . . he who makes it of prime importance produces mere diagrams, and loses sight of the greater essentials of light, colour and movement." The purpose of the Art Anatomy, therefore, would be to relieve the artist of the task of performing animal dissections, while at the same time apprising him of the effects of bone and muscle structure on external form. To this work Seton brought a unique union of biological and artistic insights which far surpassed anything he had been taught by his own instructors in Paris.[96]

Once Macmillan and Company had contracted to publish the book, Seton poured heart and soul into the research and complementary illustrations. Centering his attention on those domesticated animals which appeared most frequently in the works of the masters, he prepared highly detailed diagrams, complete with systematic scientific nomenclature, of every muscle and bone. Consideration was also given to pelage and and feathering in an effort to demonstrate accurately the ways in which movement combined with lighting to produce shadows. As the study progressed, Seton took the plates to Gérôme for his opinion. "He was very kind and encouraging. Afterwards he gave me an introduction to Aimé Morot, his son-in-law," who "admired my plates, showed [me] his animals [and] told of his many discussions with Meissonier."

---

[96]Ernest Thompson Seton, Lecture No. 1, The Practical Application of Anatomy by the Great Animaliers, MS Notes, n.d., SP; Art Anatomy, pp. i-vii, 1-2. The opinion of John Swan, RA (1847-1910) was particularly important to Seton as he had studied under Gérôme at the Ecole des Beaux-Arts and had been influenced by Barye and Frémiet. Jeremy Maas, Victorian Painters (London: Barrie and Rockliff, 1969), pp. 82-83.

Heartened by their reception, he then went to see Emmanuel Frémiet, the famous sculptor who had succeeded Antoine Louis Barye as professor of animal drawing at the Natural History Museum of Paris. Seton had a more profound respect for sculptors than for painters, especially where anatomy was concerned. "A painter," he noted, "can take refuge in color or atmosphere to hide somewhat, or condone his ignorance of the form, but the sculptor must make his model absolutely, measureably, scientifically right if you will, or else stand exposed as incompetent." Unfortunately, however, Frémiet "said he didn't take any stock in talks about art or books about art. He believed only in hard work."[97] This might have been taken as an insult if Seton had been unable to reassure himself that Frémiet's was not, after all, the last word on art anatomy. No sculptor had surpassed the exemplary work of Barye himself, a man who, in Seton's eyes, "had a most exalted idea of science as a hand maiden of art." As Frémiet had been a student of Barye,

---

[97]Seton, Journal, 2:307, AMNH; Lecture No. 1, SP; Aimé Morot (1850-1913), a student of Cabanel and winner of the Prix de Rome, exhibited regularly at the Salon after 1873, eventually (1900) becoming a professor at the Ecole des Beaux-Arts. Emmanuel Frémiet (1824-1910) had received his own early experience as a scientific illustrator. He exhibited at the Salon, winning the medal of honor in 1887. Although his work is generally considered less violent than that of Barye, his Gorilla Carrying Off a Woman (1887) may well have helped to inspire Seton's Awaited in Vain. Dictionnaire Critique et documentaire des peintres, sculpteurs, dessinateurs et gravures, Nouvelle édition, s.v. "Morot (Aimé, Nicolas);" ibid., s.v. "Frémiet (Emmanuel)."

Seton felt certain that he could change his mind. Rather than fight back, as he might have done earlier, he determined to persevere in the book and to break down Frémiet's indifference with a clear demonstration of his abilities. The fact that Gérôme had become more temperate was encouragement enough for the time being.[98]

Although the Art Anatomy was novel in conception, it was doomed to suffer for at least two reasons. In the first place, Seton was virtually powerless to extricate himself from the anti-anthropocentric perspective he had espoused in Awaited in Vain. For all his belief in measurement as the science of art, he could not deny that the wilderness, which by now was truly a part of him, demanded greater expression through him. As a result his scientific purpose was continually clouded by oblique references to his love affair with the land and its inhabitants. Seton recognized the problem, but refused to apologize for it. After a long digression on the exploits of the Winnipeg Wolf, he told one art lecture audience that

> This perhaps is purely natural history, but I do not admit on that account that it is not also art. A

---

[98]Seton, Lecture No. 1, SP. Antoine Louis Barye (1796-1875), a member of the Barbizon school, was distinctive for introducing animals into his landscapes. Although he had painted at Fontainebleau, he had prepared most of his sketches for his animal sculptures at the Jardin des Plantes. Jean Bouret, The Barbizon School and Nineteenth Century French Landscape Painting (Greenwich, Conn.: New York Graphic Society, 1973), pp. 116-20, 246; Clement and Hutton, Artists of the Nineteenth Century, 1:37-38.

> sympathetic comprehension of the animal is the foundation of the animalier's art, and I shall often drift into natural history in the form of character study as I follow up my favorite animals, just as the portrait painter prefers to study the character and poses of his subject before he can picture them. . . .[99]

It must have been difficult for him to discuss measured, objective and true art as his European teachers defined it, while at the same time trying to articulate an understanding of the animal itself. But without the animal, art would have been meaningless. For Seton, the old maxim was better rendered, "Art for the Animal's Sake." Superficially, his new book was a vehicle for social self-propulsion, for acceptance and recognition, penance for past sins. However, among the servile domesticated beasts of burden analysed in the pages of the Art Anatomy we also find the Wolf. Even in attempting to foster the praise of Gérôme and Frémiet--which alone might have constituted an abandonment of principle every bit as drastic as that signalled by the controversy in Canada--Seton could not detach himself from the land and its symbols of freedom. The second problem was really a corollary of the first. Seton had hoped to carry the book on the strength of his own illustrations. Because he was not particularly interested in men, or their history, he failed to trace developments in the study of art anatomy, trusting solely to the appeals of painters like Swan and Rivière as sufficient justification for his treatise. Furthermore, his

---

[99]Seton, Lecture No. 1, SP.

first scientific interests had emerged in the field of ornithology. His contacts with Baird, Merriam, Chapman, Fuertes, D. G. Elliot and Joseph Wolf had given him the firmest possible foundation on which to build an understanding of birds. But the criticism and rebuff of Robert Ridgway had effectively frightened him away from ornithology since 1891. The Art Anatomy centered almost exclusively on mammals, a fact which suggests that Seton was using the preparation of the book as a means of projecting himself into a new scientific, rather than an artistic, field. As a novice mammalogist he had a great deal to learn about morphology. It would not be stretching a point to argue, therefore, that what on the surface was a text book for artists was in many ways the published laboratory report of an aspiring taxonomist.

In his treatment of the horse, Seton made only passing reference to George Stubbs, the controversial eighteenth-century English animal painter whose Anatomy of the Horse (1766), for which the author had prepared both the drawings and the plates, "gave him a European reputation, both in science and art."[100] But even though Seton's book lacked

---

[100]Bryan's Dictionary of Painters and Engravers, 1964 ed., s.v. "Stubbs, George," by E. R. Dibdin. George Stubbs (1724-1806) was much more important to Seton's conception of art than he seems to have realized. He had exerted a particularly strong influence on Théodore Géricault and Eugène Delacroix, painters whose work Seton greatly admired. Barye and Delacroix had been close friends and both had

historical depth, it did reflect his keen understanding of contemporary motion studies--particularly those of the celebrated photographer, Eadweard Muybridge. Under the patronage of Leland Stanford, former Governor of California and himself an avid horseman and connoisseur of the arts, Muybridge had concluded a series of experiments between 1872 and 1882 which proved, through the use of sequential photographs, that a galloping or trotting horse raises all four hooves off the ground at once. The photographs were exhibited in Europe in 1882 and at one stroke rendered inaccurate virtually all works of art containing quadrupeds in motion. This revelation was particularly devastating for self-proclaimed realists like Gérôme, Morot and Ernest Meissonier, all of whom attended the exhibition and met Muybridge and Stanford personally.[101]

---

shared Stubbs' belief in the superiority of nature over art. Géricault, Delacroix, Barye and, to a lesser extent, Stubbs (who preceded the rest) held a romantic fascination for the "big cats"--lions, tigers jaguars and the like--which took the artists beyond the range of known animal behavior and gave to their animal paintings an imaginative violent quality which had little substance in biological fact. (Seton, on the other hand, stayed with subjects that he himself had observed in their natural wilderness habitat.) After the cats, all four seem to have preferred horses. Frequently both animals figure prominently in the same painting. At the time of his death Stubbs was preparing another monumental study, A Comparative Anatomical Exposition of the Structure of the Human Body with that of a Tiger and a Common Fowl (1817). His pre-eminence in the field required greater attention from Seton. See Sir Walter Gilbey, The Life of George Stubbs, RA (London: Vinton, 1898); Frank Anderson Trapp, The Attainment of Delacroix (Baltimore, Md.: Johns Hopkins Press, 1970), pp. 203-25.

[101]Eadweard Muybridge (1830-1904) had migrated to the United States from England in 1852. After serving as a

Encouraged by the humility with which these and other artists greeted his work, Muybridge returned to the United States intent on perfecting his methods. In 1884, probably through his friendship with Thomas Eakins, he was awarded funds by the University of Pennsylvania to pursue the project, the results of which were published in a multivolume study entitled Animal Locomotion: An Electro-photographic Investigation of Consecutive Phases of Animal Movements (1887). Before 1882 it had been customary to paint a running horse with outstretched fore and hind legs. But Muybridge's photographs demonstrated that the human eye and brain lied to the viewer, that what one saw did not necessarily duplicate what was. In fact, the point at which the horse's feet simultaneously left the ground occurred when they were bunched beneath the belly. The problem, of course, was whether the "brain picture" or the "sun picture," the artistic pose or the scientific pose, constituted the objective reality. Gérôme, who was

---

survey photographer for the federal government, he set up a private studio in California in 1870. In 1878 he lined up 24 cameras, each with a specially designed shutter, at 14 inch intervals parallel to a length of race track. Strings were attached from the shutters to the opposite railing so that as the horse passed it triggered the cameras in sequence. After 1884 he invented a more sophisticated, electrically operated contrivance which used highly sensitive, gelatin dry plates and which, again with multiple cameras, enabled him to photograph the subject simultaneously from three different angles. Coincident with these discoveries, he introduced the "zoopraxiscope" which projected the individual photographs on a screen in rapid succession, making Muybridge the father of modern cinematography. See Van Deren Coke, The Painter and the Photograph from Delacroix to Warhol (Albuquerque, N.M.: University of New Mexico Press, 1972), pp. 154-56.

among the first subscribers to Animal Locomotion, introduced Seton to its findings as early as 1891. The composition of Seton's early Wolf pictures reflects his reluctance to chance a motion study until Gérôme, who had advised against painting animals altogether, could digest, experiment with and accept the implications of Muybridge's work. By 1894, however, the controversies surrounding Robert Ridgway's attack on "The Birds of Manitoba" and William Brymner's refusal to accept Awaited in Vain had pressed Seton steadily toward a perception of objective reality more strictly tied to science. Absolute accuracy guaranteed security from public rebuke, functioning as a kind of protective shield. Muybridge had been able to show that the actions of striding quadrupeds effected variations in form that went unnoticed by the eye. It seemed likely that previous failings in the work of animaliers could be tied to the investigations of painters like Stubbs who knew nothing of photography and whose anatomical experiments had relevance for stationary attitudes only.[102]

As it happened, Gérôme, Morot and Meissonier eventually opted in favor of the "sun picture." In the case of Meissonier, a genre painter whose fascination for military subjects often

---

[102]Francis Galton, "Conventional Representation of the Horse in Motion," Nature, 26 (July 6, 1882), 228-29; E. Ray Lankester, "The Muybridge Photographs," Nature, 40 (May 23, 1889), 78-80; Goodrich, Thomas Eakins, pp. 65-69; Coke, The Painter and the Photograph, pp. 156-59.

demanded detailed motion studies of the horse, this was a particularly important development. Until his meeting with Muybridge, Meissonier had maintained that an understanding of history was the lone essential ingredient of good documentary art. But by adapting the findings of Muybridge, he shifted the emphasis of his work from history to science. Regardless of the narrative content of his later paintings, the horse--for its own sake--became a central focal point.[103]

Although Meissonier had died in 1891, Seton was an admirer of his work and had learned a great deal about him from Gérôme and Morot. Ironically, it was the deceased painter, not the living ones, who proved the significance of Muybridge's revelations. So powerful was the impact of the photographer on the artist that Seton wrote to Muybridge, requesting permission to duplicate several of his plates in the _Art Anatomy_. Eager to see his own work formally linked to the technical instruction of painters and sculptors, Muybridge was warmly disposed toward the plan, not only sanctioning the use of his plates, but also urging Seton

---

[103]Jean Louis Ernest Meissonier (1815-1891) was the first successful genre miniaturist. Prior to 1890 he was one of the most popular attractions at the Salon. At the time of his first meeting with Muybridge and Stanford (1881), he painted the latter's portrait. Seton, Lecture No. 1; Clement and Hutton, _Artists of the Nineteenth Century_, 2:106-09; Charles Yriarte, "E. Meissonier: Personal Recollections and Anecdotes," _Nineteenth Century_, 43 (May, 1898), 836; Boime, _The Academy and French Painting_, p. 106.

to expand upon their application.[104] In his treatment of the dog, Seton took a particular interest in fulfilling this commitment. Noting that the sequential footfalling of a dog, unlike that of a horse, was rotary instead of diagonal, he cautioned artists that the rules for quadrupeds in motion varied with the species. It was impossible to extend generalizations about one group to any other. It followed that anatomical variations in form as between two different mammal species running side by side at the same pace would be considerable. Although given muscles and bones might perform the same basic functions in both animals, the realistic representation of those animals by the artist would require an intimate knowledge of form dictated by footfall.[105]

The Art Anatomy depended for its success as much on the author's analysis of the Muybridge photographs as on his own extensive dissections. It is rather interesting to note that critics of the book were most moved by the dog studies which Seton had really undertaken in an effort to placate Gérôme (who had a fascination for greyhounds). As far as Seton was concerned, the dog was nothing more than

---

[104]Seton Eadweard Muybridge, May 4, 1895, May 8, 1895, June 4, 1895, Oct. 30, 1895; Seton to Macmillan and Co., May 8, 1895, May 12, 1895, June 4, 1895; Seton to Funk and Wagnall, May 8, 1895, June 8, 1895; Seton Letterbook No. 1, pp. 6, 9-11, 13, 15, 16, 20, 30, SP.

[105]Seton, Lecture No. 1; "Review of Studies in the Art Anatomy of Animals," Athenaeum, 110 (July 31, 1897), 165-66.

"the domestically descendant of a wild animal related to the Wolf."[106] The illustrations for the *Art Anatomy* were intended to do double-duty--at once serving the interests of the art community and contributing to his own growth as a scientist. Although the line dividing art from science was thus progressively obscured, Seton remained a zealous defender of the land and its inhabitants. Nothing was allowed to detract from the notion of nature's dominion over man. Indeed the driving force behind both his art and his science was the need to demonstrate nature's complexity and man's limitation. Even while putting the finishing touches on the book, Seton was using the dog studies as models in the preparation of a third Salon entry, *La Poursuite*, a large oil painting which "was supposed to be the view backward from a Russian sled, pursued by a pack of Wolves in the wintry woods."[107] Previous submissions had featured the animals in stationary poses, but *La Poursuite* was intended to amplify the findings of the *Art Anatomy*. All of the Wolves are in motion--some stragglers, about to give up the chase, have slowed to a saunter, while others closer to the prey are either running fast or leaping through the air. The sled itself is not pictured. The nobility, ferocity and power

[106] Seton, Lecture No. 1.

[107] Seton, *Autobiography*, p. 291. The narrative is derived from a story submitted to Grinnel. See Ernest Thompson Seton, "The Baron and the Wolves," *Forest and Stream*, 47 (Dec. 26, 1896), 504-06.

Plate 8. Leaping Wolf (Pencil, 1895)

of the Wolves are augmented not only by the attitudes in which they have been captured, but also by the ghostly white bleakness of the landscape against which they are set. Seton's experience with Awaited in Vain had taught him to tone down the sensational and to involve the viewer in sorting out the substance of the narrative content. Presumably fearing that man might once again win the sympathy of the jury, Seton left him out altogether, merely implying his presence by the title and by the faint suggestion of packed snow left in the trail of the fleeing sled. As a result, the intended polemic is not explicit. La Poursuite was supposed to satisfy the demands of objective realism, academic genre and moral verity. Of these three motifs, only the last was uniquely a product of Seton's imagination and experience. The eye of the camera was responsible for the first, a combination of Charlotte Schreiber and Jean Leon Gérôme for the second. By sacrificing the object of the pursuit in the interest of subtlety, the painting lost its moral commitment in ambiguity. In the end, nature's triumph is reflected not through the horrors anticipated by the artist, but through the grace of the animal forms themselves.

La Poursuite was rejected by the Salon of 1895, probably because of its impressionistic treatment of the snowy landscape. This time, however, Seton had included with his entry preliminary sketches for each of the twelve Wolves represented in the painting. Of these, no fewer than six were accepted and hung on the line--a truly remarkable endorsement for the work of a Canadian artist in the nineteenth century.[108] Ironically, of course, the success of the drawings was premised not upon narrative but upon the scientific exactness of the individual forms. The artist's ability shines through what might have been a mere anatomical exercise stressing the effects of motion by emphasizing the dominant lines of each animal, by supplementing grace with a clear perception of power and space. Each Wolf seems destined to pounce from frame to viewer. Yet there is no apparent moral bias. The Wolf--that is to say, the wilderness--is beautiful, awesome and forceful for its own sake. Although the drawings had been rescued by science, their inspiration and success were still rooted in Seton's view of nature's omnipotence, man's limitation.

When the Art Anatomy was published shortly after the Salon, Seton found himself at the center of attention among the artists he most respected. Gérôme and Frémiet, "who were at first disposed to ridicule so much attention bestowed on the minutiae of bird and beast anatomy," were truly excited by both the book and the Salon entries. Reflecting on the sketches, Gérôme told Seton that "One who draws like that can do all." Morot and Rivière, among others, congratulated

---

[108] Seton, Autobiography, p. 291; Edouard Detaille to Seton, n.d., SP.

him for the "painstaking and thoroughness" which had yielded such a "complete and valuable text-book for the world of artists and art students."[109] Seton seemed to have proved himself by accomplishing the impossible. The dreadful burden of failure had been lifted from his shoulders. Having suffered so many indignities at the hands of his Canadian contemporaries, he now relished a sense of release, basking unashamedly in the sun of the French triumph. Replying to a favorable review of the Art Anatomy, Seton wrote wryly to the editor of one journal:

> While it [the review] is very good as far as it goes, it is not, I think, nearly strong enough. . . . You are right in describing my work as 'valuable', 'complete', 'conscientious', etc. It certainly is all of these, but why not come right out flat and call it, as you know it is, 'great', 'magnificent', 'glorious'. As a matter of fact, if you had styled it 'the epoch-marking product of a colossal intellect', I should have been the last man to call you to task for it.
>
> One or two of the critics proclaim it the best thing of its kind. As it is the only thing of its kind, these men are mistaken if they think they flatter me.
>
> I'm not hunting for a cheap boom . . . but if in your next number you repent, and do your full and proper duty as my reviewer, and really do speak out for the world of brains, and, in so doing, if you should chance to say anything about the author's 'god-like intellect . . . and superhuman sagacity', or allude to possible 'Academic crowns' and 'agitation to secure' for him the Legion of Honour, or a Government pension . . . well, between you and me, it'll be all right.[110]

Yet despite this mask of arrogant confidence, Seton seems to have realized that his career as an artist was over. Both

---

[109] "The Art Anatomy of Animals," Quartier Latin, 2 (1897), 177.

[110] Ernest Thompson Seton, "Letter to the Editor," Quartier Latin, 2 (1897), 177.

La Poursuite and the composite sketches for it reflected the first real individual growth he had experienced. But they demonstrated as well that this very growth was grounded in science. Seton simply loved his animals more than he loved his art. The photographic experiments of Muybridge, Marey, Stillman and Eakins had already influenced the realists so strongly that their paintings were mired in a preoccupation with finish and technique. With the publication of the Art Anatomy Seton had unwittingly contributed to this very development. Although he continued to attend classes there until 1896, Seton also realized that Julian's heyday was over. His old friend, Robert Henri, now mixed with Canadians like J. W. Morrice who had bolted Julian's in disgust. Henri himself had openly revolted against faddish enthusiasms for academic genre at one extreme and for Impressionism at the other. In 1895 he set up his own art school in Paris, challenging students to assert their individuality and to destroy the monopoly over taste controlled by wealthy American tourists.[111] While Seton clearly shared these iconoclastic sentiments, he had persisted in his efforts to win acceptance abroad before attempting to assert his independence at home. In his heart he knew that the drawings for La Poursuite had been admitted by the Salon for the wrong reasons. The fact that they were rewarded by the unqualified

---

[111] Seton, Autobiography, p. 296; Homer, Robert Henri, pp. 85-87.

praise of acknowledged masters constituted satisfying proof that his Canadian detractors had underestimated his capabilities. But he had no desire to be judged successful solely on the strength of technique. As a missionary for the cause of pantheism he needed approval for the entire canvas, not merely for a few of its component parts. The whole work contained the whole expression of the whole perception. By selecting representative Wolves, the Salon divorced the forms from their role in the narrative, focussing attention on the mechanical skills of the artist while losing sight of the wilderness which had inspired the painting.

In Henri's terms, the Salon overreached its mandate, undermining the artist's individuality and endorsing its own extreme vision of objective reality. In truth, however, Seton's dual purpose was the villain of the piece. Individuality required an absolute personal commitment to the idea being expressed, not to the manner of its acceptance by others. Furthermore, the idea itself had to be rooted in truth. It was not enough for Seton to argue that he had applied science to art in order to isolate the proper effects of motion on form. As his moral intuition about man in nature was based in science, he also had a responsibility to depict accurately the behavior of his animals. In the interests of satisfying the requirements of genre, however, he abandoned biology in favor of an untenable narrative, sending his Wolves off in retributive quest of the man scent. Both Awaited in Vain and La Poursuite suffered for this inconsistency. Although it was unlikely that either the Canadian Victorians or the French academicians would have accepted his conviction that man was forever condemned to limitation by nature, both might well have responded favorably to a less sensational, if equally dramatic, artistic statement than Awaited in Vain afforded. But in attempting to answer this demand, Seton diluted the narrative of La Poursuite to such an extent that it was unrecognizable to the viewer, suggesting that he actually lacked commitment to an idea. While in fact the wilderness imperative was as strong as ever, the painting failed to give it the expression Seton had intended--irrefutable proof that integrity allowed no compromises.

Where other artists might have pursued the problem to a satisfactory resolution, Seton used this opportunity to escape both Bohemian Paris and the ambivalent world of art spawned by the photograph. He had no interest in entering, or being consumed by, the hostile debate between academicians and Impressionists. The real, if very superficial success of the drawings for La Poursuite had redeemed him in the eyes of other artists, establishing his reputation and licensing him to thumb his nose at the RCA. But as his real interests lay with the animals themselves, and as the formal demands of painting seemed to have betrayed his perception of their symbolic relationship with man, he now turned his attention to the literary mode, hoping still to unite art and science in defense of the wilderness he loved. This time, however, scientific and moral considerations (always inextricably united in Seton's mind) would not be sacrificed at the artistic altar.

VII

The motion photograph evoked different responses from different artists. The Victorians and French genre painters continued in their didacticism and anthropocentrism. Their work appeared analytical, but it depended for its substance upon subjective interpretations of recorded history which reflected the artists' emotional excitement. Despite the fact that many such painters elected to depict the animal revealed in Muybridge's photographs, nature itself remained a secondary concern. In this sense, realism was a purely technical phenomenon, an accurate rendering of the "sun picture" in the service of a higher, nobler end. Morality existed as between man and man, owing nothing whatever to animals or the land. Nature was majestic, but man was majesty. For the Impressionists and post-Impressionists on the other hand, scientific objectivity implied an impersonal record of perceived reality. The photograph had proved the inability of the brain to grasp instantaneously the complexities of the "sun picture." But on the positive side it disclosed the important effects of light on the moving form. Abandoning the academic preoccupation with history and emotionalism, the

Impressionists revelled in nature, using direct and reflected color to suggest atmosphere. They centered their attention on immediate, uninterpreted experience. As visual "impressions" their paintings recreated a human perception of the photographed moment, not of the photograph itself. There were no moral lessons to be drawn, no commentaries on the coarseness and sordid squalor equated with social realism.

The work of Ernest Thompson Seton was in some ways an amalgam of both traditions. Although his early training united with the exacting demands of scientific draughtsmanship and the prerogatives of Victorian Canadian society to push him into the arms of academic instructors, Seton shared with many Canadian artists abroad the disquieting sense that "acceptable" art had little to do with his own experience. Although his friendship with Robert Henri combined with Bohemian life in Paris to stimulate his intellectual and religious rebellion, he held on, at least in part, to a technical style and manner which he hoped would guarantee approval among the art élite at home. But in attempting to attach his moral perception (or scientific intuition) to the genre tradition, he created a unique body of work which at once eulogized nature, attacked history and limited man. His landscapes--particularly his treatment of snow--had begun to show the influence of the Impressionists, but the photographic precision of his animals suggested that he was a realist. However, it was the implied role of man which, on one hand, identified Seton's spiritual allegiance and, on the other, alienated his critics. In fact,

Seton had begun to do for the animate wilderness what another radical body of Canadian painters, the Group of Seven (some of whose members also attended Julian's), did later, and even more sensationally, for the inanimate wilderness.

Raised in a staunchly Presbyterian family whose piety repressed both wit and joy, Seton had learned early the agonies of isolation from his fellows. Rejecting the religion of his father, whom he despised, he created a religion of the land, in which he sought, and found a refuge. In an effort to understand and to communicate the power and complexity of the wilderness and its inhabitants, he vacillated between art and science, eventually attempting to combine the two. But as a missionary for his own special brand of pantheism, he overstepped the acceptable boundaries of both, leaving himself open to legitimate attack. Having spent so much of his life endeavoring to satisfy critics, Seton remained convinced that he, like the Wolf, was the victim of unwarranted promise of a better world to come. His idea of science was rooted in biology. He tended to link technology, or applied physical science, with the religion of his father, seeing both as mutually supporting denials of man's place in the animal kingdom from which he had evolved and to which he was forever bound. In taking the side of nature, Seton was unconsciously defending himself from persecution. But in the process of advocating a community of all living things, he also realized that his own intuition was not in itself sufficient to win converts. He would have to fight science with science in order to establish his credibility.

In 1896 Seton made a complete break with the world of art. Although forced to earn his living as an illustrator, he never again submitted his work to the Salon. Resigning from the OSA and the RCA, he tacitly rejected their appraisals of his competence. His family relationship having all but disintegrated, he no longer had any compelling reason to return to Toronto. Neither Britain nor France could promise the landscape or the animals his soul required. Accordingly, after marrying Grace Gallatin, an independent, young American feminist and author whom he had met in Paris,[112] Seton moved to a farm within commuting range of New York city and, at the age of thirty-six, prepared to launch an entirely new career.

[112] It is impossible to deal adequately with either Grace Gallatin Seton or her daughter, Anya Seton, in the context of this study. Both women had a strong influence on Seton's life and both are themselves well-respected authors. By leaving them out of the present work I mean to do them no injustice.

Chapter 3

FROM SCIENCE, ART

A new literature has been born. . . . Man is not the only nor even the chief, actor. . . . Not so very long ago animal stories were often told for the purpose of carrying a moral--they were self conscious. Now they are told because they are worth telling. The real moral is the interest in the animal and the way in which it contrives to live, not in some extraneous literary appendage that tries to make an application to human conduct. No longer can one write a good nature-piece until he has intimate knowledge of the animal or plant in the wild, and has tried to put himself in its place. . . . The new literature is founded on specific technical knowledge, but it embraces all the human sympathies. It is the outcome of the study of objects and phenomena. The first product was scientific literature. The second is the lucid resourceful nature-writing of the present day. There are new standards of literary excellence.

Liberty Hyde Bailey

I

Ernest Thompson Seton's twenty-odd volumes of animal stories and nature essays for children have attracted more scholarly attention than any other product of his career. Although his first book, Wild Animals I Have Known, did not appear until 1898, his interest in the genre dated from 1887 when he published "The Drummer on Snowshoes" in St. Nicholas.[1] Most critical analyses of the body of work which followed have been unsatisfactory, either because of their inordinate praise or because of their tendency to belittle.[2] There seems, indeed, to be no middle ground. The stories were intended to be a hybrid of literature and natural history. In many respects they succeeded at being neither. But the very fact of Seton's having written them commands our attention, as does the phenomenal number of readers they attracted.

The realistic animal story has been identified as

---

[1]Ernest Thompson Seton, Wild Animals I Have Known (New York: Scribner's, 1898); "The Drummer on Snowshoes," St. Nicholas, 14 (April, 1887), 414-17.

[2]See, e.g., Lionel Stevenson, "Nature in Canadian Prose," in Appraisals of Canadian Literature (Toronto: Macmillan, 1926), 160-70; "The Animal Story," Edinburgh Review, 214 (July, 1911), 94-118; Michel Poirier, "The Animal Story in Canadian Literature," Queen's Quarterly, 34 (Jan.-March, 1927), 298-312, 34 (April-June, 1927), 398-419; Henry Chester Tracy, "Ernest Thompson Seton, in American Naturists (New York: E. P. Dutton, 1930), pp. 233-43; S. E. Read, "Flight to the Primitive," Canadian Literature, 13 (Summer, 1962), 45-57; William H. Magee, "The Animal Story: A Challenge in Technique," Dalhousie Review, 44 (Summer, 1964), 156-64; James Polk, "Ernest Thompson Seton," in Wilderness Writers (Toronto: Clarke, Irwin, 1972), pp. 16-60; Patricia Morley, "Seton's Animals," Journal of Canadian Fiction, 2 (Summer, 1973), 195-98.

an exclusively Canadian genre, to be distinguished from the work of Jack London and Rudyard Kipling by its attempt to view life from the perspective of the animal itself. It bears no similarity to the fables of La Fontaine, nor is it candied folklore. It takes a biological stance which, by implication, has a deterministic quality attached. This determinism is premised on the cyclical relationship of all living things and to that extent may be said to constitute an ecological ethic which is immutable. This ethic is distinguished from anthropocentric morality *because* it embraces all of nature. Man's presumed dominion over nature is condemned as illusory.

The realistic animal story finds at least some of its inspiration in Montaigne and Rousseau, both of whom identified nature as the birthplace of truth and man's ultimate guide in his search for harmony. With Newton and Descartes it holds to an implicit belief in the unity of all life, and with Bacon it attempts to deduce the facts on which it is constructed from observations in the field. In this last respect it frequently fails, either through the limitations of its author or through the inexact state of his science. An overzealous allegiance to Montaigne and Rousseau, coupled with a sensual response to language, frequently makes of the story good literature but bad science. Alternatively, a devotion to turgid empiricism might yield reliable science but uneven literature. But whether the story leans toward science or toward literature, it reflects the Darwinian age in which it was born by leading its readers into that awesome never-never land where conventional morality cannot dictate to a determinism growing from the realities of survival.

In the realistic animal story the protagonists do not speak with human voices. Nor do they reason. Yet critics of the genre frequently confuse the author's overt sympathy for them with sentimentality. While there is often evidence to support the argument that the work is romantic, sentimentality does not overwhelm the stories nor does it necessarily function as a premeditated device of the author.[3] Writing in 1901, George Hellman argued that books which did not treat of human affairs would not last as literature.[4] For this reason he was prepared to cast his lot with Aesop and Kipling whose fables and tales, while avowedly untrue,

---

[3]Ernest Thompson Seton, Introduction to Famous Animal Stories, MS notes, n.d., SP; Introduction to the Volume of Animal Stories in the Young Folks Library, MS notes, n.d., SP; Autobiography, p. 353; Francis Dickie, "World Famous Manitoban," Winnipeg Free Press, Nov. 16, 1963. For some contemporary interviews with Seton on this subject, see, e.g., Nancy Bell, "The Work of Ernest Seton Thompson," Humane Review, 4 (April, 1903), 11-20; William Wallace Whitelock, "Ernest Seton-Thompson," Critic, 39 (1901), 320-25; James MacArthur, "Wolf Thompson and His Wild Animals," Bookman, 9 (March, 1899), 71-74; Myra Emmons, "Ernest Seton Thompson, Artist, Naturalist, Writer," Recreation, 6 (May, 1897), 315-30.

[4]George S. Hellman, "Animals in Literature," Atlantic Monthly, 87 (March, 1901), 391-92.

brought animals to the service of men and therefore became memorable. Many contemporary critics would maintain that Hellman's prophecy has been fulfilled, that the animal story is dead.[5] Even if this were so, it would reflect nothing more than the tragedy of a civilization preoccupied with its own creations. In pejorative terms to suggest that sympathy for animals--or for the land itself--is inherently romantic is to betray an unpardonable lack of humility before nature. Such a view relegates environment, which includes animals, to the role of backdrop. In this role, environment affects the affairs of men but has in itself no integral value. Critics in this tradition have often approached the animal story in a defensive posture. Because it casts man as part of the environment of animals, it is assumed, wrongly, to be opposed to man and therefore elicits an emotionally charged response that diverts the reader from the author's intention. The reputation of the animal story has repeatedly suffered because the literary critic has deprived it of its legitimacy. Yet the very things which the critic finds absurd in the course of his analysis are frequently those aspects of the animal story particularly important to the historian.

Amidst a sea of minor Canadian authors like Arthur

---

[5]Alec Lucas, "Nature Writers and the Animal Story," in Literary History of Canada: Canadian Literature in English, ed. Carl F. Klinck, et al. (Toronto: University of Toronto Press, 1966), p. 388.

Heming, W. A. Fraser and Archie McKishnie, there emerged at the turn of the century two writers whose work came to epitomize the best features of the genre. One was Charles G. D. Roberts, the other, Ernest Thompson Seton. Both men were products of the Canadian wilderness. Both were prolific and both enjoyed wide reading audiences throughout North America. Each has been called the father of the animal story, although it would be truer to say that each discovered it independently of the other at precisely the same moment. Both men wrote from the animal's viewpoint, and both demonstrated a remarkable awareness of natural history. Roberts, however, was distinguished from Seton by a superior grasp of the art of writing, while Seton was clearly more at home as a scientist.

Born near Fredericton, New Brunswick, Roberts was exactly Seton's age. Although they may have met in Toronto while Roberts was editor of Goldwin Smith's The Week, they did not become friends until at least 1899.[6] Prior to that time their influence on one another was negligible, yet Roberts always maintained that it was Wild Animals I Have Known that created popular interest in the animal story and encouraged him to pursue it seriously.[7] Unlike Seton,

---

[6]Seton to Charles G. D. Roberts, Jan. 22, 1899, Seton Letterbook No. 1, p. 389; Seton to Gertrude Pringle, Nov. 20, 1926; Gertrude Pringle to Seton, Nov. 23, 1926, SP; Gertrude Pringle to William Arthur Deacon, Nov. 23, 1926, copy in SP.

[7]W. J. Keith, Charles G. D. Roberts (Toronto: Copp Clark, 1969), p. 87.

Roberts enjoyed a happy childhood. His father, a Protestant minister and a highly cultured man, had encouraged his children to do the very things that Seton's father condemned as wasteful. Still, when the family moved to a new parish near Sackville, Roberts was required as a young boy and adolescent to share the chores necessary to the operation of a glebe farm adjacent to the parsonage. His free time was spent wandering through the Tantramar Marshes, subsequently immortalized in his poetry, and along the fringes of the dense New Brunswick forest so that in later life he could "claim to have had the intimacies of the wilderness as it were thrust upon him." He took no pleasure in hunting, but "To trail, to outwit, to ambush the wary dwellers of the wilderness,--to match my woodcraft against theirs and expose their furtive tactics . . . will never lose its thrill."[8] In 1874 the family moved to Fredericton where, in the company of his cousin, Bliss Carman, Roberts pursued a classical education at the Collegiate School. In 1876 he began his undergraduate degree work in philosophy and political economy at the University of New Brunswick. By 1881, when he completed his M.A., he had already published his first volume of verse, Orion and Other Poems. After serving as headmaster of one school and principal of another,

---

[8]Charles G. D. Roberts, The Watchers of the Trails (Toronto: Copp Clark, 1904), p. viii; Eyes of the Wilderness (Toronto: Macmillan, 1933), pp. 3-4.

Roberts moved to Toronto for an unsuccessful try at editing The Week, returning home to the Maritimes in 1885 to take up a Professorship at King's College in Windsor, Nova Scotia. Here he remained for ten years, writing, teaching, and raising a family, yet all the while attempting to maintain contact with the wilderness which inspired the best of his work. By 1890 he had been elected a Fellow of the Royal Society of Canada in recognition of his significant literary achievements, and by 1895 he felt sufficiently secure in his reputation to launch out entirely on his own. Although throughout his life Roberts remained one of Canada's most vocal nationalists, he spent the largest part of the next thirty years in England and Europe. Indeed, from 1907, the year after he was awarded an honorary doctorate at the University of New Brunswick, until 1925, he did not set foot in North America. It is probably not stretching the point to suggest that his phenomenal success with the animal stories, most of which were written during this prolonged absence, was due to a need to keep the land and its inhabitants alive inside his mind.[9]

---

[9]See Charles G. D. Roberts, Earth's Enigmas (Boston: Lamson, Wolfe, 1896); The Kindred of the Wild (Boston: Page, 1902); Red Fox (Boston: Page, 1905); The Haunters of the Silences (Boston: Page, 1907); The House in the Water (Boston: Page, 1908); Kings in Exile (London: Ward, Lock, 1909); Neighbours Unknown (London: Ward, Lock, 1910); More Kindred of the Wild (London: Ward, Lock, 1911); Babes of the Wild (London: Cassell, 1912); The Feet of the Furtive (London: Ward, Lock, 1912); Hoof and Claw (London: Ward, Lock, 1913); The Secret Trails (London:

Whatever his reasons for writing animal stories, Roberts' background ensured that they would differ in many respects from those of Seton. It must never be forgotten when discussing the latter's work that he did not enjoy the advantages of a university education nor the security of a well-paid profession. His childhood and youth had been marred by constant unhappiness and he had been forced very early in life to learn resourceful independence. Throughout his entire career Seton free-lanced in many different fields. Not once did he hold a full-time job. Indeed, his favorite practice was to create projects for himself, later selling the results to others when in need of cash. In the process he passed through alternating periods of poverty and prosperity, never allowing either condition to affect his basic principles. Without a formal training in literature or philosophy or science, he was all the more remarkable for accomplishing what he did. In effect, Seton began with his art, which was a native aptitude, developed it as extensively as possible, and then used it as a key to open other doors. One of these doors had led to the Smithsonian Institution where he had learned not only the rudiments of the scientific method, but the drift of biological thought and enquiry. This was particularly important to the substance

---

Ward, Lock, 1916); Wisdom of the Wilderness (London: Dent, 1922); They That Walk in the Wild (London: Dent, 1924).

of his animal stories and now must serve to rescue them from the scientific ignorance of aggressive literary critics who, in mocking isolated incidents of questionable veracity, betrayed their unwillingness to probe, much less acknowledge, the author's intention in writing them.

Exponents of the current vogue of literary nationalism in Canada have found in the animal story "a single unifying and informing symbol" which recurs in all Canadian writing. This is the symbol of Survival which implies "hanging on, staying alive. . . . The survivor has no triumph or victory but the fact of his survival; he has little after his ordeal that he did not have before, except gratitude for having escaped with his life." Furthermore, there is a "preoccupation with the obstacles to that survival." Margaret Atwood argues that in Canadian literature the majority of protagonists waste an inordinate part of their lives failing before apparently insurmountable obstacles--a fact which typifies Canada's victimization at the hands of the imperialist behemoth, whether British or American.[10] One of the major texts upon which this generalization is based is Seton's observation, in the preface to Wild Animals I Have Known, that "The life of a wild animal always has a tragic end."[11] But the realism of Seton's animal story

---

[10] Margaret Atwood, Survival (Toronto: Anansi, 1972), pp. 31-34.

[11] Seton, Wild Animals I Have Known, p. 12, Seton's

does not lie solely in the tragic end of its prótagonist. Nor, by implication, is mere survival the only issue. Seton's realism takes into account the entire life of his individual animals, suggesting an intricate existence which includes much more than a perpetual duel between hunter and hunted. Some of his animals are herbivores and do not hunt. Others are carnivores and are themselves not hunted. It is true that all of them endeavor to survive and that they frequently encounter obstacles to their survival. But even these obstacles do not dominate the stories, nor are Seton's tragic conclusions invariably related to them.

I make these points reluctantly because I see in Atwood's poetry and fiction, as in the work of Grey Owl, Fred Bodsworth, Farley Mowat, Wayland Drew and others, a binding link with Seton's animal story. Furthermore, I find no evidence of a similar link in any American literature. This phenomenon I would attribute, however, not to national origins but to the physical proximity of wilderness. Not the wilderness of parks, but a wilderness that knows no boundaries, is unquantified and unquantifiable, limiting man by its very immensity, commanding that the animals be seen and heard. Seton's attachment to the Canadian wilderness

---

italics; James Polk, "Lives of the Hunted," Canadian Literature, 53 (Summer, 1972), 53; Atwood, Survival, p. 74. See also, C. Lintern Sibley, "The Vogue of the Nature Story," Canadian Magazine, 38 (Jan., 1912), 287-92.

was not tinged with Roberts' nationalism. Nevertheless, Seton did have a profound sense of place, which to him meant the community of all living things. The tragic death of an individual being was often necessary to the survival of the whole community. Although he focussed his stories on the lives of individuals, Seton was no less concerned to keep that reality before his readers. His work was not preoccupied with the struggle for existence. Rather, it must be viewed as an affirmation of community. Only in this context can we understand Seton's antagonism toward his fellow man. The one force in the ecological cycle with which all living things must contend directly was also the greatest single obstacle to the survival of the community. In the wilderness community, where balance prevailed, the only autocthonous human being was the Indian. The white man, armed with the tools of his civilization, aspired not to become the land's, but to make the land his. His sense of community, unlike Seton's, encompassed only extensions of himself and necessitated tragedies which, far from sustaining the balance, guaranteed its annihilation. As the political and ideological embodiment of anthropocentrism, nationalism was not part of the solution, but part of the problem.

Generally speaking, literary critics have been kinder to Roberts than to Seton. And in purely literary terms the kindness has been warranted--"for Roberts it was art first; for Seton it was science." Both had a keen eye for natural history, but as one of Roberts' primary apologists notes, "Roberts has neither the time nor the taste for [Seton's] exacting science." He merely takes the animals he loves, "gives them a soul, makes them sharers in the common air and life of the world, and the result is scientific and yet more than science."[12] However, while Seton's "exacting science" may have detracted from his art, it adds to his historical importance. It is true that both men contributed to a frame of mind which could accept the prerogatives of conservation and preservation at the turn of the century. But Seton's work also reflected the new directions that biological science was about to take in the fields of animal behavior and ecology.

For his own part, Roberts regarded Seton not only as the father of the genre but as its best qualified practitioner. In 1913 he wrote that

> If there be one man, since St. Francis of Assisi whom all the kindreds of the wild have cause to bless, it is Ernest Thompson Seton. . . .
>
> [Seton's animal stories] aim above all to get at the psychology of their subjects. . . . From observed actions they strive to deduce motives and emotions. They are based on the conviction . . . that there _is_ an animal psychology. The old animal stories, if they went beyond mere external incident or adventure, simply

---

[12] Alec Lucas, "Introduction," in Charles G. D. Roberts, _The Last Barrier and Other Stories_ (Toronto: McClelland and Stewart, 1958), p. vi; James Cappon, _Charles G. D. Roberts_ (Toronto: Ryerson, n.d.), p. 19; Joseph Gold, "The Precious Speck of Life," _Canadian Literature_, 26 (Aug., 1965), 22-32.

> humanised their subjects, ascribing emotions and motives that would be proper, in like circumstances, to human beings. Seton has taught us to expect, in the animal story, a psychology immeasurably simpler . . . than that of man, . . . but none the less real and worthy of investigation. He spurns the theory that all animal life below the human plane is the blind and helpless slave of reflex action.
>
> As spokesman of the inarticulate kindreds, Ernest Seton is uniquely qualified. . . . And he is untainted by that excess of sentimentality towards them which too often perverts the view of the sympathetic nature lover. . . . He had seen more than other men. And he has seen with such discriminating, accurate and conscientious eyes, that I should be inclined to doubt the evidence of my own eyes if I found it conflicting with that of his.[13]

Roberts knew how diligently Seton strove to get at the truth, knew that the substance of his stories was drawn directly from the pages of his Journals, and knew that the study of animal behavior was about to become a major field of scientific endeavor. Accustomed to being attacked on a regular basis, Seton was never quite so effusive in his praise of Roberts. "Roberts has, of course, a high place as a stylist, but I think that if you inquire of the Smithsonian Institution or the Victoria Museum at Ottawa, you will find that he has no standing as a naturalist."[14]

Because they were both Canadians, Roberts and Seton brought to the American "back to nature movement"[15] certain

---

[13]Charles G. D. Roberts, "Ernest Thompson Seton," Bookman, 45 (Dec., 1913), 147-48; Kindred of the Wild, p. 28; E. M. Pomeroy, Sir Charles G. D. Roberts: A Biography (Toronto: Ryerson, 1943), pp. 173-74, 182.

[14]Seton to Gertrude Pringle, Nov. 20, 1926, SP; Seton, Autobiography, p. 353.

[15]Peter J. Schmitt, Back to Nature: The Arcadian Myth

assumptions about the land which were not current in the United States. The most important of these, contrary to the pessimism of Frederick Jackson Turner, was that a vast wilderness continued to dominate the North American continent. They themselves had grown up in intimate contact with it. For them it did not constitute some remote fantasy world to which they could escape from reality, or onto which they could project their own idealism. Having lived in the wilderness, Seton was quite firmly convinced that the wilderness now lived within him, demanding articulation. Gravely disappointed by the critical reception accorded his early ornithological papers, Seton, always the aspiring naturalist, diverted his attention to the mammals, using the animal story as a vehicle for exploring the uncharted territory of behavior analysis. Despite his abrasive exterior, his lack of formal scientific training constantly undermined his self-confidence. Here he was at least secure in the knowledge that there were no qualified experts.

Unlike Roberts, who wrote pure fiction, Seton adopted an autobiographical posture, frequently including himself in the stories. This was quite legitimate as most of the material for them was drawn directly from his own Journals. Furthermore, in condemning man's attitude to the natural world, Seton did not wish to cast himself as a righteous

---

in Urban America (New York: Oxford University Press, 1969).

prig and hoped that by being self-critical, he might encourage his readers to question their own values. Problems emerged, however, when he attempted to make psychological deductions from his recorded facts. While his generalizations were sometimes remarkably accurate, especially in the light of recent developments in ethology, others opened him to widespread criticism and had the effect of creating a minor controversy around the advisability of pursuing the genre.

II

The widespread success of both Seton's and Roberts' animal stories attracted many imitators in the United States. Chief among them was William J. Long, a native of Massachusetts, a graduate of Harvard and the University of Heidelberg and pastor of a Congregationalist Church in Stamford, Connecticut. After 1899 Long published several articles and books for children in the nature study line. He claimed to have spent much time in the woods and persistently defended the truth of his tales. In one instance he alleged that he had seen a woodcock with a broken leg fashion a splint from roots and grass fibres, fastening it round the fracture with a malleable paste of clay. After sitting in the sun for an hour, allowing the cast to dry, this remarkable bird flew away. In another of Long's stories, a fox trying to get at some chickens roosting beyond its reach ran rapidly in circles around the base of their perch. The chickens became dizzy watching this process and fell to the ground. Neither Seton nor Roberts was guilty of attributing reason to animals in this manner. Despite the obvious folly of some of their interpretations, the facts as presented almost always allow a logical explanation in the light of modern ethology. In Long's case the evidence submitted for the interpretation was itself apocryphal. Gullible, naive and, to say the least, anthropomorphic, Long accepted as God's truth any rare old hunter's tale or farmer's recollection. In his professed love for the wilderness, he abandoned science to sympathy and failed to allow the tension between them to guide his writing.[16] As far as Seton was concerned, an animal story without the backbone of science had no legitimate claim on the reader's attention.

This view was shared by others, especially the venerable old naturalist, John Burroughs (1837-1921). With his flowing

---

[16]See especially, W. J. Long, School of the Woods: Some Life Studies of Animal Instincts and Animal Training (Boston: Ginn, 1902); "The Modern School of Nature Study and Its Critics," North American Review, 176 (May, 1903), 689-90. For a good comparison, see Seton's treatment of healing in the Biography of a Grizzly (New York: Century, 1900), pp. 70-71. "[Wahb's] knowledge of healing was wholly instinctive. He licked the wound and all around it. . . . The licking removed the dirt, and by massage reduced the inflammation, and it plastered the hair down as a sort of dressing over the wound to keep out the air, dirt and microbes."

white beard and gentle demeanor, the "Seer of Slabsides" was a kind of national grandfather whose reputation as a calm and highly articulate interpreter of nature's ways was padded by an abiding friendship with Theodore Roosevelt, his most ardent defender. Repelled by the popularity of anecdotal animal books, Burroughs took it upon himself to correct the situation in a scathing rebuke of Roberts, Seton and Long, published in the Atlantic Monthly in 1903.[17] His kindest words were reserved for Roberts, but for some reason he chose to link Seton and Long together, denouncing both as "sham" naturalists. In Long's case this was so obvious that one must marvel at the amount of space Burroughs devoted to proving it. Seton clearly deserved criticism for some of his theories, notably those on suicide. But in focussing on these few shortcomings, Burroughs dismissed the possibility that anything Seton had to say was worth considering. He did not know Seton, was completely unaware of his background and interest in natural history, and had never read his scientific publications. Believing that instinct was the source of all animal behavior, Burroughs rejected learning theory as fabrication. He did not seem to realize that most of the problems in Seton's stories were attributable to interpretation rather than to a lack of factual truth. Furthermore, his entire analysis of Wild Animals I Have

[17]John Burroughs, "Real and Sham Natural History," Atlantic Monthly, 91 (March, 1903), 298-309.

Known, the book on which he centered his assault, was premised on his own observations in the Hudson River Valley and in the Catskill Mountains. As none of the stories in Seton's book took place anywhere near that particular vicinity, there was no ground at all for generalizing. The science of ethology demonstrates that the behavior of animals, whether learned or innate, is frequently so oddly complex, its stimuli so difficult to discern and its responses so apparently out of context with the immediate situation, that the truth is inevitably stranger than fiction. Burroughs' concept of instinct was limited to its most obvious survival function. His inability to recognize that this dimension had a multiplicity of extensions, while understandable, effectively denied the substance of his analysis of Seton's work. That he linked Long and Seton together at all suggests the limitations of his observations in the field and opens his own writing to renewed scrutiny. Yet Burroughs' attack is still cited by modern literary critics as evidence of Seton's dereliction.

In fact, Burroughs himself sensed that he was out of his depth. "I worry at times over my Atlantic article," he wrote to a friend, "for fear I have overdone it and showed too much temper. . . . I am reading Darwin again, and a German book called Animal Play, and I find a few things that might give some color of truth to some of [the] stories."[18] Seton was deeply hurt by the tenor of the article. He had always respected Burroughs but found his remarks "so bitter, so unfair and so untrue" that he considered replying to them publicly. He was dissuaded by William Brewster, one of the original founders of the AOU and a close friend of Merriam, who was terribly upset by the controversy. Like Burroughs at Slabsides, Brewster had created a wildlife sanctuary at Concord, Massachusetts, using it as a kind of outdoor laboratory from which to draw the subject matter for countless scientific papers on the ecology of birds and small mammals. He liked Seton's stories and felt they had an important role to play in arousing interest in animal behavior. "I am boiling over with wrath and indignation at Burroughs' . . . singularly unjust and ill-tempered article," he wrote to Seton, "but if I am not mistaken he has done himself irreparable harm by writing [it]." Although virtually all of his friends rallied to his defense, all were unanimous in advising him to maintain a "dignified silence"--a difficult task for Seton at the best of times.[19]

---

[18] John Burroughs, quoted in Clara Barrus, The Life and Letters of John Burroughs (2 vols.; Boston: Houghton, Mifflin, 1925), 2:49. The book to which he referred was probably Karl Groos, The Play of Animals (New York: Appleton, 1898).

[19] William Brewster to Seton, March 4, 1903, March 23, 1903, SP; William Hornaday to Seton, Feb. 19, 1904, Hornaday Letterbook No. 2, p. 18, William T. Hornaday Papers, Library of Congress, Manuscript Division, Washington, D.C.; hereafter cited as LC WTHP; Seton, Autobiography, p. 367.

Within a month of the controversy Seton found himself seated next to Burroughs at a huge dinner given by Andrew Carnegie. In his autobiography Seton makes their first meeting sound like the ultimate chastening experience for Burroughs, with the old man actually dissolving in tears when confronted with the evil he had wrought. But in Carnegie's pontifical assessment,

> Burroughs and Seton both enjoyed the trap I set for them. True it is, we only hate those whom we do not know. . . . Most quarrels become acute from the parties not seeing and communicating with each other and hearing too much of their disagreement from others.

In the course of the evening the two men made up for lost time, Seton making sure that Burroughs understood the extent of his experience in the wilderness and attempting to win his friendship. They left the party on good terms and within a short time were visiting and corresponding on a regular basis. Once Seton had been able to show Burroughs his scientific library, his collection of animal photographs, mammal and bird skins, drawings, "and last of all my journals, . . . detailing my travels and observations during thirty years, he broke down and surrendered."[20] This melodramatic elaboration of Burroughs' change of heart was not far from the truth. By 1905 the old man had

---

[20] Andrew Carnegie, The Autobiography of Andrew Carnegie (Boston: Houghton, Mifflin, 1920), p. 281; Seton, Autobiography, p. 371; William Brewster to Seton, April 1, 1903, May 20, 1905, SP.

turned increasingly more of his attention to the study of behavior and had mellowed enough to write to a friend that Seton "easily throws all other animal story writers in the shade."[21] Although still refusing to admit the possibility that learning had any significant part to play in the animal kingdom, he began to recognize that instinct was a multidimensional phenomenon which evolved with each species.

But the debate on nature faking did not end with the reconciliation of Burroughs and Seton. Another naturalist whose acceptance Seton coveted was Theodore Roosevelt. Although he believed the stories to be "based on accurate observation," Roosevelt felt that Seton went "too far in emphasizing the scientific value of his tales." As most of his own wilderness experience had been gleaned in the pursuit of big game and in reliving the imagined existence of the pioneer, he was negatively disposed to the sympathies of the animal story writers. He distrusted any story that

---

[21] John Burroughs, quoted in Barrus, Life and Letters of John Burroughs, 2:88-89; John Burroughs, "The Literary Treatment of Nature," Atlantic Monthly, 94 (July, 1904), 42. After 1903 there is abundant evidence to suggest that Burroughs frequently asked Seton for advice; John Burroughs to Seton, Aug. 1, 1908, March 25, 1920, June 10, 1920, July 5, 1920, Oct. 11, 1920; Seton to John Burroughs, March 21, 1920, March 27, 1920, June 7, 1920, June 14, 1920, June 30, 1920, Oct. 7, 1920, SP; Barrus, Life and Letters of John Burroughs, 2:54-55, 87, 97; John Burroughs, Ways of Nature (Boston: Houghton, Mifflin, 1905); Leaf and Tendril (Boston: Houghton, Mifflin, 1908); The Summit of the Years (Boston: Houghton, Mifflin, 1913).

had been related to an author by an Indian because Indians lived "in a world of mysticism" and, with a "childlike faith," attributed all manner of supernatural traits to animals. He latched onto Burroughs' early approach, labelling all the writers "yellow journalists of the woods" who had learned nothing of "the real secrets of the wilderness." He was disturbed mainly by the authors' claims to veracity. "If the child mind is fed with stories that are false to nature," he charged, "the children will go to the haunts of the animal only to meet with disappointment. The result will be disbelief and the death of interest." While Long and Jack London had earned this reproach, Seton's work for the most part stood well outside the range of any criticism Roosevelt was qualified to level. Here was one writer whose wilderness experience far outdistanced that of the President. And as for Teddy's concern about the effects of the animal story on children, there could have been no single author in America more intent upon reaching young minds with the truth than Seton. The problem was that his truth and Roosevelt's truth parted company in morality.[22]

---

[22] Theodore Roosevelt, quoted in Hamlin Garland, Companions on the Trail (New York: Macmillan, 1931), p. 205; Theodore Roosevelt to Caspar Whitney, June 7, 1901, The Letters of Theodore Roosevelt, ed. Elting Morison (8 vols.; Boston: Houghton, Mifflin, 1951), 3:89; Theodore Roosevelt, "Nature Fakers," Everybody's Magazine, 17 (Sept., 1907), 428; Theodore Roosevelt, quoted in Edward B. Clark, "Roosevelt and the Nature

Seton was a polemicist for nature. He believed that he had to stimulate an attitude toward it as well as to describe the activities occurring within it. As his work with boys was already beginning to demonstrate, he did not share Roosevelt's view about the Indians' "childlike faith" in the supernatural traits of animals. In The Trail of the Sandhill Stag he had introduced his readers to the Cree, Chaska, whom he had known in Manitoba during the eighties and whose detailed knowledge of woodcraft had provided an incentive to learn the habits of the animals. From Chaska he had learned the art of tracking, of reading the sign language of the snow on which were written many of the incidents transcribed into his stories. Seton adopted the Indian credo that nature, in both its harshness and its goodness was a priori right and necessary; anything that interfered with the biological clock was a priori evil. The pantheistic vision that emerged from this perspective relegated man to the lowest level of creation. Wholly dependent upon the animals and plants for his survival, he was the weakest and least essential link in the great chain of being, a position demanding of him the utmost humility. Anthropocentrists like Roosevelt, whose respect for nature was measured out in trophy heads, had no claim in Seton's value system to a corner on the market for truth. Seton's "sympathy" for nature, while it may have

---

Fakirs," Everybody's Magazine, 16 (June, 1907), 771, 774.

evoked sentimentality, was nothing less than an ecological conscience deprived of an ecological science. The intuition of the Indian, as accurate a gauge as could possibly have been hoped for, provided the moral and spiritual context for his stories.[23]

The nature faker controversy dragged on for several years. However, both Seton and Roberts were all but forgotten as the attacks on William Long grew more and more strident. William Morton Wheeler, who two years earlier had published his first article on ethology in Science, wrote to the same magazine in 1904 to congratulate Burroughs for having exposed Long. Carefully omitting any reference to Seton, he condemned the American public for failing to distinguish "sound and critical observations on animal behavior from the drivel in which animals are humanized beyond all recognition." E. W. Nelson of the Biological Survey joined Edward B. Clark of Everybody's Magazine in conducting and publishing the results of a symposium of nature faking. Among those whose opinions they solicited were William Hornaday and C. Hart Merriam. Although Hornaday's later work was to be scarred by excessive anthropomorphism, he jumped on the anti-Long bandwagon with a vengeance, arguing

---

[23] Ernest Thompson Seton, The Trail of the Sandhill Stag (Scribner's: New York, 1899), pp. 47-53; Autobiography, pp. 261-65. See also, Diamond Jenness, "The Indians' Interpretation of Man and Nature," Transactions of the Royal Society of Canada, 24 (1930), 57-62; Irving A. Hallowell, "Ojibwa Ontology, Behavior and World View," Culture in History: Essays in Honor of Paul Radin, ed. S. Diamond

that an "unqualified approval" of his books was "a sure index of profound ignorance regarding wild animals, their mental capacities and their ways." Merriam also focussed his entire attack on Long. Indeed, not one contributor to the symposium mentioned Seton by name or alluded to his stories. Robert Yerkes, the famous Harvard psychobiologist, continued to send his graduate students to Seton for advice. Writing in 1909, Yerkes assured him that "I believe as thoroughly in your stories as I do in the scientific research into the laws of animal behavior that I am trying to further." Nature faking, therefore, gradually became synonymous with the name of William J. Long who had clearly rearranged the wilderness to accommodate his own fantasies. As Frank Chapman was quick to point out, "no one who knew Seton would think of doubting the soundness of his observations or of classing him with his would-be emulators."[24]

(New York: Columbia University Press, 1960), pp. 19-52.

[24]William Morton Wheeler, "Woodcock Surgery," Science, 19 (Feb. 26, 1904), 348; "'Natural History', 'Oecology,' or 'Ethology?'" Science, 15 (June 20, 1902), 971-76; Edward B. Clark, "Real Naturalists on Nature Faking," Everybody's Magazine, 17 (Sept., 1907), 423-27; Chapman, Autobiography, pp. 182-83; Robert M. Yerkes to Seton, Jan. 7, 1909; Henry R. Carey to Seton, Nov. 17, 1912, SP. Long chose to defend himself rather than admit to any wrongdoing; W. J. Long, "Science, Nature and Criticism," Science, 19 (May 13, 1904), 760-67. See also, Frank M. Chapman, "The Case of William J. Long," Science, 19 (March 4, 1904), 387-89; William Harper Davis, "Natural and Unnatural History," Science, 19 (April 22, 1904), 667-75; John Burroughs, "Roosevelt as a Nature Lover and Observer," Outlook, 86 (July 13, 1907), 547-53; Paul Russell Cutright, Theodore Roosevelt: The Naturalist (New York: Harper, 1956),

III

Clearly then, the controversy surrounding the nature fakers had nothing at all to do with simple literary problems. The real issue was the state of the scientific study of animal behavior, or ethology, still in its infancy, but after 1903 gradually winning new converts. Among the believers was Seton himself, among the nay sayers, Theodore Roosevelt. Somewhere in the middle, and at least partly because of Seton's influence, stood John Burroughs. For all three of these men, two important concepts--instinct and learning--functioned as the pivot of debate. In his belief that man must limit himself, Seton argued in defense of a biological morality exemplified in the animal kingdom. He used his animal stories as a vehicle for the expression of this moral perception, which, he maintained, had grown naturally from his own observations in the field. The concepts of instinct and learning are central to an understanding of Seton's interpretation of those observations.

Although it is generally assumed that ethology was introduced by Darwin's hypothesis, recent work by historians of zoology has demonstrated that the science has much deeper roots. As early as the seventeenth century Francis Willughby (1635-1672) and John Ray (1627-1707) abandoned the tradition of investigating the sensational exploits of animals in

---

pp. 132-50; Keir B. Sterling, Last of the Naturalists: The Career of C. Hart Merriam (New York: Arno Press, 1974), pp. 322-32.

favor of consistent observation over time. Willughby is credited with the discovery of territoriality and is believed to have been the first to suggest that birds can count the number of eggs laid in a small clutch, a phenomenon described by modern ethologists as "subitizing." The German naturalist, Johann Ferdinand Adam Perneau (1660-1731), carried out remarkable experiments with birds in natural aviaries, leaving behind deductive accounts covering territorial habits, migration, call notes, reproductive behavior and so on.[25]

Greek and Roman philosophers had distinguished between man and animal on the grounds of rationality. Man possessed a creative imagination; the animals were prisoners of instinct. Christianity fortified this assumption in theological terms by denying animals the prospect of redemption and immortality. But beginning in the eighteenth century the attitudes of some began to change. David Hume (1711-1776) suggested that animals might actually be endowed with thought and reason. Julien de la Mettrie (1709-1751) challenged the Cartesians, arguing

---

[25]Philip Howard Gray, "The Early Animal Behaviorists: Prolegomenon to Ethology," _Isis_, 59 (1968), 374. (In the following pages I am deeply indebted to the remarkable work of Prof. Gray. Without the aid of his historiographical analyses my own study would have suffered immensely.) See also, C. E. Raven, _John Ray, Naturalist: His Life and Works_ (Cambridge: Cambridge University Press, 1942); F. Willughby, _The Ornithology of Francis Willughby, Wherein All the Birds Hitherto Known are Accurately Described_, ed. John Ray (London: J. Martyn, 1678); E. Streseman, "Baron von Perneau, Pioneer Student of Bird Behavior," _Auk_, 64 (1947), 35-52.

that man himself was a machine that operated according to rigidly prescribed natural laws; man was distinguished from the other animals only in terms of the complexity of his machinery. If man could learn, it followed that animals could learn. Charles Georges Leroy (1723-1789), a disciple of la Mettrie, asserted that animals learned from their own experience; their behavior was modified by the externalities they faced in the course of a lifetime. The relative complexity of the animal would affect the amount or the quality of the learning, but no animal could be said not to have learned merely because it could not think or act like a man. Leroy also maintained that an understanding of the behavior of animals would throw light on the behavior of man. Scientists, however, would require "a complete biography of every species, which would include the species' individual character, natural appetites, and its particular way of life." After observing animals in their natural habitat, the environment should be altered in different ways so that through empirical comparisons behavioral generalizations could be deduced.[26]

---

[26]Philip Howard Gray, "The Problem of Free Will in a Scientific Universe: René Descartes to John Tyndall," Journal of General Psychology, 80 (1969), 57-72; "The Nature of Man or Man's Place in Nature? Bruno Spinoza and la Mettrie," Journal of General Psychology, 76 (1967), 183-92; "Early Animal Behaviorists," pp. 376-78; David Hume, A Treatise of Human Nature (Garden City, N.Y.: Doubleday, 1961), p. 161; J. O. de la Mettrie, Man a Machine, trans. Gertrude C. Bussey and Mary W. Calkins (Chicago: Open Court, 1912), pp. 100-04; C. G. Leroy, The Intelligence

John James Audubon (1785-1851), an American of French ancestry, pursued Leroy's suggestions in his own ornithological monographs. In 1826, for example, he experimented with Black Vultures to determine whether they could smell. By manipulating the carcass of the prey at intervals, he concluded that the Vulture awaited preliminary decomposition only because its beak could not otherwise penetrate the skin. The limelight in America was shared with Lewis Henry Morgan (1818-1881) whose ethnological studies on the Iroquois have tended to obscure his behavioral research. Morgan shared Hume's enthusiasm for animal intelligence but based it largely on acquired learning stimulated by rewards, later explained by Pavlov as "conditioning." With Leroy, Morgan also believed in the desirability of animal biographies, centering his own attention on the American Beaver. Even though Darwin dismissed him as inconsequential, Morgan is today regarded as the man through whose efforts ethology "matured into a research discipline." Seton particularly admired the work of Audubon and Morgan, both of whom he cited as authorities in much of his own research. His fascination with Morgan's ideas also carried over into his later work with youth.[27]

---

and Perfectibility of Animals from a Philosophic Point of View with a Few Letters on Man, trans. Mrs. Richard Congreve (London: Chapman and Hall, 1870), pp. 7-14, 88, 184-85.

[27] Gray, "Early Animal Behaviorists," pp. 379-83; J. J. Audubon, Ornithological Biography, or an Account of

Evolutionary theory, which contradicted Descartes' man-animal dichotomy, remains the most effective scientific tool for demonstrating that "man is not essentially different from animal and that any actual difference . . . is only a difference of degree of development."[28] Nevertheless, American psychology became increasingly preoccupied with an emphasis on learning theory which, it has been suggested, "was historically inspired, and . . . remained appealing, because it somehow maintained the image of man as 'higher' or 'better.'" In other words, American psychology was guilty of "over humanizing" man by de-emphasizing Darwin. But by definition psychology was aimed at the investigation of human behavior. This, in turn, demanded that considerable attention be paid to the effects of environment and experience, with the result that inherited determinants of behavior--or the same instincts man shared with the animals--were substantially overlooked. Courageous scholars like William McDougall (1871-1938), who viewed instinct as central to human behavior, were frequently guilty of

---

the Habits of the Birds of the United States (5 vols.; Edinburgh: Adam Black, 1831-1839), 2:33-52; L. H. Morgan, The American Beaver and His Works (Philadelphia: Lippincott, 1868); Ernest Thompson Seton, Life-Histories of Northern Animals (2 vols.; New York: Scribner's, 1909), 2:1224-25, 1250; Seton, Autobiography, p. 126. See also, John James Audubon, Audubon and His Journals, ed. Maria R. Audubon and Elliott Coues (2 vols.; New York, Scribner's, 1897); and John Bachman, The Quadrupeds of North America (3 vols.; New York: V. G. Audubon, 1849-1854).

[28] H. Misiak and Virginia S. Sexton, History of Psychology: An Overview (New York: Grune and Stratton, 1966), pp. 65-66.

Lamarckism (the belief in the inheritance of acquired characteristics), a shortcoming which fortified the case against them.[29]

While instinct did not enjoy a good press in American psychology, it dominated the attention of many Europeans. The famous Austrian biologist, Konrad Lorenz, was severely critical of McDougall. But instead of abandoning instincts, he concentrated on the formulation of a body of theory which would explain them in scientifically acceptable terms. The modern science of ethology, drawn from Darwin and amplified by Lorenz, Niko Tinbergen, Julian Huxley, W. H. Thorpe and others, recognizes five primary instincts: nourishment, reproduction (including rearing of the young), sleep, care of the body surface and companionship within the species. There are, however, other acts which are not considered instincts but action patterns, or behavior patterns, through which the instincts find expression. Behavior patterns in animals demonstrate an evolution precisely the same as that of organs. In the words of Lorenz,

> the adaptation of the behavior patterns of an organism to its environment is achieved in exactly the same manner as that of its organs, that is to say on the basis of information which the species has gained in

---

[29] R. L. Eaton, "An Historical Look at Ethology: A Shot in the Arm for Comparative Psychology," _Journal of the History of the Behavioral Sciences_, 6 (1970), 177-78. See, William McDougall, _Body and Mind: A History and a Defence of Animism_ (London: Methuen, 1911); _An Introduction to Social Psychology_ (London: Methuen, 1908); "The Use and Abuse of Instinct in Social Psychology," _Journal of Abnormal Social Psychology_, 16 (1922), 285-333.

> the course of its evolution by the age-old method of mutation and selection. This is true not only for rigid patterns of form or behavior, but also for the complicated mechanisms of adaptive modification among which are those generally subsumed under the conception of learning.

Behavior patterns are now recognized to be highly conservative or tenacious. Even after they have outlived their original functions they remain, either becoming vestigial or developing new functions. This means that the physical manifestations of instinct fulfillment may appear absurd and inexplicable. One form of learning, "imprinting," still occupies a kind of grey area between instinct and association, raising special problems. But the basic question of learned behavior has been severed from Lamarckism by the ethologist's insistence that it is not bred into the species but is acquired during an individual's lifetime and/or communicated culturally from one generation to the next, in degrees depending upon the social sophistication of the species. Ethology, therefore, has been successful, where McDougall was not, in demonstrating an evolutionary continuity in behavior as between animals and man, on one hand reasserting the importance of instinct in American psychology, on the other crediting animals with the capacity to learn.[30]

---

[30]Konrad Lorenz, "Preface," in Charles Darwin, The Expression of the Emotions in Man and Animals (Chicago: University of Chicago Press, 1965), pp. xii-xiii, "McDougall's Instinct Theory," in Instinctive Behavior: The Development of a Modern Concept, ed. Claire H. Schiller (New York: International Universities Press, 1957), pp. 156-63; Sally Carrigher, Wild Heritage (Boston: Houghton, Mifflin, 1965),

The animal stories of Ernest Thompson Seton were written during a period when biologists in the United States and Europe were bridging the gap between Audubon and Morgan

---

p. 31. Learning is still the most controversial issue in ethology. In 1872 an Englishman, D. A. Spalding, invented the concepts of "behavioral maturation" and "behavioral critical periods" which saw instincts at the root of animal learning; D. A. Spalding "Instinct, with Original Observations on Young Animals," Macmillan's Magazine, 27 (1873), 282-93. Seton probably knew of this important work. Lorenz tested it in his experiments with waterfowl and Jackdaws, concluding that most birds do not instinctively recognize their own species. A bird raised in captivity in isolation from its own species will become "imprinted" on its human keeper. It will retain its innate behavior patterns but will focus them on the object to which it is imprinted. The imprinting process is restricted to a short critical period in the early life of the individual. Once it has occurred it is irreversible. As a form of early, unrewarded learning, imprinting is distinguished from associated and conditioned learning. As the individual ages, imprintability decreases while associated learning increases. In imprinting, that which is learned first is remembered best; in associated learning that which is learned last is remembered best. Conditioned learning, synonymous with Pavlov and B. F. Skinner, is generally linked with experimental psychology and need not concern us since it removes the animal from its natural environment and implies "training." Factors like imprinting and associated learning ability play a large part in determining the extent to which the cultural inheritance of the group will be transmitted from generation to generation. Eventually a point is reached where natural selection favors an animal on the basis of what it has learned rather than what it has inherited. There is even evidence to suggest that animals exhibit "ideational processes" even though they cannot conceptualize or manipulate abstract symbols. Konrad Lorenz, "Imprinting," Auk, 54 (1937), 245-73; King Solomon's Ring: New Light on Animal Ways (New York: Signet, 1972), pp. 143-92; Philip Howard Gray, "Spalding and His Influence on Research in Developmental Behavior," Journal of the History of the Behavioral Sciences, 3 (1967), 168-79; John Paul Scott, Animal Behavior (2nd ed.; Chicago: University of Chicago Press, 1972), pp. 171-73, 290-300; W. H. Thorpe, Biology and the Nature of Man (London: Oxford University Press, 1962), p. 68. For an excellent review of modern ethological literature, and an attempt at synthesis, see, C. G. Beer, "Ethology--The Zoologist's Approach to Behaviour, Part 2," Tuatara, 12 (1964), 16-39.

at one pole, Lorenz, Tinbergen and Thorpe at the other. The early animal behaviorists, whose work was not premised on Darwin, were motivated largely by a sympathy for the animal kingdom and by the radical, if frequently esoteric, insights of anti-Cartesian European naturalists and philosophers. Audubon and Morgan hoped to credit animals with greater intelligence, rescuing them from the theory that they were mere automatons. Hostile critics of Seton's animal stories usually shared the traditional anthropocentrism of American psychologists, overemphasizing (or misrepresenting) learning, denying the evolutionary continuity of instinct in animals and man, and implicitly defending Descartes against the revelations of Charles Darwin. Several biologists, many of whom Seton knew personally, were grappling with the problems of instinct and intelligence for the first time on a Darwinian plane. Seton aspired to be a part, however small, of the investigation, and his work must be viewed as a conscious attempt to acquaint the public with exciting developments in the field. If one adopts this perspective, it becomes clear that the failings of Seton's books were frequently the failings of the science. This, however, takes nothing away from the courage and imagination of a man who recognized what revolutionary ideas were emerging in biology and who made a concerted effort, first to understand them, then to give them popular expression. In the progressive dawning of the expert, the scientific popularizer came constantly under attack.[31] But had it not been for men like Seton, Americans generally might well have lost touch with biological realities. The ascendancy of the physical sciences, coupled with their offspring, technology, had created an atmosphere in which a superficially biological social phenomenon like the conservation movement gained credibility only from the fact that it promised to sustain material progress in the long term. Although he aided the cause of conservation, Seton was at heart an ecologist, which meant that he opposed anthropocentrism. Through his animal stories he attempted to raise serious doubts about man's presumed supremacy and thereby to teach him humility before nature. Before we look at the animal stories, however, we should know what the ethologists were doing while he wrote them.

Several Americans emerged before the end of the nineteenth century as potential ethologists. Among the most important of them was William Morton Wheeler (1865-1937), who studied at Clark under another pioneer behaviorist, C. O. Whitman (1842-1910), in 1892 joining him as a colleague in the newly formed department of biology at the University of Chicago. In 1898 Whitman delivered his seminal lecture on "Animal Behavior" in which he stressed four basic points:

---

[31]Ronald C. Tobey, _The American Ideology of National Science, 1919-1930_ (Pittsburgh: University of Pittsburgh Press, 1971), pp. 3-19.

(1) that instincts evolve naturally and are not improvised, (2) that they can be performed by the animal without learning, (3) that comparative study is the only means of achieving a reliable phyletic history, and (4) that flexibility of instinct is not itself a manifestation of intelligence but "the open door through which the great educator, experience, comes in and works every wonder of intelligence."[32] Wheeler followed this in 1902 with his article for Science in which, citing European developments in the field, he extracted a biological meaning for the term, ethology.

> The only term hitherto suggested which will adequately express the study of animals, with a view to elucidating their true character as expressed in their physical and psychical behavior towards their living and inorganic environment, is ethology. . . . Wasmann defines ethology . . . as 'the science of the external conditions of existence which pertain to organisms as individuals and at the same time regulate their relations to other organisms and to the inorganic environment.' It therefore embraces in its restricted sense: 'first, a knowledge of the mode of life of animals and plants, their nourishment, dwelling, mode of propagation, the care of offspring and their development, in so far as these present external manifestations; hence also, second, a knowledge of the life-relations that obtain between individuals of the same and different species . . ., [and] third, a knowledge of the conditions of existence which are essential to the life and maintenance of animals and plants.'[33]

Both Whitman and Wheeler followed Leroy's directive regarding

---

[32] C. O. Whitman, "Animal Behavior," 1898, quoted in Mary Alice Evans and Howard Ensign Evans, William Morton Wheeler, Biologist (Cambridge, Mass.: Harvard University Press, 1970), pp. 211-12.

[33] Wheeler, "'Natural History,' 'Oecology' or 'Ethology,'" pp. 974-75; "The Obligations of the Student of Animal Behavior," Auk, 21 (1904), 251-55.

"animal biography," the former working almost exclusively with Pigeons, the latter with Ants. From his investigations of the behavior of Ants Wheeler reached new conclusions about vestigial instincts, including the fact that they may remain dormant for several generations, resurfacing under the pressure of uncommon environmental stimuli. He also recognized that vestigial instincts frequently acquired new functions and suggested the desirability of comparing the behavior of related organisms, "for we know of no case in which an instinct is peculiar to a single species . . . and no case in which two species manifest an instinct in precisely the same manner."[34]

Wheeler shared Leroy's sympathetic intuition that the animal was much more subtle and complex than man would like to believe. He also recognized a "fundamental identity of the animal and human souls, the differences between which . . . are differences of degree of development." For Wheeler this implied that intelligence was not exclusively human "and that the animal mind is to be interpreted in the light of human psychology." The dangers of attributing to lower animals a greater degree of intelligence than they possessed

---

[34]William Morton Wheeler, "Vestigial Instincts in Insects and Other Animals," American Journal of Psychology, 19 (1908), 11-13. See also, C. O. Whitman, Posthumous Works of Charles Otis Whitman (3 vols.; Washington: Carnegie Institution of Washington, 1919); William Morton Wheeler, Ants: Their Structure, Development and Behavior (New York: Columbia University Press, 1910).

would be eliminated, he believed, by scientific methods of observation and experiment. But from another side, the psychological perspective led Wheeler off on a Lamarckian tangent where he argued that "instinct is essentially inherited habit." This belief in the inheritance of acquired learning opened him to the charge of vitalism, linked him with William McDougall and polarized him from biological determinists like Jacques Loeb (1859-1924), a colleague at the University of Chicago, who maintained that all actions, even ethics, were mechanical instincts determined by the inherited structure of the nervous system. Wheeler's defence of the psychological (really the anthropomorphic) approach to behavior must not, however, be misconstrued as a defence of psychology, a discipline he patently abhorred.[35]

In holding that the living animal "is not a mere mechanical system but a creative organism, a being that cannot be isolated from its environment," Wheeler believed that the most effective ethological investigations would occur in the field, not in the laboratory. This explains his faith in scientific natural historians, without whose work the new biology would not have been possible. Indeed,

---

[35] William Morton Wheeler, "On Instincts," _Journal of Abnormal Psychology_, 15 (1921), 300-03; Evans and Evans, _William Morton Wheeler_, p. 223; Jacques Loeb, _The Mechanistic Conception of Life_, ed. Donald Fleming (Cambridge, Mass.: Harvard University Press, 1964).

Wheeler's qualifications for the successful naturalist read like an itemized description of Ernest Thompson Seton.

> The naturalist is mentally oriented toward and controlled by objective, concrete reality, and probably because his senses, especially those of sight and touch, are highly developed, is powerfully affected by the esthetic appeal of natural objects. . . . He is primarily an observer and fond of outdoor life, a collector, a classifier, a describer, deeply impressed by the overwhelming intricacy of natural phenomena and revelling in their very complexity. He is, therefore, more or less irrational, intuitive, receptive and passive in his attitude toward natural objects, synthesizing rather than analyzing [them]. . . .[36]

Seton and Wheeler were both inquiring about the extent to which animals informed the experience of man. To have denied their sympathy *for* the animals would have been to reject their identification *with* them. Consciously to have humanized them would have meant eliminating their own credibility as scientists. In Seton's animal stories the mixture of sympathy and science reflected the tensions to which Victorianism and a profound love for the wilderness condemned him. Although born with a Calvinist heart, he had replaced the Bible with the *Origin of the Species*; the Darwinian maxim that man derived from animal had become a literal article of faith. A careful reading of Wheeler and others like him merely gave it an important new dimension.

Well before 1915 British and European scholars had been wrestling with complex theories related to imprinting.

---

[36]William Morton Wheeler, "What is Natural History?," *Bulletin of the Boston Society of Natural History*, 59 (1931), 3-12; "On Instincts," p. 305.

Konrad Lorenz's most influential teacher, Oskar Heinroth (1871-1945), identified the concept of homology, or inheritance from a common ancestor, in 1910. He observed that many of the acts of birds are stereotyped, executed in precisely the same manner by all members of a species. Frequently these acts were homologous with similar acts in related species. It became clear that stereotyped behavior patterns, including, for example, the social displays attached to the instinct of reproduction, were matched by stereotyped responses which had evolved reciprocally. Clearly, the behavior of one animal could function "as the releasing stimulus for behavior in another of the same species" as long as the "signal . . . or posture [could] be 'understood' by all members of the species." In order for the display to be understood, it must be conspicuous.[37] While observing the bizarre courtship habits of the Great Crested Grebe, Julian Huxley discerned three distinctive actions--"habit preening," "discovery ceremony" and "penguin dance"--none of which had any obvious function. He therefore labelled them "rituals," suggesting that they were "self-exhausting" and relating them to Darwin's mistaken notion of sexual selection which credited all animals with an aesthetic taste for pleasing colors and sounds. Still,

---

[37] C. G. Beer, "Ethology--The Zoologist's Approach to Behavior, Part 1," Tuatara, 11 (1963), 172; supra, n. 30.

Huxley's concept of ritualization is probably the most important single contribution to modern ethology. The rituals are derived from neutral behavior patterns associated with preening or feeding, of which the most conspicuous features have been exaggerated or stylized. Although apparently irrelevant acts, they serve a signalling function. And because these patterns are performed perfectly the first time they are attempted, it is clear that they, like the instincts to which they give expression, are products of evolution. In Huxley's concept of ritualization also lie the roots of the modern study of animal communication, or zoosemiotics, which begins from the premise that there are subsystems of non-verbal communication within <u>and between</u> all species, including man, manifested in signal codes which, in order for all to survive in a balanced ecosystem, each must learn. Although still in its infancy, the science of ethology had several ardent practitioners, at least some of whom were standing at the threshold of major discoveries.[38]

---

[38] J. S. Huxley, "The Courtship Habits of the Great Crested Grebe (<u>Podiceps cristatus</u>); with an Addition to the Theory of Sexual Selection," <u>Proceedings of the Zoological Society of London</u>, 2 (1914), 491-562; Thomas A. Sebeok, "Semiotics and Ethology," in <u>Approaches to Animal Communication</u>, ed. Thomas A. Sebeok and Alexandra Ramsay (The Hague, 1969), pp. 200-31; H. R. Hays, <u>Birds, Beasts and Men: A Humanist History of Zoology</u> (Baltimore: Penguin, 1973), pp. 325-27, 337; Beer, "Ethology, Part 1," pp. 172-73. Later seeking to explain the equivalent of ritualization for other instincts, Tinbergen coined the phrase "displacement activity." This concept has striking

Prior to the nature faker debacle of 1903, Seton published a tremendous number of animal stories. Most of these appeared first in magazines like *Scribner's*, *Century* and the *Ladies' Home Journal*, but were later compiled and presented in book form as was customary with other popular authors. Because the stories were written expressly for children, they were subject to obvious limitations. As a polemicist for the wilderness, Seton frequently overstated his case. But from his reading and from his observations in the field he had grasped, in however primitive a manner, certain hypotheses about learned and innate behavior. In

---

implications for aggression which, within species, and outside the realms of natural predation, it invariably serves to curb. In his research on Herring Gulls, Tinbergen noted that when one male trespassed on the territory of another, there were several potential actions open to both--the two most extreme being escape and pitched battle. Most commonly the impulses of fight and flight were activated simultaneously, resulting in apparently irrelevant movements such as "grass pulling" (associated with nest building) which functioned as a threat signal from the owner to the trespasser. The threat was also accompanied by a sideways thrust of the bill, known as an "appeasement gesture," which functioned to disavow an aggressive intention by exposing the most vulnerable part of the anatomy, the neck. An aggressor will never attack an opposite which has proferred an appeasement gesture at the instant of meeting. Displacement activities usually belong to the behavior pattern of an instinct other than the one which has been activated. While they will usually stimulate a reciprocal response in members of the same species, especially when ritualized, they may not always do so. But when actual fighting does occur, it seldom ends in the death of either bird. Ibid., pp. 173-76; N. Tinbergen, "'Derived' Activities; Their Causation, Biological Significance, Origin and Emancipation During Evolution," *Quarterly Review of Biology*, 27 (March, 1952), 1-32; *Social Behaviour in Animals: With Special Reference to Vertebrates* (London: Methuen, 1965), p. 71; Konrad Lorenz, *On Aggression* (London: Methuen, 1966), pp. 112-17.

the lives of individual animals, to which, for purposes of brevity, he ascribed the adventures of several of the same species, he hoped to convey the complexity of nature and to argue convincingly for the preservation of wildlife:

> My chief motive . . . has been to stop the extermination of harmless wild animals; not for their sakes, but for ours, firmly believing that each of our native wild creatures is in itself a precious heritage that we have no right to destroy or put beyond the reach of our children.
> I have tried to stop the stupid and brutal work of destruction by an appeal--not to reason: that has failed hitherto--but to sympathy and especially the sympathies of the coming generation.[39]

This was not a simple case of anthropocentric yearning, but a plea for the recognition of the unity of all living things:

> I hope some will find herein emphasized a moral as old as Scripture--we and the beasts are kin. Man has nothing that the animals have not at least a vestige of, the animals have nothing that man does not in some degree share.
> Since, then, the animals are creatures with wants and feelings differing in degree only from our own, they surely have their rights.[40]

Sitting beneath the light shed by W. H. Thorpe, Konrad Lorenz and Niko Tinbergen, we must recognize Seton's stories as the product of a mind well ahead of its time.

Although he had already written several nature essays for children,[41] Seton did not publish his first serious

---

[39] Ernest Thompson Seton, Lives of the Hunted (New York: Scribner's, 1901), pp. 9, 12-13.

[40] Seton, Wild Animals I Have Known, pp. 12-13.

[41] Ernest Thompson Seton, "Tracks in the Snow," St. Nicholas, 15 (March, 1888), 338-41; "The Pintail," St. N., 15 (Sept., 1888), 826-27; "The Western Meadow-Lark," St. N.,

animal story until 1894. Shortly after seeing Awaited in Vain safely into the Canadian collection at Chicago, and before returning to Paris, he went to New Mexico as the guest of Louis Fitz-Randolph, a wealthy New Jersey businessman whose daughter he had met in the Latin Quarter. In the northeastern reaches of the state Fitz-Randolph owned a huge cattle ranch which was being ravaged by Wolves. Although it is difficult to imagine him in the role, Seton may actually have been hired by Fitz-Randolph to participate in an experimental predator control program. There is no doubt at all that he joined the party of exterminators whose avowed purpose was to reduce financial losses at the ranch. It is most likely, however, that he saw this as an opportunity to expand his comparative data on Wolves. He devoted an entire volume of the Journal to an analysis of the behavior and habits of the local species, using the results as building blocks for his famous story of "Lobo, the King of Currumpaw," which appeared first in Scribner's in 1894 and which later introduced Wild Animals I Have Known. The experience itself lasted from October, 1893 until February, 1894 and appears to have been almost traumatic. Seton confesses

---

16 (Nov., 1888), 63-64; "The Screech-Owl," St. N., 17 (March, 1890), 432-33; "The Ovenbird," St. N., 17 (April, 1890), 520-21; "The True Story of a Little Gray Rabbit," St. N., 17 (Oct., 1890), 953-55; "Why the Chickadee Goes Crazy Once a Year," Our Animal Friends, 21 (Sept., 1893), 17-18; "The Birds that We See," Scribner's Magazine, 13 (June, 1893), 759-76; "The Wood Rabbit or Hare," Our Animal Friends, 22 (Nov., 1894), 65.

on one occasion that the deaths of Lobo, the leader of the pack, and of Blanca, his mate, "proved one of the turning points of my life."[42]

The story of Lobo was heralded as a landmark in the writing of animal stories--not so much by the literati as by the scientists. Rudyard Kipling wrote words of encouragement, but C. Hart Merriam, Mabel Osgood Wright, Theodore Roosevelt and David Starr Jordan wrote words of praise, urging Seton to pursue the genre.[43] The reasons were clear.

Even in death Lobo triumphs over his human killer, making a mockery of sophisticated metal traps and diabolical poisons. Although in comparison with Awaited in Vain the roles of victor and victim are apparently reversed, the intention of Seton the artist is duplicated in the work of Seton the author. The Wolf, while portrayed in its traditionally cunning posture, ceases to be the merely avaricious beast of legend. Hated by the ranchers for their diet of warm red meat Lobo and his pack are nevertheless respected for their obvious intelligence. Lobo poses no physical

---

[42]Seton, Autobiography, pp. 303-04; Journal, 5:210, and passim, AMNH; Four Months in New Mexico, MS Notes, SP; General Report, MS Notes, SP; "The King of Currumpaw: A Wolf Story," Scribner's Magazine, 16 (Nov. 7, 1894), 618-28; Julia Seton, By a Thousand Fires, pp. 111-115; Caroline Fitz-Randolph to Seton, March 13, 1895, SP.

[43]Seton to Rudyard Kipling, June 22, 1899, Feb. 11, 1899, Seton Letterbook No. 1, pp. 388, 392; C. Hart Merriam to Seton, Feb. 10, 1895, Dec. 15, 1898; Mabel Osgood Wright to Seton, Nov. 11, 1898; Seton to David Starr Jordan, Nov. 25, 1898, Seton Letterbook No. 1, p. 370, SP; David Starr Jordan to Charles F. Lummis, Copy in SP.

threat to man. But his gradual devastation of the livestock is symbolic of a strike at the economic system. The indolent, docile steers, fattening on the lush grasses of northern New Mexico, ringed round by watchful cowboys, are protected from death, for death. As the arbiter of death, man will suffer no challenge to his supremacy. Juxtaposed with the cattle are the Wolves that patiently await their chance to kill, not to challenge man's assumed mastery, but merely to survive. The cattle that do not land up in the stockyard are sacrifices to the laws of nature. Dead Wolves are sacrifices to the laws of economics. Economics determines whether the life of a Wolf is worth living; economics determines whether nature's laws will prevail or be denied.

But Seton does not leave his readers standing alone with irony. He demands that they comprehend the complexities of the animal itself--the better to understand the irony and the better to sense its injustice. We learn that Wolves run in packs and that each pack has a leader atop a social heirarchy. The lead Wolf in each pack has a mate. Lobo and Blanca are not sentimental aberrations, mere products of an artistic imagination bent on drawing moral analogies from nature. We learn that Wolves in fact are monogamous for life and that an "intense personal affection" exists between the dominant male and female. It comes as no surprise that the loss of one mate will result in a search by the other; that the death of one mate will result in mourning by the other. This is not nature faking but scientific fact.[44]

In an effort to win his bounty the author is forced to understand the Wolves on their own terms. Day by day he works at accumulating, through careful observation, a detailed knowledge of their behavioral traits. And gradually the traits themselves come to occupy the reader's attention; the story grows into a lesson in ethology, tied together by a surface plot in which man is the arch predator.

> Skin deep, man is a brute. Just a little while ago we were mere hunting brutes--our bellies were our only thought; that telltale line of dots [tracks] was the road to food. No man can follow it far without feeling a wild beast prickling in his hair and down his spine.45

The Wolves do not talk to him, nor do they attack or befriend him. They do, however, touch his life significantly, just as he does theirs. The fact that Seton makes an understanding of their behavior central to the story suggests his conviction that man must recognize the limitations he shares with the animals. Man has the paradoxical responsibility of using his reason to find a way out of his own skin, at least long enough to comprehend the intricacies of the interrelationship that binds all living things in one community. The laws of economics, under which the animals suffer injustice,

---

[44]W. H. Thorpe, <u>Science, Man and Morals</u> (London: Methuen, 1965), p. 124; <u>Biology and the Nature of Man</u>, p. 68.

[45]Ernest Thompson Seton, <u>The Trail of the Sandhill Stag</u> (New York: Scribner's, 1899), pp. 70-71.

are symbolized by the whole theory of predator control. Indeed, as he points out in another Wolf story, "Badlands Billy," man himself had created the need for predator control in the first place.

> In pristine days the Buffalo herds were followed by bands of Wolves that preyed on the sick, the weak and the wounded. When the Buffalo were exterminated the Wolves were hard put for support, but the cattle came and solved the question by taking the Buffaloes' place. This caused the Wolf-war. The ranchmen offered a bounty for each Wolf killed, and every cowboy out of work was supplied with traps and poison for Wolf-killing. The very expert made this their sole business. . . .[46]

Ignorance of nature's laws creates a vicious circle. And once man's values become attached to purely economic considerations, he imagines himself released from the animal bond.

But for Seton man seems condemned to predation of one kind or another. If he does not kill for food, he must kill for sport or kill in war, both animals and members of his own species. The wherewithal to mete out death is man's measure of power and supremacy. Reason does not dominate but merely services the will to power. Economics, the apparent tool for releasing man from his dependence upon nature, is really rationalized, if indiscriminate, predation, masquerading in mechanical laws. These laws know no morality and constitute, for Seton, an abject denial of justice. "Have the wild things no moral or legal rights?" he asks. "What

---

[46] Ernest Thompson Seton, Animal Heroes (New York: Scribner's, 1905), pp. 112-13.

right has man to inflict such long and fearful agony on a fellow-creature, simply because that creature does not speak his language?"[47] Seton was not alarmed by natural predation. "No wild animal dies of old age. Its life has soon or late a tragic end. It is only a question of how long it can hold out against its foes."[48] But he understood "foe" as a relative term. Death at the hands of a natural predator is tragic for the animal killed, but essential for the maintenance of the entire community. It is certainly no less tragic when the natural predator is deprived of its prey. The trophy hunter, a human foe like Scotty in "Krag, the Kootenay Ram," is doomed to reap the whirlwind of senseless, unnecessary predation. Nature strikes back at those who fail to heed her laws.[49] Man's treatment of both the wilderness and its inhabitants, therefore, has traditionally had very little to do with survival, but very much to do with the exercise of power through violence. The ultimate end of this confusion between strength, which is natural, and power, which is not, is alienation for man, annihilation for animals.

In none of his stories, then, is Seton kindly disposed toward man. His adult characters are particularly insensitive, petty people who take joy from tormenting a captive coyote, as in "Tito," or who measure their manhood

---

[47] Seton, Wild Animals I Have Known, p. 357.

[48] Ibid., p. 120.

[49] Seton, Lives of the Hunted, pp. 100-04.

by the size of the horns on a trophy ram, as in "Krag," or who think of Domino, in The Biography of a Silver Fox, as a walking pelt that might "lift the mortgage off our farm."[50] In one or two instances we meet better signs, either among the very old or the very young. In The Biography of a Grizzly Seton reflects on the Sisyphean lives of old prospectors whose redemption lies not in their fruitless occupation but in their ability to meld with the landscape:

> Miners came in pairs and wandered through the peaks . . . grizzly old men mostly, that had their lives in the mountain and were themselves slowly turning into Grizzly Bears; digging and grubbing everywhere, not for good, wholesome roots, but for that shiny yellow sand that they could not eat; living the lives of Grizzlies, asking nothing but to be let alone to dig.[51]

When the young boy, Yan, finally corners the great buck in The Trail of the Sandhill Stag, his mind flashes back to a doe, its mate, recently killed by hunters "when the snow was red with crime." At length, after gazing at the beautiful animal he has tracked for years, the desire to kill suddenly vanishes, and he thinks aloud:

> [W]e stand face to face, fellow creatures looking in each other's eyes, not knowing each other's speech--but knowing motives and feelings. Now I understand you as I never did before; surely you at least in part understand me. For your life is at last in my power, yet you have no fear. . . . We are brothers, oh bounding Blacktail! . . . I may never see you again. But if only you would come sometimes and look me in the eyes and

---

[50] Ibid., pp. 75-76, 274-77, 281-82; Ernest Thompson Seton, The Biography of a Silver Fox; or Domino Reynard of Goldur Town (Century: New York, 1909), pp. 172-73.

[51] Seton, Biography of a Grizzly, pp. 88-91.

> make me feel as you have done to-day, you would drive the wild beast wholly from my heart . . . and I should know more of the things that wise men have prayed for knowledge of.[52]

This reflects Seton's faith in the young, who cherish mystery, and in the old, who have learned limitation. The spectres of reason and power grip only those in between. They dictate the laws that subvert the laws. The instincts of the child are closest to the instincts of the animal. There is a unity there. But the old prospectors, who carry their futile cultural baggage out of the past, are also finding their way back into the land from which they grew.

All of Seton's animal stories are vehicles for understanding the behavior of various species. In tracing the lives of individuals the author who claims that his work is scientific is under considerable pressure to ensure that his stories are not fiction. His animals must be accurately described. The extent of their activities is limited automatically by numberless biological phenomena. He must, for example, know the geographic distribution of the species, as well as their reproductive periods. He must know whether they are social animals or whether they travel alone. He must also understand the subtle interrelationships that bind different species into a single community. It is impossible to discuss one animal without considering the others that share its territory. The author must be aware of the ground

---

[52]Seton, Trail of the Sandhill Stag, pp. 87, 91-93.

cover of a given range, not to mention the cyclical food chains that dictate survival to its indigenous species. He must understand the topography of each habitat, knowing how changes in altitude will affect the vegetation, and therefore the activities, of the individual. Given the dismal state of the science of zoology at the turn of the century, we cannot help marvelling at Seton's exactness in these matters. He did make errors, some of them glaring in the light of present knowledge, but for the most part his work reflects a rigorous attention to detail.

In the story of "Tito" Seton introduces his personalized theories of learned and innate behavior under the heading of the three sources of wisdom:

> First, the experience of its ancestors, in the form of instinct, which is inborn learning, hammered into the race by ages of selection and tribulation. This is the most important to begin with, because it guards him from the moment he is born.
>
> Second, the experience of his parents and comrades, learned chiefly by example. This becomes most important as soon as the young can run.
>
> Third, the personal experience of the animal itself. This grows in importance as the animal ages.
>
> The weakness of the first is its fixity; it cannot change to meet quickly changing conditions. The weakness of the third is the danger in acquiring it. But the three together are a strong arch.[53]

The first category represents instinct, the second would encompass imprinting and the learning of group culture, and the third would constitute individuality, explaining aberrant behavior, other than displacement activity, in response to

---

[53]Seton, Lives of the Hunted, pp. 284-85.

a given stimulus. In "Redruff," the biography of a Ruffed Grouse, Seton demonstrates how the young chicks follow their mother out of the nest almost immediately upon hatching, suggesting imprinting and the critical period. If we transpose "learning" for Seton's use of the word "reason" we get a remarkably accurate description of this behavior:

> Their start in life was a good mother, good legs, a few reliable instincts, and a germ of reason. It was instinct, that is, inherited habit, which taught them to hide at the word from their mother; it was instinct that taught them to follow her, but it was reason which made them keep under the shadow of her tail when the sun was smiting down, and from that day reason entered more and more into their expanding lives.[54]

Tito survives a pack of greyhounds by employing appeasement gestures.[55] The entire biography of "Raggylug" is given over to learning. Here Seton tells us that "Rabbits have no set time for lessons, they are always learning; but what the lesson is depends on the present stress, and that must arrive before it is known."[56]

Seton's understanding of dominance of relationships and social hierarchy was clear and unequivocal. In the story of "Krag, The Kootenay Ram," he describes these phenomena in sufficiently general terms that his young readers will understand their application to Spikerdoe, a ewe who is to

---

[54]Seton, Wild Animals I Have Known, p. 315.

[55]Seton, Lives of the Hunted, p. 281.

[56]Seton, Wild Animals I Have Known, pp. 93-143 passim, p. 113.

become the dominant female among a flock of Mountain Sheep.

> It is quite common to hear conventionality and social rules derided as though they were silly man-made tyrannies. They are really important laws that, like gravitation, were here before human society began, and shaped it when it came. In all wild animals we see them grown with the mental growth of the species.
>
> When a new Hen or Cow appears in the barnyard, she must find her level. She must take rank exactly according to the sum of her powers. Those already there have long ago ranged themselves in a scale of precedence; no one can climb in this scale without fighting all those over whom she would go. Somewhere in this scale there must be a place for the newcomer, and until this is settled, her life is one of battles.
>
> No doubt strength, courage, and activity fix her standing in most cases, but sometimes wisdom and keenness of sense are of greater importance. Which one is the leader of a band of wild animals? Not necessarily the strongest or fiercest. That one might drive the rest, but not lead them. The leader is not formally elected, as with man, but is rather slowly selected, thus, that individual who can impress the rest with the idea that he or she is the best one to follow becomes the leader, and the government is wholly by consent of the governed. The election is quite unanimous. For if in the herd are some who do not care to follow, they are free to go the other way. In many kinds of animals that go in herds, the leader whose courage and prowess have so often stood all tests, and who has inspired all the rest with confidence in his sagacity, is usually not the strongest male, but an elderly female. This is especially the case with Elk, Buffalo, Blacktail, and the summer bands of Mountain Sheep.[57]

When, in the fall, the rams join the ewes to mate, they too will duel among themselves--not, however, to establish dominance. None of the rams, which also travel in bands and which acknowledge a dominant male, will attempt to dislodge the dominant female. Their duelling and head-butting, Seton makes clear, is mere ceremony, ritualized reproductive behavior involving

---

[57]Seton, Lives of the Hunted, pp. 35-36.

the selection of mates. Because the Sheep are promiscuous, the process is viewed regularly and might lead the casual observer to believe he is witnessing constant intraspecific aggression. In fact this kind of sparring seldom results in death, except by accident. Once the rutting season is over, the ritual ends and the males and females split, the latter taking the young.[58] In the story of "Silverspot," a Crow, Seton elaborates the parallel form of dominance in birds, what Lorenz and Tinbergen would call the pecking order.[59]

Seton seems also to have had a remarkably clear understanding of the theory of territoriality.[60] In The Biography of a Silver Fox, he brings together various forms of reproductive behavior, including dominance, territoriality and ritualization. He begins by noting that "All animals have a home-region, a hunting ground that they consider theirs. For this they will fight, and on this they resent the coming

[58] Ibid., pp. 50-52, 61-65.

[59] Seton, Wild Animals I Have Known, p. 63.

[60] The behavioral concept of territoriality was not even fully articulated until 1920 when British ornithologist Eliot Howard observed that migratory birds divided their habitat into parcels of land of approximately equal value. He noted that males usually arrived before the females in early spring, fighting among themselves for possession of territory and then choosing mates. Howard assumed that this behavior pattern was associated primarily with the instinct of reproduction. Seton's speculations in this particular area were, therefore, astonishingly perceptive. See H. E. Howard, Territory in Bird Life (London: Murray, 1920); Beer, "Ethology, Part 1," pp. 176-77.

of any kindred stranger." In recognizing the individual's territory as a "hunting ground," Seton tacitly acknowledges animal dispersion, the natural law whereby species maintain population equilibrium and prevent the exhaustion of limited food resources. Defeat in battle simply forces an animal to leave the territory claimed by the victor and to locate elsewhere--one of the clearest demonstrations of the correlation between ethology and ecology. As this very problem remains the focal point of the modern research of scientists like David Lack and V. C. Wynne-Edwards, we must appreciate that Seton's insights were extremely important. Taken in conjunction with his earlier remarks about predator control and about individuals bolting from a band of Sheep, we have clear evidence that his theories on geographic distribution were in the process of breaking down the all too rigid definitions of C. Hart Merriam. Although in his formal scientific writing he was afraid to disavow Merriam's absolute reliance upon temperature and atmospheric conditions, he used the animal story to suggest other possibilities, including behavior, as determining factors. Seton also linked territoriality and ritualization in the Biography of a Silver Fox. When the protagonist, Domino, notices two strangers on his territory, he lights out after them. The time is February and Domino has been barking his mating "lonesome cry" at the moon for several days. One of the two strangers is a male, the other a female in heat. The object is to banish the male from the territory and to win the lady's favor.

> The rivals closed in fight. Blazor went down and lay there snapping his teeth. Domino stood over him, but did him no great harm. The snow lady ran off. The two pursued again. . . . As they cantered across the field, she swung away from Blazor and a little nearer the Silver Fox. All three pulled up and faced, not now as three, but a pair and a one, and the tall black Fox of the pair stood very high upon his legs. He fluffed out his ruff and raised his great tail. Thus towering he growled deeply, showed the gleaming rows of perfect teeth, walked stiffly toward Blazor, while Snowyruff came close behind, and Blazor knew it was finished. He turned and sullenly glided away.
>
> This was the mating of Domino; this was the wedding.61

Reproductive behavior does not always end so peaceably. In "Raggylug" a large buck arrives on the scene to challenge the young Cottontail for the favors of a female. Their sparring has the effect of taking them off their guard, making both easy prey for passing carnivores. Here again it is the behavior of the animals that functions indirectly as a vehicle for selection, resulting in tragedy for one individual, in survival for another.[62]

Seton clearly appreciated the subtleties of communication in mammals and birds and in some instances tried to work out for himself the meaning of specific actions and calls. As there was in individual species a pattern for

---

[61] Seton, Biography of a Silver Fox, pp. 71-72, 74-80; V. C. Wynne-Edwards, Animal Dispersion in Relation to Social Behavior (New York: Hafner, 1962), pp. 8-9, 183-87, 223-28; David Lack, The Natural Regulation of Animal Numbers (Oxford: Clarendon Press, 1954).

[62] Seton, Wild Animals I Have Known, p. 133.

each mode of communication, he assumed that if one juxtaposed the behavior of the animal with the presence of particular environmental conditions or predators or members of the same species, he would be able to arrive at some conclusions about the content of the communication. In other words, he recognized the importance of stimulus and response. The modern method of gauging bird songs involves recording them and then measuring the results on a sound spectrograph. Without this technology, Seton reduced each song to its musical scale and accompanied it with phonetic symbols in an effort to duplicate in print the tones he had heard. Thus, in "Silverspot," we find several pages given over to the various sounds made by Crows, complete with bars of music and the circumstances in which the sounds are heard. Some sounds are associated with reproductive behavior, other with danger, and so on.[63] Afraid that his readers would find his theories about communication foolish, he was constantly at pains to explain and to justify them.

---

[63]Ibid., pp. 65-70, 82. See also, Ernest Thompson Seton, "The Song of the Prairie Lark," American Magazine, 7 (April, 1888), 717-20. For the modern study of bird song, see W. H. Thorpe, Bird-Song: The Biology of Vocal Communication and Expression in Birds (Cambridge: Cambridge University Press, 1961); R. A. Hinde, ed., Bird Vocalizations: Their Relations to Current Problems in Biology and Psychology (Cambridge: Cambridge University Press, 1969); Edward A. Armstrong, A Study of Bird Song (London: Oxford University Press, 1963).

> Those who do not know the animals well may think I have humanized them, but those who have lived so near them as to know somewhat of their ways and their minds will not think so.
>
> Truly rabbits have no speech as we understand it, but they have a way of conveying ideas by a system of sounds, signs, scents, whisker touches, movements and example that answers the purpose of speech.[64]

In "Redruff" Seton links communication with ritualization in describing the "drumming" behavior of the Ruffed Grouse.[65] And in "Badlands Billy" we get a splendid digression on communication linked to territoriality.

> Wolves have no language in the sense that man has; their vocabulary is probably limited to a dozen howls, barks, and grunts expressing the simplest emotions; but they have several other modes of conveying ideas, and one very special method of spreading information--the Wolf-telephone. Scattered over their range are a number of recognized "centrals." Sometimes these are stones, sometimes the angle of cross-trails, sometimes a Buffalo-skull--indeed, any conspicuous object near a main trail is used. A Wolf calling here, as a Dog does at a telegraph post, or a Muskrat at a certain mudpie point, leaves his body-scent and learns what other visitors have been there recently to do the same. He learns also whence they came and where they went, as well as something about their condition, whether hunted, hungry, gorged, or sick. By this system of registration a Wolf knows where his friends, as well as his foes, are to be found.[66]

From these few examples, then, it must be clear that Seton took extreme care to interpret communication in such a way that he could not legitimately be accused of anthropomorphism.

Anthropomorphism was the farthest thing from his mind.

---

[64] Seton, Wild Animals I Have Known, pp. 93-94.

[65] Ibid., pp. 334-35.

[66] Seton, Animal Heroes, p. 124.

If anything, he was attempting to remind man of his underlying biological continuity with the animals. But he made many mistakes, and the mistakes came back to haunt him. Morality was central to all of the stories. As long as he equated morality with biological fact he was generally on safe ground. But when he gave in to his own assumptions, he found himself in trouble. His best stories dealt with the social animals, particularly those species demonstrating monogamy. Seton's assumption here was that monogamy was a predictable trait in all the higher animals. Monogamy was also morally good. Wolves were the supreme species because the males and females mated for life. But there were degrees of monogamy. The bull Moose, for example, was considered monogamous because, during the rutting season, he stayed with a single cow until she reached her oestrous period, then moved on to find another in the pre-oestrous stage. Seton had added difficulty here since the Moose was clearly one of the least social animals in North America. Since Moose cows are polyoestrous, he must have realized that biological laws were at work to guarantee the impregnation of all the cows in a given range in a given season. In any case, he considered Moose to be morally good, though not as good as Wolves. Brazenly promiscuous animals like Rabbits and Voles were "high on the scale of fecundity, low in the scale of general development," a probable cause for their being an easy mark for predators. They were necessary to the food chain, but few of them, with the exception of Raggylug, made good story material. Another kind of monogamy was that which existed for a single breeding season, "the male staying with the family, and sharing the care of the young till they are well grown." This behavior was observed in such admirable birds as the Hawk which stood somewhere between the Moose and the Wolf on an ascending scale of intelligence. Although Seton's interpretation was unconvincing, the subject still surfaces today, W. H. Thorpe suggesting as recently as 1965 that "natural selection, other things being equal, favours sexually aggressive polygamous males and passive discriminating females."[67]

A second instance of wishful thinking on Seton's part probably grew out of his observations at menageries (and probably accounted for his idealization of zoological parks). Wild animals could only be truly wild in their natural habitat, that is to say when they were free from man. Any sense that freedom was about to be curtailed was considered by Seton a stimulus. It elicits a chain of responses including flight or attack, but it culminates, at least in some of his animal stories, in suicide. In "The Pacing Mustang" the protagonist leaps to its death from a

---

[67]Ernest Thompson Seton, "The Natural History of the Ten Commandments," Century, 75 (Nov., 1907), 25-26; Thorpe, Science, Man and Morals, p. 125.

cliff in order to avoid capture. In The Biography of a Grizzly, Wahb recognizes that his days are numbered so instinctively wanders into a canyon filled with noxious fumes and quietly drifts into everlasting sleep. Another twist on this theme is afforded in "The Springfield Fox" when Vixen finds her one remaining offspring chained to a stake in the farmyard. She brings the young animal what she knows to be a piece of poisoned bait, ostensibly releasing it from a fate worse than death.[68]

The ethical implications of biology, so pronounced in all of Seton's work, owed much to Ernst Haeckel (1834-1919), the famous German scientist who introduced the theory of recapitulation later adopted and modified for psychology by G. Stanley Hall. Haeckel invariably stressed man's limitations within nature and is now credited with having elaborated the first scientific principles of ecology. Seton clearly shared Haeckel's view that morality was an extension of instinct. Despite their obvious shortcomings, his animal stories consistently mirrored his basic conviction that "Immorality . . . may be defined as the deflection of any natural power, member or instinct from its purpose to one that works harm for the species." Although he never denied the existence of competition in the biological order, Seton's

---

[68] Seton, Wild Animals I Have Known, pp. 10-11, 224, 270; Biography of a Grizzly, p. 3.

work is marked by an overwhelming emphasis on mutual aid, reflected particularly in the behavior of social animals. This behavior, he maintained, was innate in man as well, but had been denied expression by the power of reason. Through his overemphasis on reason, man had alienated himself not only from his environment but from his ancestors.[69]

IV

Despite the shortsightedness of Burroughs' 1903 attack, Seton was crushed to find his credibility challenged again by a man he deeply revered. Although the two resolved many of their differences, Burroughs had clearly sown seeds of doubt in the public mind, not only about the reliability of Seton's behavioral interpretations but also about the authenticity of his data. Fortunately, Seton followed the advice of his friends and declined to engage in an open debate with Burroughs. But he was determined to set the

---

[69] Seton, "Natural History of the Ten Commandments," p. 27. Although Seton had read, and been influenced by, Haeckel's work, it must be emphasized that he never succumbed to the latter's reactionary monist faith. For this, and other excesses of Haeckel's thought, see Daniel Gasman, _The Scientific Origins of National Socialism: Social Darwinism in Ernst Haeckel and the German Monist League_ (London: Macdonald, 1971). Not all of Haeckel's work appeared in translation. See Ernst Haeckel, _The History of Creation_ (London: H. S. King, 1876); _The Riddle of the Universe at the Close of the Nineteenth Century_ (New York: Harper, 1900); _The Wonders of Life_ (London: Watts, 1905); _The Evolution of Man_ (2 vols.; New York: H. L. Fowle, 1897).

record straight. Accordingly, from 1904 to 1909 he turned his scientific attention almost exclusively toward the completion of a mammoth research study on the mammals of Manitoba.

Seton was not prepared to admit wrongdoing where the animal stories were concerned. He had, after all, observed most of the events recorded in them. Nevertheless, he did realize that he would have to demonstrate a comprehensive understanding of biology if he were to gain recognition by the scientific fraternity and restore the confidence of his critics. In writing The Life-Histories of Northern Animals, therefore, he had two central objectives in mind. First, he hoped to synthesize all the available published data on the mammals known to occur in Manitoba, blending in his own field observations and revising obsolete conclusions. Second, he planned to restate his behavioral premises, anticipating that in their new context they would prove more palatable to the skeptics. However, before examining the treatise itself we must turn our attention to Seton's relationships with the two men--C. Hart Merriam and William T. Hornaday--who, more than any other living scientists, influenced its direction.

V

In 1887 Theodore Roosevelt had established the élitist Boone and Crockett Club in New York. Membership was restricted to one hundred men whose eligibility was determined by their having "killed with the rifle, in fair chase, at least one adult male individual of each of three of the various species of American large game." Although there were a few token westerners, the club was dominated by eastern Republicans like Owen Wister, Madison Grant, Gifford Pinchot, Elihu Root, and Caspar Whitney. The official mouthpiece of the organization was George Bird Grinnell's Forest and Stream, a magazine to which Seton himself was a regular contributor. Pledged to "promote manly sport with the rifle," to foster exploration, to work for the preservation of big game and "to encourage making the results of sport available for scientific study in museums," the club was a monument to social pretension disguised in woodsy causes.[70]

As early as 1889 the Boone and Crockett Club supported legislation leading to the creation of the National Zoological Park at Washington. Under the leadership of Madison Grant it was also responsible for lobbying through the state legislature at Albany a measure endorsing the construction of a still grander zoo for New York City. Indeed the New York Zoological Society, organized in 1895, contained no

---

[70] Hunting and Conservation, ed. George Bird Grinnell and Charles Sheldon (New Haven: Yale University Press, 1925), pp. 534-35.

fewer than nine Boone and Crockett members on its board of managers. In 1897 the City of New York turned over to the Society 261 acres of land in the Bronx, while an act of the state legislature provided a substantial grant for the implementation of preliminary plans. Because the Society literally owed its existence to the club, it was expected in return to uphold the compromising principles of the parent body. It began by securing entire herds of North American game animals and by placing them "as far as possible in surroundings identical with or closely resembling their natural habitats," the purpose being "to preserve to future generations some remnant of the heritage which was our fathers'."[71] (To more cynical observers the New York Zoological Park was really a Noah's Ark for hunters, insurance against game hogs and a long-term guarantee of the "fair-chase.") While the Park itself was to be maintained by the city, the Society was responsible for supplying the collections and employing the scientific expertise necessary to service them.

Scientific expertise came in the person of William Temple Hornaday (1854-1937). A native of Indiana and a graduate of the Iowa State Agricultural College, Hornaday had been chief taxidermist at the Smithsonian from 1882 to

---

[71]Madison Grant, "The Origin of the New York Zoological Society," in Trail and Camp-Fire, ed. George Bird Grinnell and Theodore Roosevelt (New York: Forest and Stream Publishing Co., 1897), pp. 317-320.

1890. Invited by Madison Grant to accept the directorship of the New York Zoological Park in 1896, he held the position until his retirement in 1926. Although eventually becoming an associate (i.e. a non-voting)[72] member of the Boone and Crockett Club, Hornaday remained a militant preservationist who frequently embarrassed his employers with outspoken attacks on rifle and shotgun manufacturers. He brought to his job an incisive mind which later won him, among other things, honorary doctorates from Yale and the University of Pittsburgh. Active in founding several conservation organizations, including the American Bison Society, the National Audubon Society and the American Game Protective and Propagation Association, he also administered the Permanent Wildlife Protective Fund established by Mrs. Russell Sage. Viewed by many as a kind of peevish sentimentalist, Hornaday was not a man to compromise his principles on wildlife issues. Nevertheless, he also realized the lobbying power of the Boone and Crockett Club in Congress, especially after its founder became President, and was prepared to go along with the measures it supported as long as they contributed to the fulfillment of his own objectives.

An adequate gauge of Hornaday's ultimate faith in the Boone and Crockett Club was his own initiative in founding

---

[72]Other associate members included Vernon Bailey, C. Hart Merriam, E. W. Nelson, Aldo Leopold, Henry Cabot Lodge, J. A. Allen, Francis G. Newlands, Warburton Pike, Carl Schurz.

a rival organization, the Camp Fire Club of America, in 1897. To be fair, it should be pointed out that the membership of this new club was made up partly, if not largely, by men who aspired to join, but who were deemed unacceptable by, the more prestigious body. Despite invitations to speak to the Roosevelt group, and even to illustrate its publications,[73] Seton himself was never recruited. He was, however, one of the first members of the Camp Fire Club which (although it too affected social airs) played down manly hunting exploits and stressed natural history and conservation. In launching this new enterprise Hornaday also enlisted the support of George O. Shields, another wildlife muckraker, whose Recreation magazine became the official organ of the club and effective competition for Grinnell's Forest and Stream.[74] Early adherents of the Camp Fire Club included A. A. Anderson and Dan Beard, both of whom, with Seton, Hornaday and Shields, served at one time or other as its President.[75] After 1904, when the club was reorganized,

---

[73]See, e.g., George Bird Grinnell, "Wolves and Wolf Nature," in Trail and Camp-Fire, ed. Grinnell and Roosevelt, pp. 152-203.

[74]When Shields' magazine collapsed, the Camp Fire Club found a new voice in Field and Stream. E. F. Warner to Seton, July 20, 1910, July 15, 1910, Oct. 21, 1910, SP.

[75]Other members included Carl Rungius, the animal painter who illustrated most of Hornaday's books, E. W. Deming, also an artist, Hamlin Garland and Dillon Wallace, both authors. Important businessmen on the roster included Frank Presbrey and George D. Pratt, both of whom would play leading roles in the Boy Scout movement.

prominent Boone and Crockett men like Roosevelt and Pinchot, whose contributions to conservation and whose prestige disguised their attitudes toward big game, were welcomed as honorary members. In some instances, notably in the formation of the American Bison Society in 1905, the two organizations were able to unite in common cause. Nevertheless, Hornaday's leadership guaranteed that the Camp Fire Club would be the more radical group. Its membership was less obviously political and clearly more intellectual. Its views on conservation issues of the day were passionate and uncompromising, serving no particular interest group and, at least where wildlife preservation was concerned, looking to the needs of the animals first. The Camp Fire Club was almost wholly responsible for funding the creation of the Montana National Bison Range.[76] It also joined Merriam in the battle to save the Northern Fur Seal in the Pribilof Islands. In 1910, the year in which Seton was President, the club lobbied Congress, through the Senate Committee on the Conservation of Natural Resources, demanding the abolition of the leasing system for the killing of Seals, the negotiation of international

---

[76]The preservation of the Buffalo was Hornaday's pet project. He was instrumental in bringing about the creation of the Wichita (1907) and Wind Cave (1913) National Bison Ranges, projects which attracted financial assistance from the Boone and Crockett Club, the New York Zoological Society and the American Bison Society. William T. Hornaday, <u>The American Natural History: A Foundation of Useful Knowledge of the Higher Animals of North America</u> (4 vols.; New York: Scribner's, 1914), 2:13.

treaties with Britain, Russia and Japan for the elimination of pelagic sealing, and a five-year injunction on the land slaughter to allow for the rehabilitation of depleted stock. Between 1911 and 1912 Congress responded by meeting all of these demands.[77] The Club pressed for the creation of both Glacier and Mount McKinley National Parks and was particularly vocal in discussions leading to the passage of the Migratory Birds Act of 1913. In all of these activities Seton and Hornaday worked together tirelessly, often contributing substantially from their own pockets.

Members of the Camp Fire Club were expected to subscribe to a rigid code of ethics inspired by the Christian doctrine of stewardship. It maintained that in "the settled and civilized regions of North America there is no real necessity for the consumption of wild game for food purposes," and that "a game-butcher or a market-hunter is an undesirable citizen and should be treated as such." Hunting was unnecessary to a successful outing; indeed, a "photograph of a large wild animal in its haunts is entitled to more credit than the dead trophy of a similar animal." Although couched in Hornaday's predictably puritanical turn of phrase, the code bore the indelible stamp of anti-anthropocentrism--a conviction that animals had a right to live, regardless of their

---

[77]Ibid., 1:123-36; Sterling, Last of the Naturalists, pp. 386-413.

service to man.[78]

Seton's formal conservation work was restricted to causes championed by the Camp Fire Club. His greatest service to the movement as a whole came through the medium of his published writing, little of which was about conservation, all of which was radically conservationist--intellectually miles ahead of progressive utilitarianism, but offering few practical alternatives. Seton was probably most conscious of creating a receptive frame of mind for the kinds of plans that men like Hornaday were better qualified to articulate. On isolated occasions he did address himself, either through a story or a brief reflection, to the conservation

---

[78]William T. Hornaday, Our Vanishing Wild Life: Its Extermination and Preservation (New York: Scribner's, 1913); Kenneth Fowler, "The Story of the Camp Fire Club," Field and Stream, 15 (Nov., 1910), 601-15. Seton to Theodore Roosevelt, Jan. 31, 1897, Feb. 4, 1897, March 10, 1897, Seton Letterbook No. 1, pp. 129, 134, 159; Theodore Roosevelt to Seton, March 23, 1897; Seton to W. T. Hornaday, Dec. 9, 1909; W. T. Hornaday to Seton, Dec. 10, 1909, Jan. 12, 1910; Jan. 31, 1910, Aug. 18, 1913, Oct. 9, 1913; Glen Buck to Seton, Dec. 18, 1912; Henry Oldys to Seton, Jan. 6, 1913, SP. W. T. Hornaday to Seton, June 8, 1904, June 17, 1904, July 29, 1904, Aug. 5, 1904, Hornaday Letterbook No. 2, pp. 139, 150, 199, 200; W. T. Hornaday to Seton, March 1, 1907, Jan. 30, 1908, Hornaday Letterbook No. 4, pp. 14, 14(a), 409; W. T. Hornaday to Seton, Jan. 9, 1909, April 17, 1909, Hornaday Letterbook No. 6, p. 329, LC WTHP. The Camp Fire Club "Code of Ethics" is reprinted in full in C. Gordon Hewitt, The Conservation of the Wild Life of Canada (New York: Scribner's, 1921), pp. 298-99. See also, Frank Graham, Man's Dominion: The Story of Conservation in America (New York: Evans, 1971), pp. 179-88; James B. Trefethen, Crusade for Wildlife: Highlights in Conservation Progress (Philadelphia: Stackpole, 1961), pp. 1-178, passim; Barrus, Life and Letters of John Burroughs, 2:187.

issue, but even here his focal point remained the preservation of birds and mammals.[79]

Seton had met Hornaday while the latter was still engaged by the Smithsonian as chief taxidermist. Each man was an artist in his own right and each had a comprehensive understanding of animal anatomy. On one occasion Hornaday had asked his friend to make some detailed illustrations to support a monograph on the extermination of the American Buffalo. Seton, who liked to draw from life, went to the Central Park Zoo, completed the work and dispatched it to Washington. Hornaday was dissatisfied with the results, complaining that the hindquarters were not sufficiently massive to suggest the animal's great weight. However, this observation necessitated an alteration, not in the drawing but in the analysis. As Seton was quick to point out, an animal in captivity lost the weight in its hindquarters through lack of normal exercise. The progressive weakening of the captive Buffalo merely accentuated a naturally tapering

---

[79]See, e.g., Ernest Thompson Seton, "How Bull-Fighting Was Suppressed in France," Our Animal Friends, 24 (Feb., 1897), 128-29; "The Protection of Birds in France," OAF, 26 (Nov., 1898), 56-58; "The New Sportsman," Recreation, 10 (Jan., 1899), 39-40; "Chanticleer vs. Egret," Forest and Stream, 49 (Oct. 9, 1897), 283. Although Seton did function as an intermediary between the Canadian and American governments on conservation matters of mutual concern, his role was never strong. He seems to have lobbied alone only in Manitoba where he fought consistently for the establishment of parks, "the great thing being to rescue the land." Seton to Hugh John Macdonald, Jan. 13, 1900, June 17, 1901, Seton, Letterbook No. 2, pp. 123, 183; Seton to R. M. Simpson, Nov. 7, 1924, SP.

tendency, having the effect of making the animal appear smaller. Hornaday was forced to note in the final draft of his paper that "the changes in the captive animal bid fair to become so great that in a few years naturalists who have only seen captive animals will study them and declare that buffaloes with such outlines as those have in our mounted group [at the Smithsonian] never existed!" An apparently simple morphological observation had made it clear that one did not preserve a species from extinction merely by putting it in safekeeping at the zoo. It was essential to think of zoos in terms, not of cages, but of parks where, as far as possible, the natural habitat could be duplicated. This would allow the animals the kind of exercise they required and would be more likely to encourage them to breed.[80]

Such were the considerations that moved Hornaday and his colleagues at the Smithsonian to press the creation of the National Zoological Park at Washington. Seton, who shared many of Hornaday's enthusiasms, was particularly eloquent in defense of this one. The pioneers, "with their horses, their deadly rifles . . . and their taste for murder," had joined forces with the railroad to decimate the huge

---

[80]W. T. Hornaday to Seton, May 4, 1889, May 7, 1889, May 22, 1889, May 24, 1889, SP; William T. Hornaday, "The Extermination of the American Bison, with a Sketch of Its Discovery and Life History," Annual Report of the Board of Regents of the Smithsonian Institution, Part II, 1887 (Washington: Government Printing Office, 1889), pp. 394-96, Plates VI, XI, XII, XIII.

herds of indigenous quadrupeds which had once ranged far into the East. It was now man's responsibility to restrain himself, to foster the building of depleted stocks, not merely because the animals had a right to live, but because man killed a bit of himself each time he killed one of them, renewed himself by renewing them.

> Evidently one can not comprehend the nature of a wheel in a machine by the study of that wheel alone; one must consider the whole machine or fail. And since it is established that man is merely a wheel in the great machine called the universe, he can never arrive at a comprehension of himself without study of the other wheels also. Therefore, to know himself man must study not only himself, but all things to which he is related. . . . It is possible that the study of 'man' would suffer more by the extinction of some highly organized animal than it did by the burning of the Alexandrian library.

On its 167 acre site the Washington park afforded the observer unique opportunities to view animals in conditions approximating their true environment. Each species had its own range and the freedom to roam, giving the naturalist an opportunity to reflect on the social behavior of individuals in the group, measuring it against that which he had observed in the field. European menageries failed to serve any useful purpose. Isolated in boxed cages, individual animals were totally separated from anything remotely resembling their natural habitat. Many species refused to breed in cages and most were totally alienated, forever condemned to look out upon a sea of inquisitive human faces. "Their bellies are cared for," wrote Seton, "but how few keepers have learned that each animal has a mentality, large or small, that ought to be considered!" Menageries merely demonstrated the extent to which man had been alienated from nature. By reuniting entire herds, the Washington park reopened the observers' eyes to the behavior of animals. Even though it was impossible to allow natural predation, for example, Seton noted that dominance relationships were among the first to assert themselves within species, guaranteeing a certain amount of aggression, but also ensuring reproduction.[81]

Seton's interest in zoos grew naturally out of his aptitude for painting animals. He had spent many hours in the London Zoo and in the Paris Jardin des Plantes Menagerie preparing sketches for later contribution to the Salon or to illustrate the Art Anatomy. For this reason his opinion on the Washington park was solicited by S. P. Langley, Secretary of the Smithsonian, who had the responsibility for seeing it through to completion after Hornaday's departure in 1890. When Hornaday assumed the direction of the New York Zoological Park in 1896, he too looked to Seton for advice. After the preservation of animals, Seton gave priority to the activities of artists and zoologists. The public ranked

---

[81] Ernest Thompson Seton, "The National Zoo at Washington, a Study of Its Animals in Relation to Their Natural Environment," Ann. Rep. Bd. Reg. Smith. Inst., 1901 (Washington: GPO, 1902), pp. 697-99, 710-16; "The World's Most Wonderful Zoo," London Daily Express, 16 Aug., 1901; S. P. Langley, "Report of the Secretary of the Smithsonian Institution," Ann. Rep. Bd. Reg. Smith. Inst., 1890 (Washington: GPO, 1891), pp. 34-41.

last. Consequently we find among his letters passionate pleas for "portable cabin studios and also a large studio fully equipped with casts, from works by the masters and from recent dead animals, as well as a collection of pictures and photographs and conveniences for the photographer."[82] Both he and Dan Beard hoped that the New York Zoological Park might become the center for an American school of animal artists whose work would rival that of Barye, Landseer, Swan, Rivière and Frémiet. It would have the added advantage of recording accurately for posterity the vast diversity of animals indigenous to the country, displacing the traditional emphasis on lions, tigers, elephants and other so-called "exotic" species which had no place in the North American experience. Hornaday welcomed any suggestions that would augment the relatively small body of existing knowledge on native mammals. Even though the day of the animal painter was over, he acted on Seton's recommendation and pressed for substantial artistic facilities. Seton, who wanted to guarantee the representation of at least one "master" animalier, graciously loaned the original drawings from the Art Anatomy to the Society.[83]

---

[82]Seton to S. P. Langley, April 3, 1899, SP.

[83]Seton to Madison Grant, Jan. 18, 1897; Seton to W. T. Hornaday, June 29, 1898, June 10, 1899, Seton Letterbook No. 1, pp. 117, 345, 463; Seton to Madison Grant, Nov. 21, 1899, Nov. 22, 1899; Seton to W. T. Hornaday, July 25, 1899, Seton Letterbook No. 2, pp. 85, 105-06; Seton to S. P. Langley, April 3, 1899, SP; Ernest Thompson Seton,

As Director of the New York Zoological Park, Hornaday had ample time to observe the habits and behavior of animals. The subject fascinated him, although as late as 1914 even he seemed cautious in discussing it.

> Judging from all that I have seen and heard of wild creatures of many kinds, . . . I believe that practically all their actions are based upon natural, inborn instinct--nearly all of it in the line of self-preservation, and the exceptions are due to the natural tendency to imitate leaders. Of hereditary knowledge--another name for instinct, some animals have an abundance. Of special knowledge, acquired by systematic reasoning from premise to conclusion, most animals have very little, and very few ever exhibit powers of ratiocination.

Hornaday was an outspoken critic of nature fakers and all who idealized or humanized the animal. But he did not include Seton in this category and openly stated that the latter's animal stories "deserve to live forever." He knew that Seton would not knowingly credit animals with an intelligence they did not possess, that his stories were based upon factual observation unless their author noted otherwise, and that behavior which appeared rational could easily be explained as instinctive. Even if Seton's interpretations of behavior warranted further investigation, the fact that the behavior described had occurred was not open to question.[84]

"Communication Regarding the Needs of Artists in the Zoological Park," <u>First Annual Report of the New York Zoological Society</u> (New York: New York Zoological Society, 1897); "A School of Animal Painting and Sculpture in the New York Zoological Park," <u>2nd Ann. Rep. N.Y. Zoo. Soc</u>. (New York: N.Y. Zoo Soc., 1898), pp. 69-75.

[84]Hornaday, <u>American Natural History</u>, 1:xxvii-xxix.

Plate 9. Franklin Groundsquirrel (Pencil, 1892)

As time passed, however, Hornaday gradually abandoned the role of skeptic, allowing preservationist zeal to usurp his scientific good sense. By the 1920's he had all but reversed his previous stand, carrying the levelling process to such extremes that even William Long might have had cause to smile. Hornaday was familiar with Romanes, Loeb, Thorndike and others working on the issues raised by animal intelligence. He understood the debate between vitalists and behaviorists and he had investigated the laboratory experiments being undertaken by the latter. But in his desire not to be identified with the problem box and the maze, he made several grave errors. On the positive side he encouraged Seton and the budding school of ethologists in their belief that

> The proper place in which to study the minds, manners and morals of wild animals is in the most thickly populated haunts of the most intelligent species. The free and untrammeled animal, busily working out is own destiny unhindered by man, is the beau-ideal animal to observe and to study. Go to the plain, the wilderness, the desert and the mountain, not merely to shoot everything on foot, but to SEE animals at home. . . . See what normal wild animals do as 'behavior,' and then try to find out why they do it.
>
> The next best place for study purposes is a spacious sanitary and well-stocked zoological park.[85]

He allowed for the existence of learned and innate behavior, describing intelligence as "the possession of knowledge either inherited [as instinct] or acquired [during a lifetime]."

---

[85]William T. Hornaday, The Mind and Manners of Wild Animals (New York: Scribner's, 1922), pp. 4-7.

He appreciated that learning could not be transmitted by evolution but was best secured by and within a social population. Nevertheless, in attempting to simplify and popularize these complex concepts, Hornaday identified acquired learning with "education," concluding that the "adjustment of a wild animal mind to conditions unknown to its ancestors is through . . . logical reasoning from premise to conclusion." Animal intelligence could be measured by taking into account such factors as "comprehension under tuition," the "powers of independent observation and reasoning," and "accuracy in the execution of man's orders."[86] This was anthropomorphism in its worst form.

The crux of this problem lay in Hornaday's ranting puritanism. Although Seton shared his friend's obsession with morality and clearly followed his lead in attempting to link exemplary ethics with the behavior of animals, he never fell victim to Hornaday's excesses. The latter's growing tendency to attribute to animals the power of reason was matched by a corresponding inclination to reduce the role of instinct. For example, Hornaday regarded the lack of warfare within species as evidence of an ethical "code" to which "free" animals willingly adhered. He identified dominance relationships, ritualization, territoriality and other reproductive behavior patterns as conscious moral choices.

[86] Ibid., pp. 9-10.

> Leaving out of account the carnivorous animals who must kill or die, all the wild vertebrate species of the earth have learned the logic that peace promotes happiness, prosperity and long life. This fundamentally useful knowledge governs not only the wild animal individual, but also the tribe, the species, and contiguous species.[87]

Hornaday attempted to argue that while the human intelligence was superior to its animal equivalent, the animal intelligence was possessed of reason and was therefore capable of achieving degrees of human intelligence. In effect, animal morality actively promoted the growth of reason. By contrast, human morality was progressively abandoning reason in a debased orgy of greed and hedonism which, through the medium of warfare, would ultimately bring about the extinction of the species.

The nature faker controversy made Seton much more cautious in dealing with such generalizations. Although urged by Hornaday to investigate the moral superiority of biological laws, he never implied that the end of animal learning was the achievement of human intelligence. For Seton, acquired behavior was merely a tool for survival. It grew in the natural animal, living in its natural environment, coping with its natural experiences. True morality was instinctive. Human reason was villainous only when it denied or suppressed instinct. (Indeed, the best laws of men were predicated on instinct.) Animal species did not consciously

---

[87] Ibid., p. 222.

decide not to fight. The biological laws of nature--in other words, natural selection--simply prevented intraspecific warfare and fostered mutual aid. The levelling process did not require the exchange of abstract ideas between animal and man. But it demanded the acceptance of instinct and the power of learning in both.

If Hornaday pressed Seton to amplify the interpretative ethological dimension of his work, C. Hart Merriam was equally influential in stressing the importance of descriptive accuracy in the technical sphere. Seton clearly appreciated that his speculations about animal behavior would carry little or no weight in the scientific community without evidence to support his basic grasp of zoological theory. In turning to Merriam for guidance, he guaranteed that his own perspectives on speciation and geographic distribution would, at least temporarily, enjoy the sanction of the United States Biological Survey.

Revolutionary changes in direction were developing within zoology at the turn of the century. In purely practical terms, the central problem facing mamalogists revolved around taxonomy and speciation. In the eighteenth century Swedish botanist Carl von Linné (1707-1778) had identified species as the fundamental biological taxa and had devised a binomial nomenclature which included both genus and species in an effort to distinguish the degrees of similarity and difference in external cranial and dental features (morphology). Thus, in the family Canidae (dogs), the Coyote is rendered Canis latrans, the Wolf, Canis lupus. Darwin's Origin of the Species, however, created problems for the binomial system by raising the spectre of organic speciation--that is, by suggesting that new species are constantly being created through evolution. It became clear that systematic classification would also have to evolve to take into account the changes wrought by selection and adaptation. Before 1870, when intermediates between species were discovered they were frequently assumed to have inherited acquired characteristics related to environment. Besides its inherent Lamarckism, this tendency to believe that most species were monotypic, and that structural variation implied aberration from the norm, effectively prevented subspecific designations. But after 1870 it became clear that the Linnaean system would have to be revised to accommodate the multiplicity of colors, sizes and skeletal characters gradually being discovered. Thus emerged, by painful steps, the trinomial nomenclature known today. This new trinomial system resolved most of the problems encountered by monotypists by adding a descriptive category to the names of genus and species. For example, a dark grey subspecies (or race) of Coyote located in southern British Columbia and Alberta is rendered Canis latrans lestis.[88]

[88] Sterling, Last of the Naturalists, pp. 199-206.

The scientific careers of C. Hart Merriam and Ernest Thompson Seton reached maturity in the midst of the debate on trinomialism. The technical limitations of their work reflect the tensions which divided mammalogists at the turn of the century over the problem of what constituted species and subspecies.

For Merriam the distinction was clear:

> A species is a group of individuals which resemble one another in all essential respects, varying only within certain definable limits, and which is separated from all similar groups by a well-marked hiatus. A subspecies or geographical race differs from a species in one respect only; namely, that intergrades exist connecting it with the parent stock: in other words, a subspecies is . . . an incipient species.[89]

This explanation falls somewhat short of the modern theory of speciation which argues that

> in sexually reproducing animals a new species develops when a population, which is geographically isolated from its parental species, acquires during the period of isolation characters which promote or guarantee reproductive isolation when the external barriers break down.[90]

In his official capacity Merriam was forced to spend much of his life behind a desk. Consequently he made many of his taxonomic pronouncements on the evidence of dead specimens. Most of his criteria for distinguishing between species were, of necessity, morphological. Nevertheless, Merriam shared

---

[89] C. Hart Merriam, "Species and Subspecies: A Reply to Mr. Conn," Science, 12 (Dec. 7, 1888), 279.

[90] Ernst Mayr, Animal Species and Evolution (Cambridge, Mass.: Harvard University Press, 1966), p. 482.

with ethologists of the same period a distrust of the laboratory, holding tenaciously to the value of direct field observation. This prejudice against the laboratory worked negatively to prevent him from grasping the role of heredity in mammalogy. While contemporary scholars like Hugo De Vries, William Bateson and T. H. Morgan were rediscovering Mendel's laws, Merriam rejected the theory of the origin of species by mutation. Today, of course, we take for granted the importance of heredity. To the modern geneticist, a species is "any group of sexually reproducing individuals which does not and cannot interchange genes with any other group."[91] That Merriam did not accept these new developments must not be taken to mean that he had no role to play in advancing the study of taxonomy.

---

[91]Lincoln P. Brower, "Speciation," in The Encyclopedia of the Biological Sciences, ed. Peter Gray (New York: Van Nostrand, Reinhold, 1970), p. 881. Because morphological discontinuity is central to geographic and genetic speciation, scientists have now been enabled to tighten their definitions, especially with respect to time and place. Allochronic speciation describes new species which evolve over time in a direct line of descent, adapting to constantly changing environmental conditions. Allopatric speciation describes the fragmentation of a given species into two or more geographically isolated groups which proceed to diverge genetically from the parent. Primarily a spatial phenomenon, allopatry will obviously foster greater changes over longer periods of time. If sufficient time has elapsed and the previously fragmented groups are reunited, they will be incapable of reproducing--that is, of exchanging genes--and may therefore be designated distinct species. As an incipient species, a subspecies is still capable of reproduction with its parent. The diversity of living species is directly dependent upon allopatry which, although allochronic speciation is implicit in it, stands alone in dividing the common ancestor. Ibid.; Mayr, Animal Species, pp. 424-28, 481-88, 516-18. See also Sterling, Last of the Naturalists, pp. 237-44.

In ornithology, Audubon's Birds of America (1827-1838) had isolated 506 forms of birds. Baird's Birds of North America (1860) raised this figure to 738. By 1884, the year after the creation of the American Ornithologist's Union, 878 forms were recognized. As one of the founders of the AOU, Merriam undertook the chairmanship of its Committee on Migration which launched a nationwide bird count of such ambitious proportions that the organization could not afford to subsidize it independently. In an effort to get adequate funding the Union lobbied in Congress, not only for an appropriation to support the census, but also for the creation of an independent branch in the Department of Agriculture devoted exclusively to the collection of scientific data on all aspects of North American bird life. This action was rewarded in 1885 by the establishment of an ornithological office in the Division of Entomology, elevated in 1886 to a separate Division of Economic Ornithology and Mammalogy, with Merriam himself as its Chief. In 1896 this section was renamed the Division of Biological Survey. Indirectly, therefore, the problems associated with taxonomic revision had a great deal to do with the origin of one of the most important government agencies in the country.

In his new capacity, Merriam enlisted the aid of field collectors like Vernon Bailey (who later joined the Biological Survey and married Merriam's sister, Florence), Frank M. Chapman and Seton, as well as that of countless farmers and private citizens. Bird skins sent by such sources from all over the continent, coupled with the more professional investigations of Merriam's own research parties, supplemented the work of the AOU and helped to demonstrate the phenomenal range of bird forms which had previously eluded classification. The AOU's first Check-List of North American Birds (1886) formally introduced trinomial nomenclature to scientific ornithology. By 1893 the number of known bird forms had jumped to 968.[92]

Developments in taxonomy had significant implications for all areas of zoology. Accurate systematization was central to questions of paleontology, geographic distribution, morphology, physiology, embryology, genetics and ethology. In his new post Merriam was to investigate the classification and distribution of mammals as well as birds. Although his own interests had always focussed on the latter, he soon discovered that mammalogy had been sadly neglected and required considerably more attention. The work of Audubon, Baird, J. A. Allen, Frederick True and others had brought the known number of mammal forms to 363 by 1885. By 1901, largely through the work of Merriam and his colleagues in the Survey, this number had quadrupled to about 1450. Forms, of course, do not represent species, but the total of species and subspecies. Modern zoology, influenced by Julian Huxley's

---

[92]Sterling, Last of the Naturalists, pp. 91-123, 207-08; Palmer, "Clinton Hart Merriam," p. 130; Frank M. Chapman, Handbook of Birds of Eastern North America (New York: Dover Books, 1966), pp. xxix-xxxvi.

"New Systematics" (which lays greater emphasis on lower taxonomic units and local populations), is marked by a tendency to reduce the number of species. Therefore, while the number of known forms has continued to rise, the number of species has declined.[93]

Merriam encountered serious difficulties in attempting to apply trinomial classification to mammals. He accepted the principle of intergradation, widely used in ornithology, which held that subspecies manifested intermediate links with their parent species. In ornithology these links were relatively easy to establish. In mammalogy, however, intergradation was often impossible to determine and, by many scientists, was merely assumed to exist where Merriam was certain it did not. His own inclination was to argue that there were many more species than taxonomists believed. In an effort to solve the problem he therefore posited what he perceived to be a more accurate criterion for isolating species and subspecies: an observed "degree of difference," he maintained, would not only distinguish one form from another but would also eliminate the subjective interpretations of amateur naturalists. Not surprisingly, however, it had precisely the opposite effect. Indeed, Merriam's own work on the Brown Bears of North America demonstrated the absurd lengths to which the principle could

---

[93]Sterling, Last of the Naturalists, pp. 209-11; Julian Huxley, "Toward the New Systematics," in The New Systematics, ed. Julian Huxley (New York: Oxford University Press, 1940), pp. 1-40.

be taken. In 1918 he claimed to have identified eighty-six distinct forms. It has been argued quite recently that all eighty-six can be reduced to one species with several geographical races.[94]

While modern taxonomists are, for obvious reasons, called "lumpers," Merriam's concept of speciation made him a "splitter." After 1900, as more and more specialists invaded the field of mammalogy, many of his conclusions were dismissed. Nevertheless, he persisted in his antipathy to the laboratory, arguing that real discoveries were made only in the field. He was like Seton in many respects--tyrannical, patronizing, an outspoken critic of the herd mind--but, as his penchant for taxonomic splitting demonstrated, he shared and fortified Seton's awe for the overwhelming diversity of nature. And despite the limitations of his work, Merriam is well remembered for the number of mammal forms which he discovered and which continue to bear his name.

Geographic distribution also occupied Merriam's attention, not only because it touched the thorny question of speciation, but also because it served the real mandate of the Biological Survey--to discover the economic impact of birds and mammals on agriculture. (In this field, as in others, his conclusions have since become valuable chiefly

---

[94]Sterling, Last of the Naturalists, pp. 212-235; C. Hart Merriam, Review of the Grizzly and Big Brown Bears of North America (Genus Ursus), with a Description of a New Genus, Vetularctos (Washington: United States Department of Agriculture, North American Fauna, No. 41, 1918).

as refutable points of departure.) He noted that the study of distribution had been dominated prior to 1850 by botanists following the lead of Alexander von Humboldt (1769-1859) who had been interested in the interaction of vegetation and environment and in the contributions of plants to the composition of landscape. After 1850 the field was appropriated by zoologists in the tradition of Louis Agassiz (1807-1873). The pre-Darwinian school had identified temperature as the most significant factor in distribution, a premise which Merriam adopted uncritically. J. A. Allen, a former student of Agassiz, supported his teacher's assumption that atmospheric conditions, particularly humidity, constituted the next most important factor, another theory to which Merriam also adhered. Allen also believed that mammals had originated in the arctic, later dispersing southward, a view which Merriam did not share, but one which is still debated today. Allen applied to North America Gloger's Rule "that mammals and birds are darkest in color where rainfall is greatest, and palest where it is least," and introduced Allen's Rule "which held that, in order to preserve body heat, appendages grew smaller as animals move northward."[95] Merriam's most controversial contribution to the discussion was his Life Zone Theory, first elaborated in 1890, wherein he suggested that northern (Boreal) and

[95]Sterling, Last of the Naturalists, p. 270.

southern (Sonoran) zones constituted the two major life regions of North America. Between them existed a neutral, Transitional zone which demonstrated features of both. While most of the United States fell within the Sonoran and Transitional regions, and while Canada occupied the Boreal, the Boreal clearly dipped south along the Alleghany, Rocky and Cascades/Sierra Nevada mountain systems. He considered the Boreal region circumpolar, the Sonoran unique to the United States. Assuming that temperature was the primary factor in distribution, Merriam hypothesized that his life zones were duplicated on a vertical plane, allowing altitude to perform the function of latitude. Geographic distribution would, therefore, duplicate "vertically, and in a smaller area, what might be found in a north-south strip of land several thousand miles long stretching from the cold arctic tundra of northern Canada into the tropical Florida Keys."[96]

Ornithologists had long subdivided the eastern part of the continent into seven "faunas," or zones, ascending latitudinally from south to north. These were designated Floridian, Louisianian, Carolinian, Alleghanian, Canadian, Hudsonian and Arctic. Merriam accepted these categories, but as most of his own research had been done in the West it did not necessarily follow that his three regions could

---

[96]Ibid., p. 278; C. Hart Merriam, "The Geographic Distribution of Life in North America, with Special Reference to the Mammalia," Ann. Rep. Bd. Reg. Smith. Inst., 1891 (Washington: GPO, 1893), 381-82.

absorb the ornithologists' seven. After further work, however, he proved that the Alleghanian fauna coincided with his own Transition zone. Dividing the Sonoran zone into two parts, he paralleled the upper half with the Carolinian and the lower with Louisianian and Floridian faunas.[97] He continued this process northward, concluding that

> all the major and minor zones that have been established in the East are found to be uninterruptedly continuous with corresponding zones in the West, though their courses are often tortuous, following the lines of equal temperature during the season of reproduction, which lines conform in a general way to the contours of altitude. . . .98

While the northernmost (Arctic) fauna of the Boreal zone would yield the fewest families, genera and species, the southernmost (Floridian) fauna of the Sonoran would yield the most. The warmer the temperature, the longer the reproductive season. This logic demolished A. R. Wallace's contention, in his Geographical Distribution of Animals (1876), that mountains posed an impassable barrier for many species of mammals. Because each fauna functioned vertically as well as latitudinally, species indigenous to them could be found anywhere within them, including the opposite sides of mountains. Furthermore, since mammals of the same species could travel back and forth over innumerable passes, mountainous

---

[97]In an effort to make his categories correspond more obviously with geographic and climatic features, Merriam finally restricted the Sonoran to the arid West, defining a new zone, the Austral, to parallel it in the East. The upper and lower parts of the Austral zone complement their Sonoran equivalents.

[98]Merriam, "Geographic Distribution of Life," p. 381

barriers could not possibly delimit distinct species.

These small modifications which, with a few others, Merriam felt had put the cap on a definitive Life Zone Theory were inadequate to immortalize all the work he had done. By emphasizing temperature and humidity he ignored altogether the ecological and ethological dimensions which dominate the modern study of geographic distribution. Failing to appreciate the subtle interrelationships of vegetation, soil, water, topographical features and predation, he was powerless to register the importance of local communities. This same tendency to look at the whole picture rather than its component parts had also affected his taxonomic work. Merriam was impatient to arrive at broad, general conclusions. Unfortunately, he reached too many conclusions before he had evidence to support them. In the end, his chief importance lies in the multiplicity of questions he asked rather than in the solutions he proffered.[99]

As Chief of the Biological Survey until 1910, Merriam

---

[99]Sterling, Last of the Naturalists, pp. 279-313, 436-37; P. J. Darlington, Zoogeography: The Geographic Distribution of Animals (New York: Wiley, 1957), pp. 364-71. See also, C. Hart Merriam, Results of a Biological Survey of the San Francisco Mountain Region and Desert of the Little Colorado, Arizona (Washington: USDA, NAF, No. 3, 1890); "Laws of Temperature Control of the Geographic Distribution of Terrestrial Animals and Plants," National Geographic Magazine, 6 (Dec., 1894), 229-38; "The Geographic Distribution of Animals and Plants in North America," Yearbook of the Department of Agriculture, 1894 (Washington: GPO, 1895), pp. 207-14; Life Zones and Crop Zones of the United States (Washington: USDA, Division of Biological Survey, Bulletin No. 10, 1898).

was perceived by many as the living authority on mammalogy in the United States. Seton's lack of scientific training, combined with his weakness for the glamor of association with the mighty, precluded any possibility that he would be seriously critical of Merriam's research. Shortly before the latter's resignation from the Survey, Seton joined Roosevelt in pressing him to publish a synthesis of his work on mammals, this despite the fact that Seton knew such a study would likely reduce significantly the pre-eminence accorded his own recently issued Life-Histories. Merriam was considerably more anthropocentric than Seton--a reflection of the dictates of the Department of Agriculture--but this fact took nothing away from his sense of awe at the diversity of nature. Interestingly, after leaving the Survey he joined the staff of the Smithsonian Institution where, fortified by a liberal endowment from the widow of railroad magnate E. H. Harriman, he diverted his attention to the ethnology of remote California Indian tribes, another clear indication of a kind of psychic union with Seton, whose interests inclined the same way at precisely the same time.[100]

As far as Seton was concerned, science was worthless unless everyone had ready access to its tools and could

---

[100]Sterling, Last of the Naturalists, pp. 339, 368-74, 438; Seton to Theodore Roosevelt, Jan. 20, 1909; Theodore Roosevelt to Seton, Jan. 22, 1909; E. A. Preble to Seton, Nov. 10, 1909; C. Hart Merriam to Seton, Nov. 30, 1909, Nov. 27, 1910; E. W. Nelson to Seton, June 15, 1910, SP.

understand its fundamental principles.[101] In his more technical studies he even followed Sir John Richardson's practice of identifying species by their Indian, as well as by their Latin names. "Experience shows," he wrote, "that a record of Indian names may be of great service to travellers and historians," a concern not shared by most professional zoologists.[102] But even though he had no compelling interest in taxonomy or speciation, he was painstaking in his own classifications, taking care to send ambiguous specimens to Merriam for proper appraisal. Not surprisingly, therefore, many of the errors in his work find their source in Merriam's monographs, of which Seton considered himself the people's interpreter. He subscribed to much of Merriam's Life Zone Theory, holding that temperature was a crucial factor in geographic distribution. Dividing North America into three regions--Tropical, Temperate and Arctic--he used the freezing point of water, the temperature "at which all life conditions are changed suddenly," as his base line of measurement. Thus, in the Arctic region

---

[101]Ernest Thompson Seton, "Letter to the Editors," Auk, 2 (July, 1885), 316-17; "On the Popular Names of Birds," Auk, 36 (April, 1919), 229-35; "English Names of Mammals," Journal of Mammalogy, 1 (Feb., 1920), 104-05.

[102]Ernest Thompson Seton, "A List of the Mammals of Manitoba," Transactions of the Manitoba Historical and Scientific Society, 23 (May, 1886), 1; "Birds of Manitoba," p. 458; Life-Histories, 1:ix, 23. Sir John Richardson was the naturalist and surgeon who accompanied John Franklin on both his overland journeys (1819-1822 and 1825-1826). He published the zoological results of these expeditions in his Fauna Boreali-Americana (1829-1837), one of Seton's major source books.

"water is always more or less frozen," while in the Tropical region "water never freezes." The north point of the Temperate region was "the limit of trees," its south point "the limit of frost."[103] He shared Merriam's view that the northern distribution of animals and plants was "governed by the sum of the positive temperatures for the entire season of growth and reproduction," and that southern distribution was "governed by the mean temperature of a brief period during the hottest part of the year."[104] Seton subdivided the two northernmost regions (with which his own work was wholly concerned) into faunas, the Arctic embracing the Alaskan, Barren Ground, Greenland, Alpine and Aleutian faunas, the Temperate encompassing the Hudsonian, Canadian, Sitkin, Pacific Coast, Campestrian and Alleghanian faunas.[105] Most of these were again partitioned into subfaunas with varying geographic areas, each containing either distinct species or distinct forms of broadly distributed species. For example, the Labrador subfauna of the Hudsonian fauna was "characterized chiefly by very dark races of species that are widely spread over several faunas." His humid Sitkan

[103] Seton, Life-Histories, 1:11-12.

[104] Merriam, Life Zones and Crop Zones, p. 54; Seton, Life-Histories, p. 14.

[105] Ibid. Here Seton worked some new variations on some old themes. For example, the Aleutian fauna was normally contained in the Temperate region. He followed W. H. Osgood in moving it north. See W. H. Osgood, A. Biological Reconnaissance of the Base of the Alaska Peninsula (Washington: USDA, NAF, No. 24, 1904), p. 24.

fauna reflected J. A. Allen's influence. "It is characterized . . . by exceptional rainfall, dense forests, and heavy-colored races of birds and mammals that have a wide distribution; but it has very few distinctive species. . . ."[106] All of Seton's categories fit perfectly into Merriam's life zones, including the Alleghanian fauna which he also called the Transition zone. Like Merriam's, Seton's faunas functioned vertically as well as latitudinally. Most of southern Manitoba fell within the Alleghanian, but "Riding, Duck and Porcupine Mountains may be considered islands . . . surrounded by Alleghanian conditions, but so elevated as to be Canadian in flora and fauna."[107]

In the work of men like Audubon and Lewis Henry Morgan there was evidence of a desire seriously to probe the behavior of North American animals and to sweep aside Cartesian mythology. Leroy's dream of a huge collection of "animal biographies" had found numerous supporters of varying degrees of competence.

---

[106] Seton, Life-Histories, pp. 14-17. Merriam himself had followed Allen's lead in the distribution question. Seton had clearly read Allen first, but employed Merriam's revision to bring his own interpretation up to date. See, for example, J. A. Allen, "The Geographic Distribution of the Mammalia Considered in Relation to the Principal Ontological Regions of the Earth, and the Laws that Govern the Distribution of Animal Life," Bulletin of the United States Geological and Geographical Survey of the Territories, 4 (May 3, 1878), 313-77; "Geographic Distribution of North American Mammals," Bulletin of the American Museum of Natural History, 4 (Dec. 29, 1892), 199-243.

[107] Seton, Life-Histories, pp. 18-22.

At the Smithsonian Institution, Coues, Baird and Ridgway, among the greatest living ornithologists in America, had launched Seton into the realm of rigorous and exacting science. Interestingly, it was their biological specialization which, through C. O. Whitman's work with Pigeons, introduced William Morton Wheeler to the study of ethology in the late 1890's. Indeed, ornithology was to prove central in Huxley's work on ritualization, Howard's research on territoriality, and Heinroth's and Lorenz's investigations of imprinting. However, Seton's inability to take criticism, which surfaced with Ridgway's rebuke of his monograph on "the Birds of Manitoba," had the effect of hastening his departure from the ornithologists' camp. After 1891, fortified by the international success of his Wolf paintings and by discoveries he had made while preparing the Art Anatomy, he turned increasingly more of his attention to mammalogy, where authoritative research was at a much more primitive level.

William T. Hornaday and C. Hart Merriam represent polar touchstones in Seton's work on mammalogy. Merriam was the analyst and systematizer, giving some, but proportionately very little, time to conservation and preservation. Hornaday was both conservationist and preservationist. Supporting Seton throughout the nature faker controversy, he allowed his sympathy for animals to overcome his scientific judgment. Although they began cautiously, his own behavioral interpretations ultimately snowballed into absurdity. Unhampered by a full-time occupation, Seton enjoyed the leisure to keep pace with much of the important current literature in diverse areas of the field, ensuring that he would avoid Hornaday's extremes. His research for the Life-Histories of Northern Animals began in 1897. Its development paralleled the regular publication of his animal stories. Because he had no formal scientific qualifications, Seton was cautious in his generalizations. He never claimed to have interpreted behavior correctly, and he left open to his readers the option of finding alternative explanations. He was not consciously anthropomorphic, but he, like many of the scientists whose work he read and synthesized, was guilty of errors. With Hornaday and Merriam he believed that legitimate observations could be made only in the field. For this reason he carried his Journal with him wherever he went, taking each available free moment to note down the natural activities transpiring in his midst. Under Merriam, for whom he collected specimens, he learned the art of proper description, something at which he always excelled. But with Hornaday he shared an underlying sense of unity with the very nature he was investigating. Unlike Merriam, who remained coolly objective about mammals, even while revelling in their complexity, Seton was submerged in both the landscape and its inhabitants. His were the eyes of the earth, the eyes of the persecuted. No merely mechanical explanations would ever eradicate the responsibility he felt for and the limitations he shared with the animals he studied.

He was caught in the crossfire between science and sympathy.

VI

Seton's understanding of morphology can be attributed in part to the vast number of scientific illustrations he was asked to make for Merriam, Chapman, Hornaday and others who recognized the superiority of his draughtsmanship. Before the success of Wild Animals I Have Known, he really depended upon commissions of this kind to sustain himself. He seems to have kept a fairly specific clientele, although he branched out upon occasion to prepare drawings for popular books, magazines and school texts. He joined Louis Agassiz Fuertes in illustrating an influential study of birds by Florence Bailey, Merriam's sister. The Macmillan Company, publishers of the Art Anatomy, sent some large contracts in his direction, including one for Mabel Osgood Wright's Four-Footed Animals and Their Kin, which required him to spend the summer of 1897 drawing from life in Yellowstone Park. On this trip he conceived the idea for the Life-Histories and seriously began compiling material for it. By the turn of the century he was also much in demand on the lecture circuit, another good source of income and a sound excuse for regular trips across the continent. On all such excursions Seton used his spare time to collect specimens for the

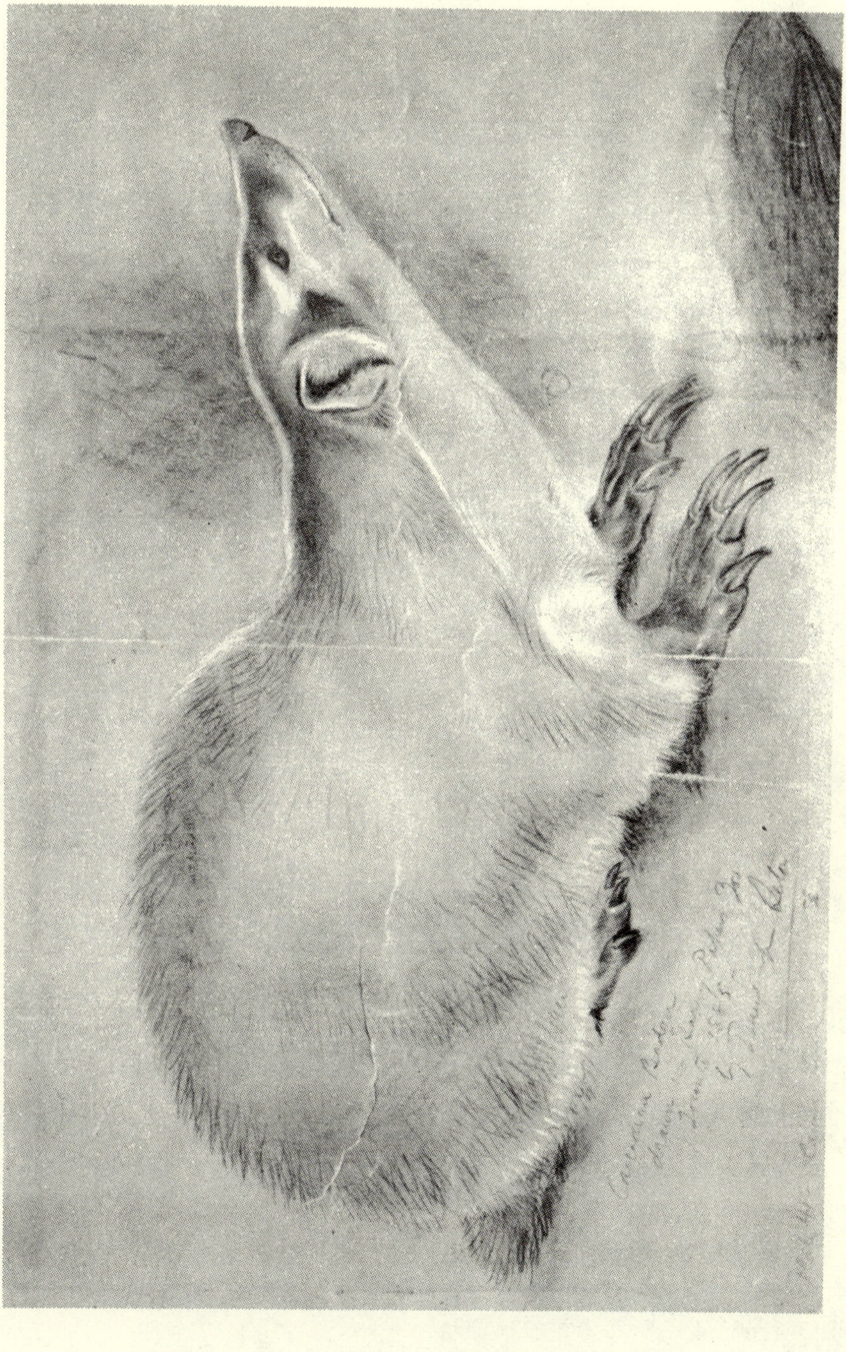

Plate 10. Canadian Badger (Pencil, 1885)

Smithsonian Institution and for the Biological Survey. In other words, the demands on his art not only contributed to the depth of his understanding of animal anatomy, but also created situations in which he could carry out original field work, measuring at first hand the interrelationships of various species and laying the groundwork for his skill as a descriptive naturalist.[108]

After 1903 Seton cautiously avoided animal stories. Until 1909 he withdrew almost exclusively into the popular essay format, only occasionally venturing forth with some children's fantasy.[109] His earlier work, by this time appearing in

---

[108] Seton to C. Hart Merriam; May 1, 1897, May 11, 1897, May 15, 1897, Nov. 1, 1898; Seton to Elliott Coues, Nov. 18, 1897; Seton to Frank Chapman, n.d.; Seton to George P. Brett, May 26, 1897, May 31, 1897, Dec. 27, 1897, Dec. 29, 1897; Seton to Florence Merriam, Nov. 6, 1896, Nov. 7, 1896, Feb. 14, 1898, March 19, 1898; Seton to the Art Director, Youth's Companion, Feb. 18, 1899, April 3, 1899, Oct. 12, 1900; Seton to Robert N. Martin, May 10, 1898; Seton to Edward F. Bigelow, July 21, 1898; Seton to Charles S. Fee, March 19, 1898, Nov. 1, 1898; Seton to J. M. Caldwell, May 5, 1899; Seton to the Southern Lyceum Bureau, May 3, 1899, Seton Letterbook No. 1, pp. 75-76, 182, 186, 189, 207, 219, 247, 252, 277, 279, 298, 316, 318, 331, 351, 354, 356, 397, 420, 437, 456, SP. C. Hart Merriam to Seton, Sept. 4, 1908, Sept. 9, 1908; Florence Merriam to Seton, March 27, 1898; W. M. R. French to Seton, May 13, 1899, SP. Mabel Osgood Wright, Four-Footed Americans and Their Kin, ed. Frank M. Chapman (New York: Macmillan, 1898); Florence Augusta Bailey, Birds of Village and Field (Boston: Houghton Mifflin, 1898); J. H. Stickney, Bird World (Ginn: Boston, 1898). Ernest Thompson Seton, "Elkland, 1," Recreation, 7 (Sept., 1897), 199-201; "Elkland, 2: The Beaver Pond," Rec., 7 (Oct., 1897), 286-90; "Elkland, 3: Old Timers," Rec., 7 (Nov., 1897), 369-72; "Elkland, 4: Flies and Weather," Rec., 7 (Dec., 1897), 456-57; "Elkland 5: Puss and the Bear," Rec., 8 (Jan., 1898), 33-34; "Elkland, 6: Duels," Rec., 8 (Feb., 1898), 117-19.

[109] See, e.g., Ernest Thompson Seton, Woodmyth and Fable (New York: Century, 1905).

several editions, in several languages, brought in substantial royalties. In an effort to discover which way the wind was blowing, Seton also undertook to print in family magazines the results of his research on fragments of the forthcoming Life-Histories. Proper care in the presentation of these, he was sure, would provide evidence of his documented knowledge, easing all doubt about previous claims to veracity. Another advantage of this pre-publication technique was the volume of correspondence it elicited from the readership, yielding new insights into distribution, diet, climatic conditions and so on.[110] The scientist whose work would be popular must involve his public in the writing of it.

Seton's real skills lay in descriptive work. He had begun in the early 1880's to publish rather amateurish ornithological descriptions and instructions on how to locate, kill and stuff specimens. While these gave hints of an interest in behavior, they generally stayed close to the path of accepted

---

[110] Ernest Thompson Seton, "The Wapiti and His Antlers," Scribner's Magazine, 39 (Jan., 1906), 15-33; "The Moose and His Antlers," Scr. Mag., 39 (Feb., 1906), 157-78; "The Caribou and His Kindred," Scr. Mag., 39 (April, 1906), 426-43; "The Prong-Horned Antelope," Scr. Mag., 40 (July, 1906), 33-49; "The White-Tailed (Virginian) Deer and Its Kin," Scr. Mag., 40 (Sept., 1906), 321-41; "The American Bison or Buffalo," Scr. Mag., 40 (Oct., 1906), 385-405; "The Habits of Wolves: Including Many Facts About Animal Marriage," American Magazine, 64 (Oct., 1907), 636-45; "The Snow-Shoe Rabbit," Everybody's Magazine, 16 (May, 1907), 599-608; "The Merry Chipmunk," Success Magazine, 10 (May, 1907), 328-31, 368-70; "Dogs of Song," Suc. Mag., 10 (Aug., 1907), 537-39, 562-63.

knowledge. He was a regular contributor to The Auk, the official organ of the American Ornithologists' Union, after its introduction in 1884.[111] The mere fact that his sightings were published in such an important journal gave him the confidence to extend, sometimes to overextend, his reach. After 1888, in a series of articles on the Song Sparrow and English Sparrow, he laid out a detailed methodology for describing the life histories of birds in general. In some of these he stated his belief that birds possessed "minds" that could be measured through data accumulated by the observation of behavior in the natural habitat. Any proper descriptive investigation, he maintained, required attention to several headings: migration, habitat, the "voices" of male and female (including variations, height from the ground when singing, time of day, noticeable causes for alarm, environmental influences), habits (including evidence of social behavior such as play and communication), the physical condition of specimens, food (including contents of mouth, gullet and gizzard), the form, color and measurement of plumage (noting differences of sex, season, age, moult and locality), reproductive behavior (with special attention to courtship), the measurement and composition of nests, the size, color

---

[111]Long after he had immersed himself in mammalogy, Seton continued to send some original material to the AOU. See, e.g., Ernest Thompson Seton, "Hybrid Pinicola enucleator X Carpodacus purpureus," Auk, 10 (Jan., 1893); "Bird Records from Great Slave Lake Region," Auk, 25 (Jan., 1908), 68-74; "Recent Bird Records for Manitoba," Auk, 25 (Oct., 1908), 450-54.

and number of eggs, parasites, nearest congeners, competitors and predators. His choice of the English Sparrow and the Song Sparrow was itself important. The first species (really a Weaver Finch) was not indigenous to North America but had been artificially introduced at Brooklyn, New York, Quebec City and Halifax after 1850 to help keep down caterpillar populations. By the 1880's it had multiplied and dispersed so broadly as to become a pest, driving away native birds that fed naturally on caterpillars, and itself acquiring a taste for grain. Information on the species was in constant demand by various government agencies concerned with agriculture so that Seton could, as it were, kill two birds with one stone. The Song Sparrow was indigenous, ate only weed seeds, and was a positive economic asset, affording a nice comparison with its less desirable neighbor.[112] Other minor methodological articles appeared throughout the 1890's and early 1900's, discussing everything from coloration to the proper methods for keeping a journal. As late as 1895 Seton devised the key to species used in Frank Chapman's definitive Birds of Eastern North America.[113]

---

[112] Ernest Thompson Seton, "Notes on the English Sparrow," Forest and Stream, 30 (April 5, 1888), 204-05; "The Song Sparrow: An Investigation of His Life History," Forest and Stream, 30 (April 19, 1888), 244; "The Myth of the Song Sparrow," Bird-Lore, 1 (April, 1899), 59; "Notes for Observation of Habits of Birds," Osprey, 4 (Sept., 1899), 6-8.

[113] Ernest Thompson Seton, "Directive Coloration of Birds," Auk, 14 (Oct., 1897), 395-96; "The Recognition Marks of Birds," Bird-Lore, 3 (Nov.-Dec., 1901), 187-89; "On Journal Keeping," Bird-Lore, 3 (Nov.-Dec., 1902), 175-76.

Reflecting on the responsibilities and rewards awaiting naturalists interested in studying mammalogy, Seton concluded that

> no man can possibly gather, even in a lifetime, from his own observation, facts enough to . . . map out the faunal areas of a few square miles. Obviously, then, if the work is ever to be done, it must be by hundreds, or thousands . . . of observers uniting their efforts, i.e. each contributing his little mite of ascertained fact, unmarred by generalisation, and put down in such a way that it will always mark a step in advance and may ultimately be used, without revision, as material for preparing the fullest and most accurate account; and then, after a sufficiency is gathered, the whole plan, with its many complex reasons, will unfold itself to our eyes like a magnificent pattern, so intricate that no one could have imagined it, and so beautiful that its mere contemplation will be found more elevating than many a noble work that is credited with shaping the human mind. . . .[114]

He believed that naturalists were indispensable to the success of all zoological inquiry. Natural history was a progressive accumulation of data which, in proportion as it increased in volume, opened itself to new interpretations. But the accuracy of the interpretations depended directly upon the accuracy of the data. Before 1891, Seton had merely dabbled in mammalogy. In 1886, with the help of Macoun, Merriam and J. A. Allen, he had prepared an immature, anecdotal "List of the Mammals of Manitoba" in which he identified and described fifty-two forms.[115] But by 1898, using his ornithological tools as

---

[114]Ernest Thompson Seton, "On the Use of Faunal Lists," Proceedings of the Canadian Institute, 7 (1888-89), 278-79.

[115]Ernest Thompson Seton, "A List of the Mammals of Manitoba," Transactions of the Manitoba Historical and Scientific Society, 23 (May, 1886), 1-26.

guides, he had devised an intricate schedule for the ordered investigation of all species of animals, which he applied consistently to the rest of his scientific research and writing. The new schedule required the accurate measurement of newly killed specimens, including data on sex, age, physical condition, weight, color, date and location, as well as a comparative description of analagous observations made on members of the same species. Like Merriam, he stressed the importance of morphological characters, urging fellow naturalists to include measurements of eyeballs, ears, hocks, hooves, and cranium, simplifying the procedure in a series of detailed diagrams (Fig. 1). Descriptive observations implied "habits" to most scientists; but to Seton, habits were evidence of "mind"--that is to say, of learning and instinct. Behavioral traits which could be employed to assess the relative sophistication of mammals would include evidence of ritualization, sociability, communication and so on. Other observations essential to an understanding of the composite functioning of species embraced physical facts (such as the information on diet afforded by scatology, the cyclical nature of populations and dispersal), environment (with special reference to the interrelationship of plants and animals) and predation.[116] The analytical rigor of this schedule accounts for the accuracy

---

[116]Ernest Thompson Seton, Schedule for the Horned Ruminants, A Paper Read Before the Camp Fire Club of America, 1898, Unpublished MS, SP; "For a Methodic Study of Life-Histories of Mammals," Journal of Mammalogy, 1 (Feb., 1920), 68-69.

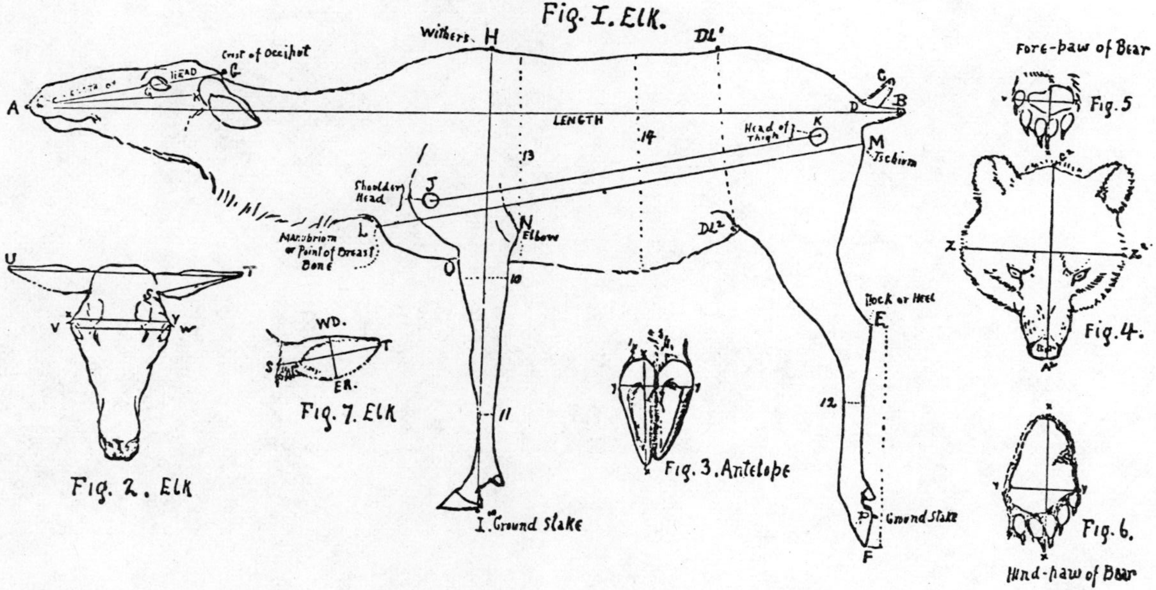

Fig. 1. Ernest Thompson Seton's plan for measuring an animal.

Source: Ernest Thompson Seton, "How to Measure an Animal," Recreation, 8 (April, 1898), 271.

of the Life-Histories and helps to explain why it is a landmark in the development of North American natural history.

In 1893 Seton had been appointed naturalist to the government of Manitoba. After 1896, although he lived in New York and Connecticut, he retained his British citizenship and continued to call Manitoba, the heart of the wilderness, his home. In 1900, partly to observe diplomatic niceties, partly to maintain his one legitimate title, but largely to give the proposed book added credibility, he wrote to the provincial Premier, Hugh John Macdonald, to ensure that his office was still recognized. Assured that it was, Seton could feel secure in the knowledge that Manitoba, at least, had not forgotten its illustrious stepson. Whether Americans realized it or not, the author they were calling one of their own was telling them stories primarily about his experiences in Canada, where, by the time of the nature faker controversy, he had spent well over half of his forty-three years. Furthermore, with the exception of some men like J. H. Fleming, W. E. Saunders and John Macoun, all of whom Seton knew well, Canada could boast relatively few competent naturalists of international stature. Among the members of the American biological fraternity there were many, like C. Hart Merriam, who needed information on animals whose patterns of migration and geographic distribution knew nothing of political boundaries. In most instances they were forced to send their own representatives to carry out field work on Canadian species.

But in Manitoba they had their own resident expert. Seton, therefore, was in the somewhat ironic position of being a British citizen, holding a Manitoba government appointment, living in the United States and performing a function of international significance recognized only by Americans.[117]

Seton's Life-Histories may be viewed as a scientific, yet imaginative trip back to the land of his youth, where he had experienced the "golden days." Although it also included comparative data from every possible extension of the range of each species, the two volume study dealt exclusively with mammals found within Manitoba. The extensions gave the work greater relevance for Americans since the habitats of many mammal forms in Manitoba reached well south of the border. By employing Merriam's Life Zones, he was able to make these extensions comprehensible to the public as well as to the scientific fraternity. In 1898 Seton had proposed to investigate only the horned ruminants, thirty species in all. But as the scope of the work was initially published in Forest and Stream, we may assume that this was an attempt to win favor with, and perhaps membership in, the Boone and Crockett Club.[118] The final product, published in 1909, analyzed

---

[117]Seton to Hugh John Macdonald, Jan. 13, 1900, Seton Letterbook No. 2, p. 123, SP; L. L. Snyder, "In Memoriam: James Henry Fleming," Auk, 58 (Jan., 1941), 1-12; P. A. Taverner, "Memories of William Edwin Saunders, 1861-1943," Auk, 61 (July, 1944), 345-51.

[118]Ernest Thompson Seton, "A List of the Big Game of North America," Forest and Stream, 51 (Oct. 8, 1898), 285-86.

sixty-four forms of mammals in Manitoba (an increase of twelve from 1886). Of these, fifty-one were identified as parent species indigenous to the province. Of the remaining subspecies, five were related to indigenous species and eight to parent species from outside the province. Seton added the fifty-one indigenous species to the last eight subspecies to arrive at a total of fifty-nine species representatives. His investigations were restricted to forms located within the pre-1912 political boundaries of Manitoba--that is to say, south of the fifty-third parallel of latitude. The boundary extension of 1912 more than tripled the size of the province, embracing a segment of the District of Keewatin already partially explored in 1900 by Edward A. Preble, a colleague of Merriam in the Biological Survey. Preble's research had turned up seventeen additional forms. As one of Seton's closest friends, and as the primary consultant on the Life-Histories, Preble guaranteed that he would have a sound knowledge of those forms lying north of the existing boundary. By uniting the compilations of both men we can safely say that the number of species and subspecies known to occur in what is now Manitoba totalled eighty-one in 1909. Modern research on the mammals of Manitoba

---

Here, Seton credits Merriam and T. S. Palmer (Biological Survey), Gerrit S. Miller (Smithsonian) and J. A. Allen (American Museum) for initial direction in the study. Being careful not to alienate the Camp Fire Club, he also published the results of his 1897 Yellowstone Park expedition in Recreation. The material from both articles was incorporated in the Life-Histories. Ernest Thompson Seton, "Mammals of the Yellowstone Park," Recreation, 8 (May, 1898), 365-71.

conducted by J. Dewey Soper has since revised this number upward to one hundred nineteen species and subspecies, an increase of thirty-eight forms.[119]

The relatively primitive state of taxonomy at the turn of the century and the proverbial controversy between lumping and splitting account in part for these discrepancies. Seton did keep a cautious eye out for new forms and is still credited with the discovery of one subspecies of Caribou in the Queen Charlotte Islands.[120] Although he never formally entered the theoretical debate on speciation, he was fastidious in his attention to detail, carefully reading all the monographs touching the forms he did locate in Manitoba. In his desire to avoid extravagant conclusions, he sometimes erred on the

---

[119]E. A. Preble, A Biological Investigation of the Hudson Bay Region (Washington: USDA, NAF, No. 22, 1902); J. Dewey Soper, "The Mammals of Manitoba," Canadian Field-Naturalist, 75 (Oct.-Dec., 1961), 171-73. Merriam declined Seton's invitation to become chief consultant, although he volunteered information and agreed to identify specimens and estimate populations. E. A. Preble to Seton, Jan. 1, 1908, July 11, 1908, July 29, 1908, Aug. 4, 1908, Sept. 15, 1908, Oct. 4, 1908, Nov. 11, 1908; C. Hart Merriam to Seton, July 16, 1908, July 24, 1908; Seton to C. Hart Merriam, July 23, 1908, SP.

[120]Ernest Thompson Seton, "Rangifer dawsoni: Preliminary Description of a New Caribou from Queen Charlotte's Islands," Ottawa Naturalist, 13 (Feb., 1900), 257-61. Now identified as Rangifer tarandus dawsoni (Seton, 1900), this subspecies is extinct. Seton claimed discovery of a second species, Rangifer montanus, later reduced to subspecific status and finally "lumped." Ernest Thompson Seton, "Preliminary Description of a New Caribou," Ottawa Naturalist, 13 (Aug., 1899), 129-30; A. W. F. Banfield, A Revision of the Reindeer and Caribou, Genus Rangifer (Ottawa: National Museum of Canada, Bulletin No. 177, 1961), pp. 88-92.

side of conservatism in the Life-Histories, excluding forms now known to have existed at the time the research was in progress. Either he failed to capture representatives of the omitted forms, or he simply neglected to send ambiguous specimens to Merriam or Preble for analysis, trusting instead to his own fairly accurate judgment and inadvertently lumping certain species and subspecies together. For each genus identified in Manitoba, Seton included a list of the species and subspecies known to exist everywhere else on the continent. This too entailed a taxing volume of reading and is a good indication of the lengths to which he was determined to go to enlarge his knowledge and to reinforce his credibility. As a result, his conclusions may be measured against modern investigations of southwestern Manitoba with highly satisfactory results. For example, in 1973 Robert E. Wrigley conducted a study of the Carberry Sandhills, the precise location where Seton did most of his own collecting. He discovered sixty-three species of mammals, merely four more than Seton. Accounting for migrations and for alterations in nomenclature and systematics since Seton's day, and admitting a certain margin of error, we must marvel at the fundamental soundness of the original work.[121]

As an illustration of Seton's attention to the complexities

---

[121] Robert E. Wrigley, "Mammals of the Sandhills of Southwestern Manitoba," Canadian Field-Naturalist, 88 (Jan.-March, 1974), 21-39.

of speciation, let us look briefly at his treatment of the Voles, comparing it with the much later research of J. Dewey Soper (who was one of Seton's specimen collectors[122]), and A. W. F. Banfield, one of Canada's foremost contemporary mammalogists. Soper's study of "The Mammals of Manitoba" (1961) is still taken as definitive for the province, although Wrigley has added new data for the southwestern region.[123] Banfield's Mammals of Canada (1974) is the authoritative work on all known forms. As the latter's distributional data are exceedingly accurate, it is a relatively simple task to extrapolate the Manitoban situation from his maps and figures.[124] From Table 1 it can be seen that in 1909 Seton could identify four different forms of Voles. Both Soper and Banfield, writing long after the boundary extension, identify the same four forms, plus an additional seven forms, on all of which they are in basic agreement. Differences in nomenclature reflect

---

[122] Seton to J. Dewey Soper, June 26, 1918; J. Dewey Soper to Seton, March 21, 1918, April 14, 1918, SP.

[123] Soper followed Merriam's Life Zone Theory with variations suggested in R. M. Anderson, Catalogue of Canadian Recent Mammals (Ottawa: National Museum of Canada, Bulletin No. 102, 1946); Soper, "Mammals of Manitoba," pp. 176-81, 196-99.

[124] A. W. F. Banfield, The Mammals of Canada (Toronto: University of Toronto Press, 1974). Banfield's concept of geographic distribution is drawn from the most recent developments in biogeography; see, e.g., Edwin M. Hagmeier and C. Dexter Stults, "A Numerical Analysis of the Distributional Patterns of North American Mammals," Systematic Zoology, 13 (1964), 125-55; Edwin M. Hagmeier, "A Numerical Analysis of the Distributional Patterns of North American Mammals, 2: Re-evaluation of the Provinces," Systematic Zoology, 15 (1966), 279-99.

	Seton (1909)	Soper (1961)	Banfield (1974)
1	Evotomys gapperi (Vigors)	Clethrionomys gapperi (Vigors)	Clethrionomys gapperi (Vigors)
2	Evotomys gapperi loringi (Bailey)	Clethrionomys gapperi loringi (Bailey)	Clethrionomys gapperi loringi (Bailey)
3	Microtus pennsylvanicus drummondi (Audubon and Bachman)	Microtus pennsylvanicus drummondii (Audubon and Bachman)	Microtus pennsylvanicus drummondii (Audubon and Bachman)
4	Microtus minor (Merriam)	Pedomys ochrogaster minor (Merriam)	Microtus ochrogaster minor (Merriam)
		Additions to Manitoba Voles Since 1909	
5		Clethrionomys rutilus dawsoni (Merriam)	Clethrionomys rutilus dawsoni (Merriam)
6		Clethrionomys gapperi hudsonius (Anderson)	Clethrionomys gapperi hudsonius (Anderson)
7		Clethrionomys gapperi athabascae (Preble)	Clethrionomys gapperi athabascae (Preble)
8		Microtus pennsylvanicus aphorodemus (Preble)	Microtus pennsylvanicus aphorodemus (Preble)
9		Microtus xanthognathus (Leach)	Microtus xanthognathus (Leach)
10		Phenacomys ungava mackenzii (Preble)	Phenacomys intermedius mackenzii (Preble)
11		Phenacomys ungava soperi (Anderson)	Phenacomys intermedius soperi (Anderson)

Table 1. The Voles of Manitoba, showing additions since 1909. Most of the Voles discovered by Preble were known to Seton before 1909 but, as they lay outside the political boundaries of the province at that date, they were excluded from the Life-Histories of Northern Animals.

Sources: Ernest Thompson Seton, Life-Histories of Northern Animals (New York: Scribner's, 1909), 1:506, 513, 515, 533; J. Dewey Soper, "The Mammals of Manitoba," Canadian Field-Naturalist, 75 (Oct.-Dec., 1961), 196-99; A. W. F. Banfield, The Mammals of Canada (Toronto: University of Toronto Press, 1974), pp. 178-85, 191-93, 201-22.

Seton (1909) North America	Banfield (1974) Canada	Hall and Kelson (1959) North America
1 Evotomys gapperi gapperi (Vigors)	Clethrionomys gapperi gapperi (Vigors)	Clethrionomys gapperi gapperi (Vigors)
2 Evotomys gapperi loringi (Bailey)	Clethrionomys gapperi loringi (Bailey)	Clethrionomys gapperi loringi (Bailey)
3 Evotomys gapperi athabascae (Preble)	Clethrionomys gapperi athabascae (Preble)	Clethrionomys gapperi athabascae (Preble)
4 Evotomys gapperi ochraceus (Miller)	Clethrionomys gapperi ochraceus (Miller)	Clethrionomys gapperi ochraceus (Miller)
5 Evotomys gapperi galei (Merriam)	Clethrionomys gapperi galei (Merriam)	Clethrionomys gapperi galei (Merriam)
6 Evotomys gapperi saturatus (Rhoads)	Clethrionomys gapperi saturatus (Rhoads)	Clethrionomys gapperi saturatus (Rhoads)
7 Evotomys gapperi rhoadsi (Stone)	no reference	Clethrionomys gapperi rhoadsii (Stone)
8 Evotomys caurinus (Bailey)	Clethrionomys gapperi caurinus (Bailey)	*Clethrionomys occidentalis caurinus (Bailey)
9 Evotomys ungava (Bailey)	Clethrionomys gapperi ungava (Bailey)	Clethrionomys gapperi ungava (Bailey)
10 Evotomys proteus (Bangs)	Clethrionomys gapperi proteus (Bangs)	Clethrionomys gapperi proteus (Bangs)
11 Evotomys wrangeli (Bailey)	no reference	Clethrionomys gapperi wrangeli (Bailey)
12 Evotomys brevicaudus (Merriam)	no reference	Clethrionomys gapperi brevicaudus (Merriam)
13 Evotomys carolinensis (Merriam)	no reference	Clethrionomys gapperi carolinensis (Merriam)
14 Evotomys idahoensis (Merriam)	no reference	Clethrionomys gapperi idahoensis (Merriam)
15 Evotomys dawsoni (Merriam)	Clethrionomys rutilus dawsoni (Merriam)	Clethrionomys rutilus dawsoni (Merriam)
16 Evotomys orca (Merriam)	no reference	Clethrionomys rutilus orca (Merriam)
17 Evotomys occidentalis (Merriam)	Clethrionomys occidentalis occidentalis (Merriam)	Clethrionomys occidentalis occidentalis (Merriam)
18 Evotomys obscurus (Merriam)	no reference	Clethrionomys occidentalis obscurus (Merriam
19 Evotomys mazama (Merriam)	no reference	Clethrionomys occidentalis mazama (Merriam)
20 Evotomys californicus (Merriam)	no reference	Clethrionomys occidentalis californicus (Merriam)
21 Evotomys nivarius (Bailey)	no reference	Clethrionomys occidentalis nivarius (Bailey)

Table 2. The genus Evotomys and the genus Clethrionomys. A comparison of the forms known by Seton in 1909 and their contemporary equivalents. Since 1909, nineteen new forms have been added to the list of Clethrionomys, 5 subspecies of the species rutilus and 14 subspecies of the species gapperi.
*Note that Banfield disagrees with Hall and Kelson on only one form.

Sources: Ernest Thompson Seton, Life-Histories of Northern Animals (New York: Scribner's, 1909), 1:506-08; A. W. F. Banfield, The Mammals of Canada (Toronto: University of Toronto Press, 1974), pp. 178-85.

E. Raymond Hall and Keith R. Kelson, The Mammals of North America (New York: Ronald Press, 1959), 2:711-18.

developments in the history of taxonomy. A careful look at the seven new forms reveals that only two, Phenacomys intermedius mackenzii and Phenacomys intermedius soperi, occur within the pre-1912 boundaries of Manitoba. The first was noted by E. A. Preble in 1902, the second by R. M. Anderson in 1942. The remaining five forms occur well north of the fifty-third parallel of latitude. Using the faunal lists and descriptions available at the time, therefore, Seton failed to locate only one known form of Vole.[125]

In the Life-Histories, however, Seton itemizes all of the known species and subspecies for each of the genera he had been able to isolate in Manitoba. If we look again at Table 1, we can see that he had located the parent species gapperi of the genus Evotomys, otherwise known as Gapper's Vole or the Gapper Mouse. He also located one subspecies, gapperi loringi. In order to give other naturalists more ground for comparison, he included a distributional map for the entire genus Evotomys, naming and locating on it the range of all known species and sub-species. In column one of Table 2 are listed all the forms of Evotomys known to Seton. In column three are listed the same forms as they are known today under the genus Clethrionomys, current since about 1930 when the lumpers took over from the splitters. In column two are listed those forms of Clethrionomys presently

---

[125]Banfield, Mammals of Canada, pp. 180, 184, 193, 211-12, 220.

known to exist in Canada. A close look at Table 2 reveals many things about Seton. First of all it is clear that he recognized every single species and subspecies of the genera known to exist in 1909. He did not discover any new forms himself although, since 1909, the number of forms of Clethrionomys has almost doubled. This at once suggests the range, detail and accuracy of his scientific reading, and the caution with which he undertook his own field investigations. As indicated earlier, Seton saw himself as the people's translator, the synthesizer of available material. The prudence which guided his own observations was dictated by the fear that originality might yield one more raking through the coals by another Burroughs or Ridgway. Accordingly, he stuck by the traditions established by Merriam and his colleagues. In 1909 Merriam was still firmly in command of the Biological Survey, splitting away to his heart's content. Little did Seton know that with Merriam's departure from the civil service in 1910 would come a generation of lumpers, bent on undoing the systematics of their former Chief.[126]

Column one of Table 2 shows exactly how strong the influence of the Biological Survey was on matters of speciation. Merriam and his brother-in-law, Vernon Bailey, were responsible for discovering fifteen of the twenty-one known

---

[126]See, Vernon Bailey, "Revision of the American Voles of the Genus Evotomys," Proceedings of the Biological Society of Washington, 11 (May 13, 1897), 113-38.

forms of Evotomys. If we add the subspecies discovered by Preble, over eighty percent of the forms of Evotomys known to Seton in 1909 can be said to have originated with the Biological Survey. Column two reveals that every single form of Clethrionomys known in Canada in 1909 was discovered by an American. Seven of the eleven forms, more than seventy percent, known in Canada were discovered by members of the U.S. Biological Survey, a clear indication of the value of American scientific records to Canadian ecological history. A comparison of column one with column three yields up all the essential information for an understanding of the lumper-splitter controversy. Seton was clearly a child of Merriam's scientific universe. He despised excessive generalization and was a devotee of the principle of diversity in all things. The complexities of nature did not admit of reductionism. There was a system, but the system was particularist. Unwittingly, Seton was in league with the splitters. He accepted uncritically their identification of the fifteen different species of Evotomys. Modern taxonomists, adopting a new name for the same genus, and lumping former species into subspecific ranks, have reduced this number to three.

In discussing Seton's treatment of the Voles I have looked at only four of the sixty-four mammal forms identified in the Life-Histories. An extended analysis embracing a broader sample would yield the same basic information, with perhaps some minor variations. The Voles were chosen, first because of the multiplicity of their forms, and second because they never appear in the animal stories as anything but prey.

With proper allowances for developments in science, most of Seton's technical information is remarkably accurate. If the trouble is taken to make conversion charts, such as those embodied in Tables 1 and 2, it is possible, in the case of virtually every species known to him at the time, to find parallels in the present. The same is also true for most of his distribution maps, although some, like that for the Wolf, are imprecise in the light of current knowledge.[127] Seton's descriptive work, based on the schedule he adopted in 1898, is fascinating. Measurements and color demarcations are perfectly exact. Morphological data are laid out in a careful, well-ordered format, satisfying scientist and amateur alike.

There is also evidence that Seton was beginning to think seriously about ecology, especially as it related to population dispersal. Even before beginning his research for the Life-Histories he had referred obliquely to the idea that vegetation

[127] In 1909 virtually nothing was known about the range or taxonomy of Wolves. He identifies the species found in Manitoba as Canis occidentalis (Richardson). Modern systematists have reduced this to Canis lupus occidentalis, locating it in Alberta, the Yukon and Northwest Territories. Its range may have altered in response to human populations. But Seton also calls this Wolf the "Buffalo-runner." The Buffalo Wolf, now extinct, is identified by modern taxonomists as Canis lupus nubilis (Say). Seton called Canis nubilus the Timber Wolf. He may well have seen nubilis in its last throes, but this is unlikely. Seton, Life-Histories, 2:749-54; Banfield, Mammals of Canada, pp. 289-95.

(as well as temperature and atmospheric conditions) was an instrumental factor affecting the geographic distribution of mammals.[128] Guided by Robert Bell of the Geological Survey of Canada he discussed the composition of soils in Manitoba and made comprehensive plant collections in all regions where he investigated mammal populations.[129] In the summer of 1907 Seton and Preble took a canoe trip to the Arctic Barren Grounds in an effort to verify the continued abundance of the Caribou and to gain some insights into its migration patterns. However, most of the interesting results of the expedition had nothing to do with the Caribou. Among other things, Seton explored and mapped Aylmer and Clinton-Colden Lakes (northeast of Great Slave Lake) for the Royal Geographical Society. He also gathered sixty species of plants which he donated to the herbarium of the Geological Survey of Canada for further analysis.[130] But he was particularly interested by local reports around Fort Smith and Fort Resolution that animal populations were in a regular state of flux, showing extreme density one year and all but disappearing the next. In 1904, for example, Snowshoe Rabbits had become a plague, but in 1907 Seton saw fewer than twenty in six months. Seeking to

---

[128] Seton, "On the Use of Faunal Lists," pp. 275-78.

[129] Seton, Life-Histories, 1:3-11; Robert Bell to Seton, Sept. 30, 1908, Oct. 14, 1908, SP.

[130] Ernest Thompson Seton, The Arctic Prairies (New York: Scribner's, 1911), pp. 328-38; R. W. Brock to Seton, Jan. 14, 1910, SP.

determine whether there was any cyclical correlation for all mammal forms, he secured the area returns of the Hudson's Bay Company for the period 1821 to 1905, and charted the seasonal volume for several species. From his tabulations he deduced three major principles:

(a) The high points for each species are with fair regularity 10 years apart.

(b) In the different species these are not exactly coincident.

(c) To explain the variations we must seek not the reasons for the increase - that is normal - but for the destructive agency that ended the increase.

In determining the factors that ended the increase, he found it necessary to break the species down into three groups. The first group comprised the Beaver and Muskrat, mammals whose food supply and enemies remained constant. Their variability Seton related to fluctuations in water level and ice conditions. The second group included the Snowshoe Rabbits, Mice and Lemmings whose numbers increased in the decacycle to the point where epidemic disease replaced natural predation and available food supply as the controlling agency. The third group took into account all the carnivores whose existence was dependent on the Rabbits, particularly the Lynx and the Fox. This schema was especially useful in working out the cycle of the Lynx which accorded almost perfectly with that of the Snowshoe Rabbit. A year after the decimation of a Rabbit population, Lynx numbers invariably reached their own apex, falling off drastically in the next year. Foxes responded in the same way, although their numbers were further affected by the fact that they also became prey to the Lynx in the absence of Rabbits. These observations have all been confirmed by modern studies and function as a clear demonstration of Seton's ability to generalize from observed data when he wanted to.[131]

Many similar conclusions reached during the Barren Grounds expedition were expanded in the Life-Histories to accommodate new observations made in Manitoba. For example, the correlation between the Lynx and Snowshoe Rabbit population cycles was retained with the added comment that Lynx numbers were not affected by disease or by distant migrations. This would have meant that the Lynx simply starved to death without an adequate local food supply--a fact for which Seton had found abundant evidence in the Mackenzie Valley. However, in the light of recent work we can say that he was correct regarding disease, but mistaken regarding migration, it having been proved that a Lynx will travel up to sixty-four miles when a scarcity occurs.[132] The important point is not necessarily that Seton discovered the correct ecological or ethological

---

[131]Seton, Arctic Prairies, pp. 95-112; Stuart Criddle, "A Study of the Snowshoe Rabbit," Canadian Field-Naturalist, 52 (March, 1938), 31-40; Banfield, Mammals of Canada, pp. 80-85, 349-51. See also, E. A. Preble, A Biological Investigation of the Athabaska-Mackenzie Region (Washington: USDA, NAF, No. 27, 1908).

[132]Seton, Life-Histories, 2:699; Banfield, Mammals of Canada, p. 350.

explanations for such phenomena, but rather that he looked for them at all. He was quite clearly on the right track and had a good grasp of the basic principles of interspecific relationships and food chains.

The Barren Grounds trip also supplemented his studies by bringing him into contact with Bison bison athabascae (Rhoads), the rare subspecies of Wood Buffalo. Here he had the added advantage of knowing personally the three living authorities on the mammal, J. A. Allen, W. T. Hornaday and S. N. Rhoads,[133] a fact which guaranteed an elaborate synthesis and a diligent effort to introduce new material. His estimates of pre-contact populations are still definitive according to F. G. Roe, although his theories on migration are again considered unsatisfactory. The major factors accounting for population regulation included blizzards, a hitherto unexamined cause, prairie fires, quicksand bogs, and rotten spring ice. Seton found that death by drowning was the predominant natural regulator and laid little stock in Wolf predation or disease. This was a courageous, if incorrect stand, reflecting some desire to reach beyond conventional explanations, relying at once upon historical accounts like Alexander Henry's Journal (1799-1814) and his

---

[133]It will be recalled that Seton illustrated Hornaday's treatment of the subject. See, also, J. A. Allen, The American Bisons, Living and Extinct (Cambridge, Mass.: Welch, Bigelow, 1876); S. N. Rhoads, "Bison bison athabascae," Proceedings of the Academy of Natural Science, 49 (1898), 498.

growing belief that the land itself must be taken into account in the explanation of all biological phenomena.[134] One other example, which does not find its origin in the Barren Grounds expedition, is afforded in Seton's treatment of the Pocket Gopher, "the master plowman of the West." This small rodent replaced the earthworm as the indirect producer of black loam on the prairies.

> The black loam in Manitoba is from one foot to two feet thick, an amount probably not exceeded over any large area elsewhere in the world. This is not a solid bed of decayed vegetation, but is thoroughly mixed with the upper formation.
>
> There is no doubt . . . that, in the absence of earthworms, this mixing is done . . . by a number of species of burrowing animals, but by far the most important of these are the Geomyidae or Pocket-gophers.

Darwin had concluded that earthworms brought up enough subsoil to cover the ground one inch thick. Seton estimated that a Pocket Gopher did as much in five months. Unlike the earthworm, it did not digest the substance of its castings. It did, however, bring subsoil to the surface and buried stubble to form humus. This recycling of the earth's surface "answers the purpose of fully disintegrating and mixing the dead vegetation with the soil" to produce the fertile earth necessary to sustain natural communities as well as agriculture. Again these conclusions still hold today, and again they reflect

---

[134]Seton, Life-Histories, 1:267-73; F. G. Roe, The North American Buffalo: A Critical Study of the Species in Its Wild State (Toronto: University of Toronto Press, 1970), pp. 3-11, 50-57, 154-58, 334-47, 536-41, 492-93, 829-41.

Seton's awareness of the essence of ecology.[135]

But the real purpose of the Life-Histories was the vindication of the animal stories. Tucked neatly away amidst the mounds of synthesized data and descriptive observations are many of the protagonists encountered in Wild Animals I Have Known, Lives of the Hunted and Animal Heroes. Most of them are acting precisely as they did in their "biographies." The information is reorganized, and new species are added, but the two volumes of the Life-Histories are replete with behavioral theorizing--this time without the blemish of the fictional form to obscure the author's intention. Urged on by Hornaday, Seton also used the Life-Histories as a platform for the expansion of his ideas on morality.

Addressing the Harvard Union in 1907, Seton argued that "five of the Ten Commandments are natural laws, namely, the ordinances against disobedience, murder, impurity, theft and falsification, the breach of which among animals entails severe punishment."[136] In light of the work of Konrad Lorenz and W. H. Thorpe, these intuitions no longer seem so absurd. In answer to literary critics who found Seton guilty of anthropomorphism, it must be argued that he was attempting quite the reverse, extracting himself from his own humanity in an

---

[135] Seton, Life-Histories, 1:581, 585; "The Master Plowman of the West," Century, 68 (June, 1904), 300-07. Cf. Stuart Criddle, "The Prairie Pocket Gopher," Journal of Mammalogy, 11 (1930), 275-80; Banfield, Mammals of Canada, pp. 147-51.

[136] Seton, Life-Histories, 1:31.

effort to ascertain its origins. In other words, he argued from animal to man, suggesting that nature's morality was superior to that of church or state. The only laws of church and state that deserved serious consideration were those which found their roots in nature. When aberrant behavior in animals seemed to deny the justice of nature's laws, Seton attributed the phenomenon to man's interference, noting that it was most pronounced among captive and domesticated individuals. He held that "the environment is the creator of the animal, the mould in which each species was cast."[137] The environment, therefore, deserved as much consideration as the animals whose habitat it constituted. In this respect Seton again drew parallels with man and demonstrated an emerging ecological consciousness which was inextricably tied to behavioral considerations. The range of the animal was part of its environment. If the ranges of animals altered, it might be concluded that the environment was itself altered by the presence of the animals in it. For example, if populations outdistanced food supply, vegetation would be directly affected. In this context Seton again acknowledged territoriality, but attributed it to the availability of proper foods before the instinct of reproduction.

> The high development of the property instinct is remarkable in some of the lowest forms of mammalian life. Beginning with food or mates, it extends to nest and range. . . . The devices used as property marks are most interesting. Some animals, as Bears, claw and

---

[137]Ibid., 1:25.

> gnaw the trees on their range--but most use the scent produced by special glands.[138]

Seton recognized sociability in certain mammals, especially Wolves, which he continued to single out for special attention. "Wolves do not den in colonies or continually move in bands, yet they unite their powers to help each other in tasks beyond the strength of one. . . ."[139] This kind of sociability, or, in Kropotkin's terms, mutual aid, was paralled by the growth of communication.

> Communication must progress with sociability. Other things equal, we find animals profiting by each other's society; that is, truly sociable in proportion as they have advanced in methods of communicating with one another. . . .
>
> While voice, gesture and touch are widely used, the most surprising [remain] the smell signals.[140]

In addition, Seton seems to have appreciated the importance of sociability to learning in animals. John Burroughs, who shared the Lamarckism of Wheeler, Merriam and countless others, believed that instinct was inherited habit. Yet he also argued that habits acquired during the lifetime of an individual "hardly have to be reckoned with at all."[141] Besides begging the question about where the inherited habits

---

[138] Ibid., 1:25-27.

[139] Ibid., 1:28.

[140] Ibid; see, also, Ernest Thompson Seton, "Intercommunication of Wolves," Forest and Stream, 48 (Jan. 23, 1897), 64-65; "Some More About Wolves," Forest and Stream, 48 (March 6, 1897), 183-84.

[141] John Burroughs, Summit of the Years, p. 157.

which constituted instinct came from, Burroughs tacitly denied the possibility that an animal could learn. Although Seton sometimes confused reason and learning, he did not do so intentionally. In modern ethology many species of social animals are known to be able to learn accumulated cultural habits in the course of a lifetime. Indeed, it is argued that those species which show the highest degree of social organization, and which transmit through learning culturally acquired habits, are the best qualified to survive.[142] Seton seems to have recognized this fact in Wolves, although he articulated the phenomenon in response to Burroughs' assumptions.

> Instinct has been defined as hereditary habit. But it is easy to cite many habits that become instinctual within the life of the individual so the modern definition of instinct as an unreasoning [unlearned] habit is better. . . .
>
> Instinct, then is the mere start in life that all animals are blest [sic] with at life. The later supervention of reason [learning] is I believe all powerful not only in man but in the beasts. And every instinct that the adult Wolf . . . retains is one that has persisted because subsequently endorsed by reason [learning].143

In this particular respect he was much closer to the truth

---

[142] Scott, Animal Behavior, p. 293.

[143] Ernest Thompson Seton, The Intelligence of the Wolf, Unpublished MS, n.d., SP. Seton was frequently ambivalent on this question, afraid to stir up a fuss. When put on the spot he tried to shift the emphasis. At one point he suggests that the training of the young "is given chiefly by means of example; whether consciously or not I do not know. I do not know what consciousness is; it may be that most human acts are not conscious. . . ." Seton, Life-Histories, 1:30.

than Burroughs.

Seton was fascinated by the observation that "a monogamous animal [like the Wolf] will beat a polygamous in the struggle for life. As a rule the higher quadrupeds in North America that hold their own against man are monogamous."[144] This was an indication of the superiority of monogamy. He did not argue from his own moral beliefs backward, but from his own empirical deductions, forward. As fellow members of the genus, Canis, the domesticated dog had rendered itself inferior to the Wolf by becoming polygamous, a fact which led Seton to believe that "domesticity is notoriously bad for the morals of animals."[145] In other words, domesticity, or the association of animals with man, had the effect of perverting nature's laws. For this reason, Seton joined Wheeler, Merriam and Burroughs in denying the value of the laboratory as the proper place for the observation of animals. But his insistence on tracing only those habits of animals, whether learned or innate, which corresponded with human morality did open him to the charge of anthropomorphism.

The crux of the problem--if it was a problem--still lay in Seton's conviction that animals were more intelligent than was customarily believed, and that man was more a creature of instinct than he cared to admit. His own field observations

[144] Ibid., 1:29.

[145] Seton, "Habits of Wolves," p. 636.

had confirmed him in this view, but the science of ethology had not reached the stage where it could provide adequate tools to investigate, let alone solve, the riddle. Merriam, whose major interests lay in systematics and geographic distribution, considered excursions into the realm of animal behavior as "twaddle" and advised Seton to keep to the guidelines he himself had set forth, steering clear of generalizations which could not be tested scientifically.[146] Merriam's attitude was mirrored in the operations of the Biological Survey where, until 1910, science came first, while conservation and preservation were always secondary.[147] Merriam considered Seton a good descriptive naturalist. As long as he stuck to monitoring the habits of animals, leaving the problems of interpretation to others, he would be safe. But to step beyond his qualifications would be to invite severe criticism.

Seton desperately needed an interpretative dimension. But without the security of a scientific training he knew he would be treading on dangerous ground. With Hornaday's prompting he resolved this difficulty in two ways. First, he created the animal story as a vehicle for carrying both objective truth and subjective interpretation. Stories did not require the same sort of rigorous analysis as science, but they could create impressions and arouse sensibilities.

---

[146] C. Hart Merriam to Seton, April 13, 1901, SP.

[147] Sterling, Last of the Naturalists, p. 385.

They induced a frame of mind which would be receptive to his general view that animals were intelligent beings. And they would also encourage a sympathetic understanding of Darwinian principles, chief among which was the kinship of man and animal. Secondly, in offering his first major scientific work, The Life-Histories of Northern Animals, to the public, he noted that "I have kept my theories apart from my facts, in order that the reader may judge the former for himself." In each case he allowed himself the luxury of the best of both worlds.[148]

VII

The Life-Histories was a monumental achievement by any standard. It was the first natural history ever to include an accurate map of distribution for a multiplicity of species. Some of these, such as the one for the Elk, which detailed pre-contact distribution, required a wide reading in zoo-paleontology, drawing Seton to the British Museum on his regular trips to England, as well as to the Library of Congress, the American Museum of Natural History and to countless other institutions in the United States. The determination of Elk distribution alone required "the looking up and copying out of passages in some 500 ancient books . . . and all of

---

[148]Seton, Life-Histories, p. vii.

this labor appears in the simple black outline which forms one element only of the map."[149] As the work contains sixty-eight such maps one gets a good idea of the staggering amount of research involved. Added to this, Seton supplied one hundred full page plates made from his own oil and water-color paintings and nearly three hundred descriptive illustrations, all done by hand. He was extremely careful not to be overly critical of the views of others, merely juxtaposing with those he opposed an exhaustive mountain of contrary fact and theory. Each of the sixty chapters constitutes a complete synthesis of all the known literature plus original observations and interpretative conjecture on the species discussed. Many--especially those on the Wapiti, the Buffalo, the Mule Deer, the Redsquirrel, the Beaver, the Snowshoe Rabbit, the Pocket Gopher, the Black Bear and the Wolf--are the earliest authorities cited in modern writing on the species.

The importance of the book was recognized immediately by the American scientific community whose membership joined in unanimous applause. Men like Gifford Pinchot and Madison Grant, with whom Seton had been trying to curry favor for years, wrote glowing tributes. William Brewster noted, with obvious reference to John Burroughs, that Seton had produced "a work on Natural History so nearly perfect at every point

---

[149] Seton to Edward F. Bigelow, Dec. 16, 1909, SP.

that any attempt to criticize it adversely would be likely to recoil, like a boomerang, upon him who made it." (This, of course, was exactly what Seton hoped.) His old friend from the American Museum, Frank Chapman, wrote that "Seton has done for mammals what Audubon did for birds but has done it better." Chapman's colleague, and the dean of natural historians J. A. Allen, found the Life-Histories "by far the best work ever written on such a subject." But Allen went even further than Brewster in acknowledging the value of the behavioral aspect.

> In this work you have built to yourself an enduring and enviable monument. . . . Your persistence and zeal through so many years is not less impressive and admirable than the affectionate interest that impels you so irresistably to this important line of research--the attempt to learn something of 'the intimate side' of the lives of animals.

Henry Fairfield Osborn, President of the American Museum, felt that the work "at once gives you a place among the leading students of the living mammals of North America," and "marks a new level in American natural history literature." But the most important congratulations of all came from Theodore Roosevelt who regarded the study "as one of the most valuable contributions any naturalist has made to the life histories of American mammals. In fact," he wrote, "I do not know anything that quite parallels your work. The book you have written is one of those . . . which combine great present interest with permanent value." Merriam and Hornaday, each for his own predictable reasons, were equally effusive.[150]

With the exception of Hornaday, Brewster and Allen, very few of those who commented publicly on the work's merits could muster the courage to discuss the behavioral question. The irony, of course, was that Seton had expanded the book to justify his intuitions about behavior. These intuitions had grown out of his sympathy for the wilderness and for its inhabitants. But sympathy was not science. In order for people fully to understand their kinship with the wild, to recognize the limitations imposed by environment, sympathy had to be allied with science in the work of an author whose credibility went unchallenged. Credibility was measured out in a formidable collection of precisely organized data. Theories of speciation and geographic distribution, widely accepted at the time but now no longer recognized, gave the book its substance. Speculations on behavior, based upon Seton's own observations in the field, were patronizingly passed over with smiles and muffled laughter. The fact was, however, that he had grasped an important trend in contemporary scientific writing. Because he was a synthesizer he had to make it his business to read widely in the field of zoology.

---

[150]Gifford Pinchot to Seton, Nov. 16, 1909; Madison Grant to Seton, Feb. 25, 1910; William Brewster to Seton, Nov. 21, 1909; Frank Chapman to Seton, Nov. 26, 1909; J. A. Allen to Seton, Nov. 16, 1909, Nov. 29, 1909; Henry Fairfield Osborn to Seton, Dec. 17, 1909; Theodore Roosevelt to Seton, Jan. 6, 1911; W. T. Hornaday to Seton, Nov. 26, 1909, Nov. 30, 1909; C. Hart Merriam, Nov. 29, 1909, SP.

His general grasp of the literature probably surpassed that of scientists and naturalists preoccupied with technical work on a single species or theoretical generalization. Ethology, the levelling science, has since suggested to man that he become more circumspect in his treatment of the animals with which he shares the environment. Ethology has also demonstrated the intricacies of animal behavior, revealing a dimension incorrectly cast off by many as anthropomorphism. In his animal stories and in the Life-Histories, Seton was attending the birth of this new science. Far from being anthropomorphic, he was trying to show what man could learn from the observation of animals. In this respect he carried on the tradition established by Leroy, Audubon and Lewis Henry Morgan, providing a popular link between them and Konrad Lorenz, Niko Tinbergen and W. H. Thorpe.

Those, like Burroughs, who had been quick to criticize Seton were equally hasty in re-orienting the direction of their thinking toward the investigation of behavioral problems. Burroughs, of course, held tenaciously to all his preconceived notions on the question and occupied himself with the refutation of new ideas. Hornaday went to the other extreme and made a fool of himself, while Merriam, who was not really interested, went off in pursuit of the Indian. After the publication of the Life-Histories, however, Seton's qualifications as a naturalist were never again seriously challenged. His

Plate 11. Black Bear (Pencil, 1885)

new animal stories[151] regained their former place atop the best-seller lists and the author enjoyed once more the glory in which he revelled. But the reasons for which the book was a success in 1909 have little bearing on the reasons for its being a classic now. In 1909 it depended upon a synthesis of the work of others for its accolades. Now we look to the originality of the author's own speculations on behavior and recognize in them a significant contribution to the science of ethology.

Seton's intuitions regarding animal behavior linked the wilderness with morality. Because it perverted wilderness, technology also perverted morality. The inhabitants of the wilderness, whether man or animal, were united with it and were, therefore, either a priori moral or in the process of evolving toward an absolute moral state of grace. Civilized man must look to his birthplace in the wilderness to find his own moral peace. For morality was a balanced combination of instinct and learning premised on the survival, not of the individual, but of the individual within the community of all living things. Nothing which ignored this concept of community could be called moral. Nothing which ignored this concept of community would survive.

---

[151]Ernest Thompson Seton, Wild Animals at Home (New York: Doubleday, Page, 1913); Wild Animal Ways (New York: Doubleday, Page, 1916).

Chapter 4

WILDERNESS INNOCENCE: INDIANS, BOYS AND BIOLOGY

> Do you really love nature, or will you remain strangers and aliens to her mighty heart? Taste and see that she is purest, truest, noblest. We sprang from her bosom and inherit vastly more than we ever dreamed of her wisdom, and to her all that is mortal of us will return. In affliction and calamity, when conscious purpose and endeavor fail, we can sink back into her arms, and when creeds and philosophies fade we know that if "our bark sink, 'tis to her larger sea." She is all law and no chaos, life abounds wherever life is possible, good will is there because the best and not the worst survive, and youth is in a peculiar sense the consummate flower of nature, more worthy than anything else on earth, of love, reverence, and devoted service.
>
> G. Stanley Hall

I

There has been in recent years a growing interest in the history of youth organizations in North America and abroad. Stimulated in part by the work of psychohistorians Erik Erikson and Robert Jay Lifton,[1] the study has since branched out in several different directions.[2] The birth of the so-called counter culture in the 1960's spawned new questions about the extent to which young people had learned, usually in defiance of parental or "establishment" values, to make decisions touching their own destinies. As most of these decisions had social, economic and political implications of material significance for the long term, and because they contravened the assumptions of existing institutions, they gave rise to analytical queries about "the individualization of moral judgments" and "the relativization of truth."[3] One of the leading exponents of this line of inquiry, Kenneth Keniston, has suggested that

the drastically altered historical conditions of

---

[1]See, for example, Erik Erikson, Childhood and Society (2nd ed.; New York: Norton, 1963); Ghandi's Truth: On the Origins of Militant Nonviolence (New York: Norton, 1969); Identity, Youth and Crisis (New York: Norton, 1968); Young Man Luther: A Study in Psychoanalysis and History (New York: Norton, 1958); Robert Jay Lifton, Boundaries (Toronto: Canadian Broadcasting Corporation, 1969); History and Human Survival (New York: Random House; Vintage Books, 1971).

[2]For a review of the literature, see Kenneth Keniston, "Psychological Development and Historical Change," Journal of Interdisciplinary History, 2 (Autumn, 1971), 329-45.

[3]Ibid., p. 344; Theodore Roszak, The Making of a Counter Culture: Reflections on the Technocratic Society and Its Youthful Opposition (Garden City, N.Y.: Doubleday, 1969).

> the twentieth century (extended mass education, widespread affluence, exposure to other cultures, threat of holocaust) have in turn changed the modal matrix of human development, "producing" on a mass scale a kind of questioning, restless youth who, if he existed at all in the past, was always part of a tiny and exceptional minority.[4]

In effect, rapid historical change had made youth "a new stage of life."[5] Superficially, dutiful obedience to parental dictates remained relatively constant; upward social mobility retained its appeal; a "good job" was assumed to be the guarantor of future choices. But, says Keniston, there was building within the innocent psyche a pressure which belied the tranquility of non-resistance. Historical change produced psychological change, whether apparent or not. It would seethe internally, gradually gaining the power of its own expression, and then burst, itself generating a political and social transformation requiring more adaptations. The first manifestation was the spontaneous youth movement, witnessed in the late 1960's. It was leaderless to the extent that it was not sponsored by the church or by the state or by the DAR. It was not the Boys' Brigade, the Knights of King Arthur, the Civilian Conservation Corps, or the Camp Fire Girls. Above all it was not the Boy Scouts. (Ironically,

---

[4]Keniston, "Psychological Development," p. 345. Cf. Thomas F. Neblett, "Youth Movements in the United States, Annals of the American Academy of Political and Social Science, 194 (Nov., 1937), 141-51.

[5]Kenneth Keniston, "Youth: A 'New' Stage of Life," American Scholar, 39 (Autumn, 1970), 631-54.

declining Boy Scout membership over the past decade might serve as an inverse indicator of the change effected by youth movements.) The leaders of youth movements were youths. Like Daniel Cohn-Bendit in France, they, for the most part, disclaimed their own leadership. Through many of them, anarchism discovered a renewed legitimacy.[6]

The psychological phenomenon of which we are speaking is called Developmentalism. Because it lies beneath the surface and because it passes largely unnoticed, it has suffered from a dearth of interest. In other contexts and with other interpretations it has been raised, mocked and dropped before now. The work of G. Stanley Hall, particularly his controversial study of adolescence,[7] suffered the rebukes of countless theoreticians after the turn of the century, but has recently become the object of renewed interest.[8] The same kind of concern is reflected in a relatively new periodical, The History of Childhood Quarterly, distinctive for its psychohistorical orientation. A burgeoning revisionist

---

[6]Daniel Cohn-Bendit and Gabriel Cohn-Bendit, Obsolete Communism: The Left Wing Alternative, trans. Arnold Pomerans (London: Penguin, 1969).

[7]G. Stanley Hall, Adolescence: Its Psychology and Its Relations to Physiology, Anthropology, Sociology, Sex, Crime, Religion and Education (2 vols.; New York: D. Appleton, 1904).

[8]Dorothy Ross, G. Stanley Hall: The Psychologist as Prophet (Chicago: University of Chicago Press, 1972).

literature has followed on the heels of Philippe Ariès' classic, Centuries of Childhood,[9] which was "founded on the important and incontrovertible assumption that much can be learned about a culture by investigating the way it regards its young."[10] The revisionists, while applauding Ariès' insights, have joined with John Demos to "find certain underlying themes in the experience of children in a given culture or period in order to throw some light on the formation of later personality." In other words, they have reversed Ariès' perspective, assuming "the ironic truth that 'the child is the father to the man,'" not simply a mirror, but a creator of culture.[11]

The connection between the history of youth organizations and the history of childhood is obvious enough. It is my intention in this chapter to mingle the two somewhat, employing the work of Ernest Thompson Seton as a vehicle for the discussion. A devotee of G. Stanley Hall's Darwinian, biological developmentalism, Seton was the father of one youth organization and the unorthodox uncle of another. The first, founded in

---

[9] Philippe Ariès, Centuries of Childhood: A Social History of Family Life, trans. Robert Baldick (London: Jonathon Cape, 1962).

[10] John Demos, "Developmental Perspectives on the History of Childhood," Journal of Interdisciplinary History, 2 (Autumn, 1971), 315.

[11] Ibid., p. 316. For a practical application of this thesis, see, John Demos, A Little Commonwealth: Family Life in Plymouth Colony (New York: Oxford University Press, 1970).

1902, was called the Woodcraft Indians; the second, established in 1910, the Boy Scouts of America. In the light of Keniston's and Demos' analyses, Seton is an anomaly. On the one hand, he believed in the spontaneity of childhood and youth and sought diligently to encourage its expression. On the other, he chose a modified organizational scheme in an effort to provide direction. I suppose "modified" is the operative word, for the Woodcraft Indians were highly decentralized and, in the end, hopelessly disorganized, incapable of becoming a potent national force. This, of course, can be considered their redeeming feature, depending upon how one views youth organizations.[12] Again, it was his disorganization which contributed, at least partially, to his declining importance within, and ultimate ouster from, the Boy Scouts of America. Throughout a long career in youth work, however, Seton sustained an abiding faith in the native instincts and ideas of children and adolescents. In this sense he never lost touch with the biological continuum weaving its way through all his other interests. While it might have been a deterministic perspective, it gloried in the powers of imagination and creativity, rejecting competition and nurturing community.

---

[12]For some analytical possibilities, see Herbert S. Lewin, "Hitler Youth and the Boy Scouts of America: A Comparison of Aims," Human Relations, 1 (1947), 206-27; John R. Gillis, "Conformity and Rebellion: Contrasting Styles of English and German Youth, 1900-1933," History of Education Quarterly, 13 (Fall, 1973), 249-60.

Had Seton lived to witness the birth of a world-wide ecological movement commanded by the young, he would have been much less concerned about being repudiated by the heirarchy of the Boy Scouts which was part of the unthinking machinery that made it necessary.

The Seton or Woodcraft Indians have enjoyed very little scholarly attention to date. Brian Morris sees Seton as a Rousseauan glorifier of the noble savage, but does not touch on the ecological question at all. He does, however, find in the woodcraft movement the foundation of the Boy Scouts and, from a British perspective, gives added weight to the contention that Robert Baden-Powell plagiarized many of Seton's ideas. Indeed, he goes so far as to suggest that Baden-Powell merely gave the woodcraft movement "a patriotic slant, made its appeal universal, and by resting its constitution on the pillars of established society ensured its survival."[13] David Macleod rejects Morris' conclusions, arguing that the "Woodcraft Indians were probably an abortive alternative to scouting, partly absorbed by the new movement, rather than a direct antecedent." Macleod suggests that "Seton tried to make woodcraft an end in itself and disdained utility," but he lays no emphasis on the ecological side of woodcraft and is content to assume that because the Boy Scouts won the

---

[13]Brian Morris, "Ernest Thompson Seton and the Origins of the Woodcraft Movement," Journal of Contemporary History, 5 (1970), 194.

numbers game, the Boy Scouts had grasped the reality.[14] Allan R. Whitmore has looked at the career of Daniel Carter Beard who, in 1905, set up an organization to rival Seton's Woodcraft Indians. Initially called the Sons of Daniel Boone, it was renamed the Boy Pioneers in about 1910, capitalizing on the growing enthusiasm for the Boy Scout movement in England. Beard was a patriot and a nationalist who glorified the frontiersman--the very symbol Seton most despised. Despite this fundamental difference, however, Seton and Beard maintained a tempestuous friendship which managed even to survive the debate over which of them originated the Boy Scout idea. Whitmore suggests that Beard was the folk hero of the Boy Scouts of America until his death in 1941.[15] Certainly he was one of the founders and, unlike Seton, managed to hold down his position as National Scout Commissioner and member of the Executive Board from 1910 onwards. This again, however, is no reflection on Seton. Beard was a reactionary while Seton was, if anything, a kind of Red Tory, with emphasis on the "Red." Seton was neither an American citizen nor a patriot, incidentals which supplied

[14] David Irving Macleod, "Good Boys Made Better: The Boy Scouts of America, Boys' Brigades and YMCA Boys' Work, 1880-1920," Unpublished Ph.D. dissertation, University of Wisconsin, 1973, pp. 215, 218.

[15] Allan Richard Whitmore, "Beard, Boys and Buckskins: Daniel Carter Beard and the Preservation of the American Pioneer Tradition," Unpublished Ph.D. dissertation, Northwestern University, 1970.

the pretext upon which he was ultimately dropped from the organization. Any comparisons about the appreciation of the needs of youth, as between Beard and Seton, will find no satisfactory evidence in the official annals of the Boy Scouts. Their respective destinies were determined in a web of political intrigue which inevitably favored the righteousness of the State over anything that might remotely have been construed as a challenge to it.

J. O. Springhall shares Brian Morris' conviction that Baden-Powell's objectives require more sensitive criticism. Springhall also demonstrates that, despite frequent protestations to the contrary, the Boy Scouts were essentially a bourgeois, nationalist and militaristic body, organized "as an instrument for the transmission of the norms and values of one generation to [those] of the next."[16] Paul Wilkinson adds his voice to the British chorus, seeing the Boy Scouts as a carefully designed attempt "to harness their energies for great causes approved by their adult leaders--imperial defence, national defence, international co-operation, national efficiency, and so on."[17] While arguing that the Boy Scouts of America rejected Seton and tried to maintain the British ideal, Macleod

---

[16] J. O. Springhall, "The Boy Scouts, Class and Militarism in Relation to British Youth Movements, 1908-1930," International Review of Social History, 16 (1971), 158.

[17] Paul Wilkinson, "English Youth Movements, 1908-1930," Journal of Contemporary History, 4 (April, 1969), 22.

is forced to admit that even Baden-Powell's model for the movement was highly decentralized by comparison. Indeed, the American movement was so bureaucratic, so preoccupied with keeping the Scouts "under constant adult surveillance,"[18] that it would have been surprising if it had reflected anything less than an absolute extension of middle class, progressive values.

This brief review of the literature makes no pretense at being definitive. It aspires merely to introduce some of the perplexing questions that will concern us in the following pages. The Woodcraft Indians were established in 1902, well before the Boy Pioneers or the Boy Scouts. As a movement it was anti-militarist and anti-pioneer in conception. It favored the spontaneous and democratic expression of youthful thought and action with a minimum of adult guidance. It differed from antecedents like the Boys' Brigade for its lack of affiliation with existing institutions, and it was highly decentralized, implicitly rejecting both the Church and the State by adopting an Indian model. Its objectives were, in modern terms, ecological, and for that reason it was misunderstood and attacked. Many of its basic principles were adopted, altered and eventually perverted by the Boy Scouts. During this process, Ernest Thompson Seton remained a controversial focal point of debate.

---

[18]Macleod, "Good Boys Made Better," p. 289.

II

Seton's autobiography tells us virtually nothing about his involvement in the Boy Scout movement. As the movement was to become such a potent force in the lives of American young people, and as the subject has been treated so poorly to date, it seems proper that newly discovered records supporting Seton's indispensable participation in it should now be brought to light. It is difficult to know exactly where to begin such an investigation because his entire creative life was inextricably tied to notions of the purity and innocence of youth. "Guided by common sense, by observation, by zoology, and above all, by the sayings of the Greatest of all Teachers," he wrote, "I began with the assumption that All children come here direct from God and are pure as God can make them. We do not have to reform them, but rather to keep them from being deformed."[19] His love for the wilderness embraced all the innocents inhabiting it--from the land itself to the animals, to the indigenous peoples. If his art and his writing had focussed on the first two categories, his

---

[19]Ernest Thompson Seton, A History of the Boy Scouts, Unpublished MS, n.d., p. 2, SP. Segments of this 62 page manuscript do appear in print; see for example, Ernest Thompson Seton, "The Spirit of the Woods: A Confession," Century, 103 (Dec., 1921), 213-24. Their sense, however, is altered in the larger context. The manuscript was prepared for inclusion in the autobiography but was withdrawn at the request of Seton's second wife, Julia, shortly before the book went to press. Maxwell Perkins to Seton, Jan. 4, 1940; Maxwell Perkins to Julia Moss Seton, May 9, 1940; Julia Moss Seton to Maxwell Perkins, May 13, 1940, SP.

educational activities would now centre primarily on the third. Seton did not love Frederick Jackson Turner's pioneer. When he spoke of retaining the old values, he meant the values of the Algonkian, the Iroquois and the Sioux. In effect, the Woodcraft Indians grew directly out of Seton's own way of life.

Although Seton maintained a studio and apartment at A. A. Anderson's Beaux-Arts Building in New York,[20] he purchased "a few small abandoned farms" in Cos Cob, Connecticut, a short commuter trip from the city. Naming the tract "Wyndygoul" after a legendary Seton estate in Scotland, he proposed, in a symbolic reversal of the Turner hypothesis, to return the West to the East, allowing the wilderness to reclaim the pastoral.

> I was now absolute owner of a tract of tangled wildwood, with hills, rocks and trees in abundance, a few little meadows, and a beautiful brook that spread into an alder swamp. The swamp was easily turned into a lake, and furnished the central beauty spot to a little paradise of birds and wildlife.
>
> I built an inexpensive cabin, and around the place put an impassable mesh wire fence ten feet high, and finished at the top with barbwire entanglements.
>
> Here I proposed to live most of the time. Here I meant to try out many experiments at conservation and restocking. Every native kind of wildlife was to be encouraged. I rejoiced in my dream come true. I gloried in my wildwood.[21]

The site, which Seton hoped would recall the varied Canadian landscapes of his youth, had been selected with the utmost care and attention to detail. "I wanted level land," he

---

[20] F. W. Halsey, "Ernest Seton-Thompson in Bryant Park, N.Y.," American Authors and Their Homes (New York: James Pott, 1901), pp. 281-292.

[21] Seton, Autobiography, p. 377.

told one visitor. "I do not like anything to obstruct my view. I regard the mountains as highly impertinent."[22] The architectural style of the newly-constructed central buildings could only be described accurately as "Tudor-Indian,"[23] reflecting the Victorian-wilderness tension that characterized the man. "We ripped off the architect's imitation timbers, his bogus rafters, his pretended beam-ends, his dummy bolts, and his tinware hammered iron," substituting the real thing in each case. No expense was spared to achieve the desired effect on the exterior:

> The basement and the first floor are of field stone, as smoothly laid and plastered as possible. In the cement was put a little paris green and yellow ochre, so that a faint olive green tone prevails in the stone work. . . .
>
> When it came to preparing the cement [for the upper storey] I had four bins mixed: one with a little red ochre, one with yellow ochre, one with a dash of lamp black, and one uncolored. Each had consisted of a shovelful from each color unmixed, and each plasterer was told to lay it on with his trowel. . . .
>
> The results are that it shows the pleasant traces of human handiwork throughout, and the color, while a faint rose terra cotta in general, has a beautiful play of several soft and agreeable tints.[24]

Seton stocked the lake with brook trout, black bass and twenty-three species of ducks and geese. To the grounds around he introduced imported rabbits, pheasants and

---

[22] Ernest Thompson Seton, quoted in Myra Emmons, "With Ernest Seton-Thompson in the Woods," Ladies' Home Journal, 18 (Sept., 1901), 3.

[23] Anya Seton Chase, Taped Interview, 17 December, 1971.

[24] Ernest Thompson Seton, "The Story of Wyndygoul," Country Life in America, 16 (Aug., 1909), 446-48.

guinea hens.[25] Wyndygoul was not without its precedents. William Brewster, the ornithologist, maintained a similar estate at Concord, Massachusetts--"about three hundred acres of woodland" with "many interesting birds and beasts (including wild deer)."[26] While Seton clearly modelled much of Wyndygoul after Brewster's sanctuary, the latter's only material assistance came in the form of birch and pine seedlings, cultivated at Concord and shipped to Cos Cob to be planted in the old pastures. These trees, the lake and local rock formations combined with both indigenous and imported wildlife to reconstitute an environment strongly reminiscent of the Laurentian Shield as it rises to form the Muskoka and Haliburton Highlands of Ontario.[27]

Wyndygoul was to be the focus of much of Seton's activity from 1901 to 1912. Close to his friends in New York, he was far enough removed to observe at first hand the lives of the animals about which he wrote. Although he continued his annual expeditions into the field, he was no longer forced to look west or north to find wilderness. Among other ingenious inventions, he constructed here his

---

[25] Ernest Thompson Seton, "The Story of Wyndygoul," Country Life in America, 16 (Sept., 1909), 505-08, 540, 542, 544.

[26] William Brewster to Seton, April 8, 1902, SP.

[27] William Brewster to Seton, April 20, 1902, April 25, 1902, May 10, 1903, May 17, 1903, SP.

famous hollow tree. The source of several articles,[28] this artificial habitat might, in the words of one observer, "have deceived a Dryad. Its bark, its tapering shape, its broken top, its rotting boughs were, if I may say so, to the life." At the foot of the tree was a small door through which a human could gain access to the interior.

> On the floor fine sand was dusted, so that you might make out . . . what visitors had entered by that way. . . . To the tree top a ladderway, alternating at stages from this side to that, was fixed roughly, and yet with much economy of space, to the inside uprights, and you could climb up at your leisure to investigate the various devices with which the tree was enriched. They consisted chiefly of boxes of various sizes with differently shaped openings. Each of the boxes, in addition to a spy hole, was fitted with a door which could be opened from the inside, enabling you to see from day to day what stuff had been brought in by what animal. Opposite the most convincing of the arms of the tree a casement opened which gave a chance . . . of photographing any bird which chose this perch. . . .[29]

---

[28]Ernest Thompson Seton, "Making a Hollow Tree and What Came Into It: Bulletin 1," Country Life in America, 15 (Nov., 1908), 47, 84; "Bulletin 2," CLIA, 15 (Dec., 1908), 226; "Bulletin 3," CLIA, 15 (Jan., 1909), 310; "Bulletin 4," CLIA, 15 (Feb., 1909), 414; "Bulletin 5," CLIA, 15 (March, 1909), 542; "Bulletin 6," CLIA, 15 (April, 1909), 658, 660, 662, 664; "Bulletin 7," CLIA, 16 (July, 1909), 354, 356; "Bulletin 8," CLIA, 16 (Aug., 1909), 455; "Bulletin 9," CLIA, 16 (Sept., 1909), 552; "Bulletin 10," CLIA, 16 (Oct., 1909), 662; "The Hollow Tree," CLIA, 20 (Sept., 1911), 88; "Further Annals of a Hollow Tree," CLIA, 22 (Oct., 1912), 64, 66; "The Hollow Tree," Bird-Lore, 11 (Jan., 1909), 1-3.

[29]W. Beach Thomas, "A Wonderful Home: How a Great Naturalist Lives," London Daily Mail, Nov. 29, 1909.

A member of the American Breeders' Association since 1898, Seton dabbled in commercial fur farming, raising Mink, Marten, Skunk, Otter and even Wildcats at Cos Cob, once more using personal experience to support a host of authoritative papers.[30]

Wyndygoul was the first of the three main estates designed and owned by Seton before his death. De Winton, the second, and Seton Castle, the third, reflected, with Wyndygoul, the latent, if thwarted, lord-of-the-manor air with which his soul was burdened.[31] He seemed subliminally doomed to bear forever the banner of Victorianism. Neither his natural history investigations nor his experience as an artist could

---

[30]J. Walter Jones, Fur Farming in Canada (2nd ed.; Ottawa: Canada, Commission of Conservation, 1914), pp. 98, 147. See also, Ernest Thompson Seton, "Raising Fur Bearing Animals for Profit," Country Life in America, 9 (Jan., 1906), 294-97; "Annals of a Fur Farm: Chapter 1," CLIA, 23 (Nov., 1912), 38-40; "Chapter 2," CLIA, 23 (Dec., 1912), 61-62; "Chapter 3," CLIA, 23 (Jan., 1913), 30-32; "Practical Fur Farming: Part 1," Field and Stream, 18 (March, 1914), 1146-41; "Part 2, The Care and Feeding of Skunks," Field and Stream, 18 (April, 1914), 1299-1302; "Part 3, The Diseases and Breeding of Skunks," Field and Stream, 19 (May, 1914), 19-24; "Part 4, Marking and Disarming," Field and Stream, 19 (July, 1914), 296-99; "Part 5., Marketing Skins," Field and Stream, 19 (Aug., 1914), 391-94; "Part 6, Mink Farming, With a Footnote on Skunk Raising," Field and Stream, 19 (Nov., 1914), 746-51; "Part 7, Marten Farming," Field and Stream, 19 (Jan., 1915), 923-25.

[31]Seton always took a particular interest in planning and architectural design. See Ernest Thompson Seton, "Achieving the Picturesque in Building," Country Life in America, 35 (Oct., 1918), 44-47; "The House That Is Mine," House and Garden, 44 (Oct., 1923), 62-63, 112, 114, 116; "If Da Vinci Came to Town," American City, 5 (Nov., 1911), 252-54.

release the grip of élitism. This was unfortunate in the sense that he desperately wanted to cast off the artificial trappings of society. In many respects, of course, he succeeded. The wilderness within compelled its own outward expression. He is invariably remembered for personal untidiness. Seton's corporeal presence was preceded through life by his smell. He seldom shaved or bathed. His wild and shaggy locks, like untended sails, luffed madly in the breeze. He had a "gleeful way of wrecking conventionality. With some unexpected . . . word, look, or prank, he [wilted] the starch out of the stiffest social drapery and [reached] straight for the heart beneath it."[32] He relished the company of strong and obvious personalities. Lincoln Steffens, who was particularly fond of Seton, and often joined him for long nature walks around Wyndygoul, described Cos Cob as an "art colony" where

> we talked art, and . . . had a contempt for people who talked business and politics. The townspeople and fishermen were all right; they fitted into the land and seascapes, but toward the rest of the world, toward the successful New Yorkers who were buying up Greenwich . . . we were such snobs that after cutting us and then discovering that we were cutting them, they used to try to get in with us. And they couldn't.

Surrounded by painters like John Henry Twachtman and Childe Hassam, both close friends of Robert Henri, by writers like Bert Leston Taylor, Wallace Irwin and Irving Bacheller, and by editors and publishers like Gilman Hall of Ainslee's

---

[32]Emmons, "With Ernest Seton-Thompson in the Woods," p. 4.

and Everybody's, and Don Seitz of the New York World, Seton did not have far to look for controversial figures.[33] Frequent guests at Wyndygoul included Juliet Wilbor Tomkins, Richard Le Gallienne, Lloyd Osbourne, Carl Lumholz, Emery Pottle, Mark Twain, Gertrude Lynch, Cecelia Beaux, William Hornaday, Hart Merriam and, after 1903, John Burroughs. Theodore Roosevelt stayed on a few occasions, as did Charles, Will and Mary (Fanton) Roberts. Havelock Ellis, with his independent wife, Edith Lees (whom Seton considered "a prophetess from the wilderness, neither wholly woman nor wholly man,") were also welcome visitors.[34] But Seton's closest friend from the world of letters, and one of the few men who really understood him, was Hamlin Garland. With his wife, Zulime, and his brother-in-law, the sculptor, Lorado Taft, Garland was in regular attendance at Wyndygoul.

Of Scotch and New England ancestry, Garland was born in West Salem, Wisconsin, in 1860. He had homesteaded in the Iowa and Dakota country just as Seton had done in Manitoba. Attached ideologically to the doctrines of Henry George, and culturally to a unique notion of provincialism, he had become one of the great Populist crusaders for artistic and

---

[33]Lincoln Steffens, The Autobiography of Lincoln Steffens (2 vols.; New York: Harcourt, Brace and World, 1931), 1:436-37, 440, 2:583; Homer, Robert Henri, pp. 102, 118, 121.

[34]Ernest Thompson Seton, quoted in Havelock Ellis, My Life: Autobiography of Havelock Ellis (Boston: Houghton, Mifflin, 1939), p. 479.

literary impressionism and political decentralism--especially after the success of the 1893 Exposition at Chicago.[35] A significant proportion of his work was, like Seton's, autobiographical, depending for its substance upon the author's multivolume diary.[36] The two men were first introduced at a New York Players Club luncheon in October, 1896, shortly after Seton's return from Paris.[37] Within months of this meeting, Seton was wrestling with Garland's idea of western localism and attempting to articulate the desirability of forming "a guild or brotherhood of the West." The composition of this body, he wrote to Garland, should be "men (or women) of mind who found their inspiration in the West." Believing that "eastern men of mark are wholly European in thought," Seton championed the cause of authors and artists like Bret Harte, Joaquim Miller, Mark Twain, Edward Kemeys and E. W. Deming, all of whom constituted part of a "virile and autocthonous" school.

---

[35]Hamlin Garland, Crumbling Idols: Twelve Essays on Art Dealing Chiefly with Literature, Painting and the Drama, Jane Johnson, ed. (Cambridge, Mass: Harvard University Press, 1960), pp. 7-17, 97-124; John Pilkington, "Fuller, Garland, Taft and the Art of the West," Papers on Language and Literature, 8 (Fall Supplement, 1972), 39-56.

[36]Donald Pizer, "Introduction and Editorial Note," In Hamlin Garland, Hamlin Garland's Diaries, Donald Pizer, ed. (San Marino, Calif.: Huntington Library, 1968), pp. xi-xii. For a complete listing of secondary sources on Garland, see, Jackson R. Bryer and Eugene Harding, Hamlin Garland and the Critics: An Annotated Bibliography (Troy, N.Y.: Whitston, 1973).

[37]Hamlin Garland, Roadside Meetings (New York: Macmillan, 1930), pp. 326-27.

> It is too soon as yet to characterize this new school. But it has shown at least one element of lasting greatness--it gets close to the naked truth. It has cast aside the tradition of . . . hedges, superb chargers and European dialects. . . . Its reckless adhesion to literal truth has already incurred for the school the reproach that it dilutes arc with science; that its writers are mere naturalists, its figure painters mere ethnologists. These gibes are . . . praises.
>
> I think . . . these men write, paint or speak because they have a story that they burn to tell, not because they are enamoured of their art.

Seton selected the Thunderbird, which he called "the central ganglion of the Western world" because it was acknowledged by all North American Indians, as the symbol for his organization.

> What better badge could there be for man aspiring to restore to the West its intellectual leadership [than] . . . the form of the mystic bird? It would symbolize all that such a guild could ever hope to be and would unite us with the learned past of this wonderful West we inherit.38

In effect, Seton had a story, beyond mere written words, that he himself burned to tell. He aspired to create an artful organization, a community of young men to duplicate the comradeship, in the spirit of the West, that Wyndygoul had afforded him and his friends. By introducing the Indian and by identifying him with a noble past, Seton carried Garland's ideas beyond geographically determined cultism. He also removed himself from the charge of nativism often levelled at other western writers.[39] Instead

---

[38]Seton to Hamlin Garland, May 15, 1897, Seton Letterbook No. 1, p. 191, SP. It is not clear how well Seton actually understood the Thunderbird symbol of creation. He had obviously begun to read the reports of the Smithsonian Institution sent to him in exchange for ornithological specimens and probably developed his interest through them. See, for example, Garrick Mallery, "Picture Writing of the American Indians," <u>Tenth Annual Report of the Bureau of Ethnology to the Secretary of the Smithsonian Institution, 1888-89</u> (Washington: Government Printing Office, 1893), pp. 483-87. See also Ellen Russell Emerson, <u>Indian Myths</u> (Boston: J. R. Osgood, 1884), pp. 32-40.

[39]Brownwen J. Cohen, "Nativism and Western Myth: The

of deriding the urban immigrant as a source of evil while extolling the pioneer as virtue incarnate, Seton distinguished between the East and the West in purely cultural terms. The East was Europe reconstituted, the West was the true America--"true" because it had an aboriginal tradition which still remained open to emulation by the white man.

Although Garland was a sensitive defender of the native cause, he was forced to admit that his aims were "not precisely the same" as Seton's.

> We both saw the red man as Catlin saw him, an animal adapted to a certain environment. Biologically he was guiltless as the panther or the eagle. We had no economic or religious prejudices concerning him. His sign language, his songs, his dances were of absorbing interest to us. . . .[40]

(Needless to say, the word "animal" bore no pejorative connotations for Seton). Garland's extensive travels among the Crow, Northern Cheyenne and Blackfeet led him to applaud the passing of the cowboy, the scout, the trapper and all the "lonesome men."

> As a novelist, I am sorry to see them go; as a well-wisher of the red men, . . . I am glad of the cowboy's diminishing hoof-beats. He carries with him something fine, but his room is better than his presence when all is said and done.

For Garland, however, and even for men like Deming and Frederic Remington, the artists who illustrated his books, magnanimity

---

Influence of Nativist Ideas on the American Self-Image," Journal of American Studies, 8 (April, 1974), 23-39.

[40]Hamlin Garland, Companions on the Trail (New York: Macmillan, 1931), pp. 14-15.

for the Indian was couched in the rhetoric of the white man's burden.

> We are answerable for them, just as we are answerable for the black man's future. As the dominant race, we have dispossessed them; we have pushed them to the last ditch--which will be their grave, unless we lay aside greed and religious prejudice and go to them as men and brothers, and help them to understand themselves and their problems; and only when we give our best to these red brethren of ours, do we justify ourselves as the dominant race of the Western continent.[41]

Seton, who regarded all white men as immigrants in perpetuity, reversed the order, identifying them as the red man's burden. The "true" American was obliged to eschew the pioneer model, to "adopt the best things of the best Indians," and to acknowledge that "those live longest who live nearest to the ground, . . . who live the simple life of primitive times."[42]

While in Paris Seton had discussed the preachments of Edward Bellamy with Robert Henri. His more recent acquaintance with Garland, and their spirited exchanges at Wyndygoul on such topics as the single tax, led him to delve still deeper into the complexities of utopian socialism. Before long he was equating _socialism_ with _sociability_, the latter a term frequently used in the same context by Kropotkin. "Sociability

---

[41]Hamlin Garland, "The Red Man's Present Needs," _North American Review_, 174 (April, 1902), 487-88. See also _Hamlin Garland, The Captain of the Gray Horse Troop_ (New York: Harper, 1902); _Book of the American Indian_ (New York: Harper, 1923).

[42]Ernest Thompson Seton, _The American Boy Scout; the Official Handbook of Woodcraft for the Boy Scouts of America_ (New York: Doubleday, Page), p. 1.

as a fundamental of human nature," wrote Seton, "is fully recognized in all Indian Tribes . . . and is the inevitable solution of many of the troubles that are harrassing the White race in America." The fact that "the Indian was a socialist in the best and literal meaning of the word" had clearly inspired Henry George to advocate the national ownership of land, condemning private property as the "foundation of all misery and poverty among the Whitemen." In Indian societies, stressed Seton,

> No man owns the wood of the forest, or the water of the rivers, or the soil of the earth. He did not make them, they are the harvest of the land that belongs to the whole people; and only so much of them is his as he can gather with his own hands and use in his own home. . . .
>
> The Tribe may give to one family the exclusive right to hunt or gather wood or forage or wild rice or fruit in one tract or range, but the family cannot sell this right, nor can they hire others to hunt or gather for them, lest the hunting be destroyed by overdrain.[43]

Seton was also aided in reaching these conclusions through his extensive reading of the work of Lewis Henry Morgan, who, in addition to his pathbreaking investigations in animal behavior, had made major contributions to the study of Iroquois ethnology.[44] (These, in turn, had strongly influenced

---

[43]Ernest Thompson Seton and Julia M. Seton, eds., The Gospel of the Redman: A Way of Life (Santa Fe, N.M.: Seton Village Press, 1966), pp. 26, 28-30; Henry George, Progress and Poverty (New York: D. Appleton, 1880).

[44]Lewis Henry Morgan, League of the Ho-de-no-sau-nee, or Iroquois (2 vols.; Rochester, N.Y.: Sage, 1851); Ancient Society, or Researches in the Lines of Human Progress from Savagery through Barbarism to Civilization (New York: Henry Holt, 1877).

Friedrich Engels, who expanded on their implications as early as 1884.)[45] In 1843 Morgan initiated at Aurora, New York, a secret society called The Grand Order of the Iroquois. Modelled after the Five Nations Confederacy, with Cayuga, Mohawk, Oneida, Onondaga and Seneca Tribes extending throughout the state, the Grand Order was made up almost exclusively of whites who viewed the Indian past as the backbone of the American heritage. With Morgan as its chief, Skenandoah, the society observed Iroquois political and religious rituals, dressed in Iroquois garb, and undertook a sophisticated research program dedicated to understanding and bettering the condition of Indians.[46] From this experiment, which Seton was soon to copy in modified form, Morgan had cultivated the seeds for a body of thought which made him the father of American anthropology.[47] Although it is probably too much to assume that Seton was swayed by Engels' extrapolations from Morgan, it is interesting to find him continually reiterating his belief that "By his life and tribal constitution, [the Indian] has shown us that the nationalization of all natural resources . . . puts a stop at once . . . to abject poverty and to

---

[45]Friedrich Engels, Origin of the Family, Private Property and the State, in Light of the Researches of Lewis Henry Morgan, trans. Ernest Untermann (Chicago: C. H. Kerr, 1902).

[46]Carl Resek, Lewis Henry Morgan, American Scholar (Chicago: University of Chicago Press, 1960), pp. 21-24.

[47]Meyer Fortes, Kinship and the Social Order: The Legacy of Lewis Henry Morgan (Chicago: Aldine, 1969), pp. 3-17.

monstrous wealth."[48]

Seton's eclectic interests were unified only by his overriding concern for the land itself. But because the Indian was the physical embodiment of a spiritual union between man and land, he very early assumed a special place in Seton's value system. As a child he had been introduced to the noble savage through the work of James Fenimore Cooper. Helen Hunt Jackson's famous Century of Dishonor (1881) had united this idealized Indian and the muckraking tradition around the theme of persecution, automatically predisposing Seton, while still weighing the relative merits of white and red societies, to favor the latter.[49] Nevertheless, it must not be assumed that he intuited all his philosophical premises from a sense of moral outrage. In fact, by 1902 Seton had read a great deal about, had travelled extensively among and knew personally many North American Indians. He had encountered them first in Lindsay and later in Manitoba. Long before going to Paris he had learned the intricate distinctions separating several linguistic groups.[50]

In 1893 the Toronto artist, George Reid, brought a young Iroquois poet from Brantford, Ontario, to view Seton's painting, Awaited in Vain. Pauline Johnson, or Tekahionwake,

---

[48]Ernest Thompson Seton, The Book of Woodcraft and Indian Lore (New York: Doubleday, 1925), p. 573.

[49]Ibid., pp. 9, 51-53.

[50]See, for example, Seton, Autobiography, pp. 262, 376.

"quickly sensing the painter's sympathy with the Wolf, claimed him as a Medicine Brother, for she herself was of the Wolf Clan of the Mohawks."[51] After 1885, when she began to contribute short pieces to periodicals and literary magazines, she managed to create a respectable collection of short stories and verse.[52] Formally introduced to Toronto society in 1892, she quickly became "one of the most popular touring entertainers" in Canada,[53] achieving sufficient stature by 1894 to be invited to read on the London stage. Here, before packed houses, she appeared for the first half of each performance dressed in native costume, "a beautiful buckskin dress trimmed with ermine skins and with silver brooches scattered over the bodice." The theatrical effect was heightened by the addition of a bear-claw necklace, bracelets of wampum beads and a red broadcloth blanket draped over her shoulders.[54] (As if to declare the cultural duality marking her own birth,

---

[51] Ernest Thompson Seton, "Introduction," in E. Pauline Johnson, The Shagganappi (Toronto: William Briggs,1913), p. 5; Autobiography, p. 290.

[52] See, for example, E. Pauline Johnson, Canadian Born (Toronto: Morang, 1903); Flint and Feather (Toronto: Musson, 1912); Ojistoh (New York: Werner, 1911); The White Wampum (London: Lane, 1895).

[53] Marcus Van Steen, Pauline Johnson: Her Life and Work (Toronto: Musson, 1965), p. 17.

[54] Walter McCraye, Pauline Johnson and Her Friends (Toronto: Ryerson, 1947), pp. 61-62. Both Seton and another young Englishman named Archibald Belaney, attempted to emulate this manner of presentation. See Lovat Dickson, Wilderness Man: The Strange Story of Grey Owl (Toronto: Macmillan, 1973).

she donned a conventional evening gown for the second part of the program.) Although her mother was a native of Bristol, Pauline proudly claimed the blood of her father, a Mohawk chief. "There are those who think they pay me a compliment in saying that I am just like a white woman," she told Seton, but "I am an Indian. . . . Forget that I was Pauline Johnson, but remember always that I was Tekahionwake,. . . the sad historian of her own heroic race."[55] Until her premature death in 1913, Tekahionwake counted Seton among her most trusted friends, frequently soliciting his advice and encouragement, always addressing him affectionately as "my dear Wolf-Man."[56] For his part, Seton had every reason to remember "the sad historian." It was she who first introduced him to the image of Tecumseh.

As an ardent Canadian nationalist, Pauline Johnson came to the attention of Charles Mair, a founder of the Canada First movement and himself a distinguished man of letters. In 1886 Mair had published Tecumseh, a rhetorical drama styled after Shakespeare, portraying the character of the great Shawnee chief whose services to the British crown during the War of 1812 had "endeared him to the Canadian

---

[55]E. Pauline Johnson, quoted in Seton, "Introduction," in Johnson, The Shagganappi, pp. 5-7.

[56]E. Pauline Johnson to Seton, Dec. 4, 1903, Grace Gallatin Seton Papers, Old Greenwich, Conn.; E. Pauline Johnson to Seton, Aug. 2, 1905, Aug. 17, 1905, SP; McCraye, Pauline Johnson and Her Friends, pp. 52-53.

heart." History had never recorded, maintained Mair, "a nobler example of true manhood and patriotism."[57] After reading the play in 1892, Pauline prepared a glowing review for the Toronto World, applauding Mair's sensitivity not only to Tecumseh but to Indians in general. "His drama," she wrote, "bears upon every page evidence of long study and life with the people whom he has written of so carefully, so truthfully."[58] While this rather extravagant praise had some (though not much) substance in reality, it was most important for the fact that it mirrored once more the cultural schizophrenia characteristic of most Victorian Canadian literati. Mair loved Tekahionwake because she had given his protagonist the Iroquois seal of approval. For both of them, however, Tecumseh was great, not only because he was a dignified and intelligent Indian, but also because he "appreciated British justice" and had "helped materially to preserve our independence." There was certainly no doubt in the mind of Charles Mair that Indians had to yield to white civilization,

> but, before quitting the stage they left behind them an abiding memory, and an undying tradition. And, thus, "Romanticism," which will hold its own despite its hostile critics, is their debtor. Their closeness to nature, their picturesque life in the past, their mythical religion social system and fateful history have begot one of the

---

[57] Charles Mair, Tecumseh, a Drama, and Canadian Poems (Toronto: Radisson Society, 1926), pp. 201-02.

[58] E. Pauline Johnson, Toronto World, March 22, 1892, quoted in Norman Shrive, Charles Mair: Literary Nationalist (Toronto: University of Toronto Press, 1965), p. 191.

> wide world's "legends," an ideal not wholly imaginary, which as a counterpoise to our Realism, our literature needs, and probably never shall outgrow.[59]

By the turn of the century Seton had lost all faith in pioneers, the realists, "the men who penetrated and eventually possessed the West." Indeed, he argued, "The best that could be said of the best of these . . . was that he was almost as good as an Indian."[60] When, therefore, he was consciously looking for a palatable aborigine to hold out as a carrot to American youth, Mair's Tecumseh, translated and legitimized for him by Tekahionwake, provided the obvious answer. Seton's own Victorianism enjoyed another momentary resurgence as he chose for his ideal symbol this Indian, who, while native to America, had apparently dreamed the Canadian Dream, yet managed to remain defiantly noble.

There can be no doubt that Seton's Tecumseh was Mair's Tecumseh--"physically perfect, wise, brave, picturesque, unselfish, dignified."[61] All the words used to describe him are found in the text of the drama. Nor did Seton, even when describing his hero to young boys in the United States, attempt to disguise the bias of his source. "The War of 1812,"

---

[59]Charles Mair, "An Appreciation," in E. Pauline Johnson, The Moccasin Maker (Toronto: William Briggs, 1913), pp.11-12.

[60]Ernest Thompson Seton, Blazes on the Trail No. 1: Lifecraft or Woodcraft, the Fourfold Way (Greenwich, Conn.: Indian Village, Little Peequo, 1928), pp. 6-7.

[61]Seton, Autobiography, p. 376.

he told them pointedly, "gave Tecumseh a chance to fight the hated Americans." In battle, "his personal prowess, his farseeing statesmanship, his noble eloquence, and lofty character" gave him "a place on the very highest plane among patriots and martyrs."[62] Seton did not, however, require the justification of Canadian nationalism. His own values were of a much more spiritual nature, linked with the biological order to which he had directed so much of his intellectual energy. Even in that context, Tecumseh "was the most Christlike character presented on the pages of American history."[63]

To answer those critics who would argue that his theories about the Indian were merely romantic extrapolations from the words of legendary chiefs and professional myth-makers, Seton buried himself in the work of contemporary historians and anthropologists. Among his major sources were scholars like F. W. Hodge, Frances Densmore and James O. Dorsey, all of whom he had known at the Smithsonian.[64] Seton was an

---

[62]Seton, Book of Woodcraft, p. 525.

[63]Seton, History of the Boy Scouts, p. 19. Seton subscribed to Tecumseh's theory of aboriginal rights, seeing it as a logical extension of territoriality in animals; Seton, Gospel of the Red Man, pp. 78-80; Seton to F. W. Hodge, Jan. 16, 1929, SP.

[64]F. W. Hodge et al., Handbook of American Indians North of Mexico (2 vols.; Washington: Government Printing Office, 1907-1910). See Charles Hofman, ed., Frances Densmore and American Indian Music; a Memorial Volume (New York: Museum of the American Indian: Heye Foundation, 1968), for a complete bibliography of Densmore's work;

original subscriber to Edward S. Curtis' North American Indian Project, referring often to the monumental study produced by it.[65] He also knew and was in regular correspondence with Natalie Curtis whose distinguished book on Indian music and folklore is still considered a classic in the field.[66] She in turn introduced him to the path-breaking work of ethnomusicologist Alice Fletcher whose translation of the "Omaha Tribal Prayer" became a standard feature of his camp fire ritual.[67] With Natalie Curtis and Frances Densmore, both he and Garland took a special interest in the music research of Frederick R. Burton and contributed financially to his important study of Ojibway songs.[68]

Frances Densmore to Seton, May 30, 1911, Sept. 20, 1918, SP; Ernest Thompson Seton, The Woodcraft Manual for Boys: the Fifteenth Birch Bark Roll (Garden City, N.Y.: Doubleday, Page, 1917), p. 73. James O. Dorsey, "A Study of Siouan Cults," Eleventh Annual Report of the Bureau of Ethnology to the Secretary of the Smithsonian Institution, 1889-90 (Washington: Government Printing Office, 1894), pp. 351-553.

[65] Edward S. Curtis, The North American Indian, ed. F. W. Hodge (20 vols.; Cambridge: Cambridge University Press, 1907-1930); Edward S. Curtis to Seton, Jan. 25, 1910, SP.

[66] Natalie Curtis, ed., The Indian's Book: An Offering by the American Indians of Indian Lore, Musical and Narrative, to Form a Record of the Songs and Legends of Their Race (New York: Harper, 1907); Natalie Curtis to Seton, Feb. 4, 1904, Jan. 15, 1910, June 17, 1911, March 29, 1916; Seton to Natalie Curtis, Jan. 8, 1910, SP; Seton, Woodcraft Manual for Boys, p. 76.

[67] Alice Fletcher, Indian Story and Song from North America (Boston: Small, Maynard, 1900); Seton, Woodcraft Manual for Boys, pp. 9, 26, 71.

[68] F. R. Burton, American Primitive Music; with Especial Attention to the Songs of the Ojibways (New York: Moffat, Yard, 1909); Garland, Companions on the Trail, p. 196;

Among Seton's warmest Indian friends was Dr. Charles A. Eastman (Ohiyesa), a Santee Sioux from Minnesota. Born in 1858, Eastman graduated from the Boston University Medical School in 1890. After serving as Government Physician at the Pine Ridge Agency, where he was responsible for injured prisoners taken at the Battle of Wounded Knee, 1890, he was appointed Field Secretary of the International Committee of the YMCA. He occupied this post from 1894 to 1897, during which time he organized forty-two Indian YMCA's in Canada and the United States. From 1897 to 1900 he represented the Eastern Sioux as Attorney before Federal Courts of Claim and Congressional Committees. He served as Government Physician at Crow Creek, South Dakota, from 1900 to 1903, when Theodore Roosevelt appointed him to revise the Sioux allotments and to establish family names so that descent of property through tribe would be protected. Through his association with the YMCA, and through his friendship with Seton, Eastman was also to play a significant role in establishing the Boy Scouts of America. An outspoken defender of his people, he found in Seton a sensitive and loyal sounding board for his own convictions. Indeed, Seton's vigorous denunciations of General George Custer and his

---

Hamlin Garland to Seton, n.d.; F. R. Burton to Seton, March 19, 1905, June 29, 1906, Jan. 29, 1908, July 3, 1908, Sept. 2, 1908, SP; Seton, Woodcraft Manual for Boys, pp. 73-74.

constant condemnation of American racism can be traced directly to his discussions with Eastman. Eastman was also responsible for introducing him to James R. Walker, a Pine Ridge Agency doctor whose studies of Siouan religion were in Seton's hands long before their formal publication.[69]

As a result of the infamous events which closed the nineteenth century, keeping the bloodied Sioux on the front page of daily newspapers in both Canada and the United States,[70] Seton harbored a special fascination for the plains Indians. During his biological expeditions to the American West and to Manitoba he made a special effort to meet the survivors of various wars. One of his favorites was White Swan, a Crow brave who had been severely wounded by the Sioux while on a scouting mission for Custer. Because he was deaf and dumb, White Swan was forced to communicate in the sign language common to all plains tribes. Seton became his willing pupil in 1897, and shortly afterwards began to prepare his own magnificent study, Sign Talk,[71]

---

[69] Charles A. Eastman, Indian Boyhood (New York: McClure, Phillips, 1902); From the Deep Woods to Civilization: Chapters in the Autobiography of an Indian (Boston: Little, Brown, 1916); Seton to William Howard Taft, Oct. 15, 1912, SP. James R. Walker, The Sun Dance and Other Ceremonies of the Oglala Division of the Teton Dakota (New York: Trustees of the American Museum of Natural History, 1917); James R. Walker to Seton, Sept. 13, 1910, Sept. 21, 1910, July 15, 1914, SP.

[70] For the Indian perspective, see, Dee Brown, Bury My Heart at Wounded Knee (New York: Holt, Rinehart, Winston, 1971).

[71] Ernest Thompson Seton, Sign Talk (Garden City, N.Y.:

published in final form in 1918, after years of research among the Yanktonnais Sioux and Piegans. With the aid of the sign language he was also able to converse at length with different linguistic groups, quickly learning from them a history of the wars that did not coincide with official Washington versions. Sifting through the accounts of military men who had conducted what he perceived to be a genocidal mission, he extracted evidence that they themselves had acknowledged their own guilt. He then reprinted it, with analytical commentary endorsing the Indian interpretation, for the edification of young boys.[72] Seton was especially disturbed by Washington's expedient justifications for

---

Doubleday, Page, 1918). The book is especially important in the light of recent research in semiotics; see, for example, William C. Stokoe, Semiotics and Human Sign Languages (The Hague: Mouton, 1972). In his work, Seton drew on other authorities; e.g., Garrick Mallery, "Sign Language Among North American Indians, Compared with that Among Other Peoples and Deaf Mutes," First Annual Report of the Bureau of Ethnology to the Secretary of the Smithsonian Institution, 1879-80 (Washington: Government Printing Office, 1881), pp. 269-552; William Philo Clark, The Indian Sign Language (Philadelphia, Pa.: L. R. Hamersly, 1885); but he was personally responsible for unearthing and preserving the now famous "Hadley Indian Sign Prints," the pictographic representation of hand signs devised by Lewis F. Hadley (Ingonompashi), a missionary at Anadarko, Oklahoma. The Seton Papers are replete with letters on this important topic; see especially, F. W. Hodge to Seton, Oct. 20, 1910, March 4, 1916, Aug. 23, 1916, Dec. 11, 1916; Seton to F. W. Hodge, March 23, 1916, May 25, 1916, Sept. 6, 1916; F. W. Hodge to Cato Sells, May 17, 1916; F. W. Hodge to J. W. Dortch, May 23, 1916, SP.

[72] O. O. Howard, Famous Indian Chiefs I Have Known (New York: Century, 1908); James McLaughlin, My Friend the Indian (Boston: Houghton Mifflin, 1910); Francis E. Leupp, In Red Man's Land (New York: Fleming H. Revell, 1914); The Indian

breaking treaties, arguing that "the record of our official dealings with the Indians is far worse than anything we can find in Russia or Spain's colonies during modern times. It is . . . 'an unbroken narrative of injustice fraud and robbery.'"[73] He maintained that the Battle of Wounded Knee, which he had carefully researched and then discussed with Charles Eastman, had no parallel in the history of human cruelty. "As sure as there is a God in Heaven," he warned, "this thing has to be met again, and for every drop of righteous blood spilled that day . . . a fearful vengeance is being stored and will certainly break on us."[74]

Seton's pantheism was also fortified by his growing interest in the Indian. Like the animals, "this great pagan" observed most of the Ten Commandments--"all but the first and third: that is, he had many lesser gods besides the one Great Spirit, and he knew not the Sabbath Day of rest." As far as Seton was concerned, only those of the Commandments that could be traced to biological roots had

---

and His Problem (New York: Scribner's, 1910); Seton, Book of Woodcraft, pp. 45-47, 55-59.

[73]Seton to the Editor of the New York Herald, Nov. 27, 1912, SP.

[74]Ibid. Seton's major source for the details of the massacre at Wounded Knee was James Mooney, "The Ghost Dance Religion and the Sioux Outbreak of 1890," Fourteenth Annual Report of the Bureau of Ethnology to the Secretary of the Smithsonian Institution, 1892-93 (Washington: Government Printing Office, 1894), pp. 641-1136. Although a "scientific" ethnologist, Mooney was, like F. W. Hodge and many others at the Smithsonian, remarkably subjective about the righteousness of the Indian cause.

any real substance. As Eastman had noted, the Indian saw "no need for setting apart one day in seven as a holy day, since to him all days were God's." Seton reminded his readers that the early missionaries and explorers who had tasted the fruit of native cultures believed, "from their worship of God, their strict moral code, their rigid laws as to foods clean and unclean, and their elaborate system of bathing and purification, that in these red men of the New World they had indeed found the long lost tribes of Israel." The condition of the body was directly related to the condition of the soul. And the strong soul was built upon stoic self-denial, moderation, and humility before all living things. Seton's isolation as a child, his experience of the empty Manitoban prairie, and his studies of the solitary animals whose lives, he had so often pointed out, ended in the irony of life-giving tragedy, had given him a unique notion of nature's mystery. Humility worshipped mystery through ritual and through silence, and, in Eastman's words, "never claimed that the power of articulate speech was proof of superiority over dumb creation." As "the great prophet of outdoor life," and "the master of woodcraft," the Indian "sought for the beautiful in everything." Having found beauty, he also felt a responsibility to acknowledge it. His songs, stories, dances, ceremonies and craftsmanship "were mere expressions of his broad creed that the Great Spirit is in everything, everywhere, all the time." The religion of these "Spartans of the West" had grown out of the land itself. The pioneer had continued to survive in North America only because he had learned and adapted the ways of the Indian. But now the First American was being pushed aside by a pious Christian, before whose alien, self-made laws he refused to compromise his soul.[75]

Wyndygoul was the milieu in which the Woodcraft Indians was organized; the Thunderbird symbolized the Indian origin of the spiritual values that Seton sought to demonstrate. The specific content of the teaching of the movement took shape, appropriately, in the context of Seton's work as a writer and artist.

With his reputation as an artist and writer of animal stories well established, with his concern for the Indian growing, and with a geographical setting like Wyndygoul in which to pursue his biological investigations, Seton was suddenly rewarded with a vehicle for the integration of his thought. Edward Bok, editor of the Ladies' Home Journal since 1889, and Vice-President of the Curtis Publishing Company since 1893, had watched his progress for some time. An astute businessman with a keen sense of family tastes, Bok had already published the work of such luminaries as Garland, Kipling and Twain, by 1900 raising the Journal's monthly circulation to 800,000,

---

[75]Seton, Book of Woodcraft, pp. 12-13, 24, 50-51, 58, 572-73; Charles A. Eastman, The Soul of the Indian: An Interpretation (Boston: Houghton, Mifflin, 1911), pp. 46, 89-90, quoted in ibid., pp. 24, 515.

larger than that of any other American magazine. It was only natural that Bok would eventually get around to Seton, given the latter's obvious popularity with children.[76]

As early as 1898 Seton had toyed with the idea of publishing a "woodcraft dictionary," stressing Indian folk tradition, for Frank Doubleday. But when Doubleday failed to follow up on the proposal, Bok (who had gotten wind of it through their mutual friend, Rudyard Kipling,) invited Seton to contribute to the Ladies' Home Journal a regular "boys' department," consisting of "a page--text and pictures--[which would] appeal in all its contents to boys in their relation to outdoor subjects." In 1901 Seton accepted Bok's offer and entered into a formal contract with the Journal. By 1902, under the heading, "Ernest Thompson Seton's Boys," the eccentric lord of Wyndygoul had commenced a series of articles destined to launch America's largest youth movement.[77]

---

[76]Edward Bok, The Americanization of Edward Bok: The Autobiography of a Dutch Boy Fifty Years After (New York: Scribner's, 1920), pp. 160-80, 190-250; Frank Luther Mott, A History of American Magazines, 1885-1905 (Cambridge, Mass.: Harvard University Press, 1957), p. 545. Circulation reached one million in 1903.

[77]Seton, History of the Boy Scouts, p. 19; Seton to F. N. Doubleday, April 1, 1899, Seton Letterbook No. 1, p. 414; Seton to F. N. Doubleday, Dec. 23, 1899, Seton Letterbook No. 2, p. 113; Edward Bok to Seton, July 18, 1901, May 25, 1903; Seton to Edward Bok, June 3, 1903, SP. Ernest Thompson Seton, "Ernest Thompson Seton's Boys: Part 1, Trailing," Ladies' Home Journal, 19 (May, 1902), 15, 41; "Part 2, The Second Chapter on Tracks," LHJ, 19 (June, 1902), 15; "Part 3, Playing 'Injun,'" LHJ, 19 (July, 1902), 17; "Part 4, Archery," LHJ, 19 (Aug., 1902), 16; "Part 5, Teepees," LHJ, 19 (Sept., 1902), 15; "Part 6, Woodcraft Indians and Getting Lost," LHJ, 19 (Oct., 1902), 14; "Part 7, Freezing," LHJ, 19 (Nov., 1902), 15.

III

Committed to the notion that the young would show the way to salvation, Seton had been following recent advances in child psychology. Given his own interest in ethology, it was not surprising that he should have centered his attention on the investigations of G. Stanley Hall, genetic psychologist, father of the American child study movement, former Haeckel sympathiser, and champion of the theory of recapitulation. Hall's insights suggested that children would wholly mature only if they were encouraged to relive the history of the human race. In a Darwinian age this meant conscious passage, within a lifetime, from the distant ancestor, animal, through the existence of primitive man, to the present advanced stage of development. "To the young child," he maintained, "there is no gap between his soul and that of the animals." Because the characteristics distinguishing man from animal were also the last to develop, it followed that "Children . . . in their incomplete stage of development are nearer the animals in some respects than they are to adults." As a result, there was a basic sympathy for animals among young people generally. Hall also related man's latent fear of animals to the "evolutionary chasm" that separated them. This factor partially accounted for the use of animal symbols in Indian totems and illustrated the "way in which the stages of a child's development repeat those through which the race has passed."[78]

---

[78]Hall, Adolescence, 2:220-21.

Hall was convinced that the theory of recapitulation could be applied to pedagogy. Indeed, he went much further and suggested that if education were "to become adequate to the needs of the soul," animals would have to be investigated in terms other than their material usefulness to man. City life, he felt, fostered only a crude understanding of the mechanics of science, "but so removes the child from the heavens and animate nature that it is pathetic to see how unknown and merely bookish knowledge of them becomes. . . ." Biology alone could not meet the new pedagogical task which wanted less discussion of structure and more about habitat, food supply, migration, folklore, myth and literature. Education must be "full of the spirit of the field naturalist, observer and lover;" and it must, above all, be rescued from professors and science teachers who had "lost contact with the nature and needs of childhood and youth." In its broadest sense, recapitulation was an extension of Whitman's and Wheeler's work. As Hall himself suggested, "the new tentatives in ethology" demonstrated that "we are influenced in our deeper, more temperamental dispositions, by the life-habits and codes of conduct of we know not what unnumbered hosts of ancestors, which like a cloud of witnesses are present throughout our lives, and that our souls are echo-chambers in which their whispers reverberate."[79]

---

[79] Ibid., 2:60-61, 228-29.

Hall's work was both prophecy and calling, requiring all of Seton's special gifts. Nevertheless, the latter's unique understanding of ethology, when combined with the theory of recapitulation, yielded results which probably could not have won Hall's endorsement. Seton's investigations of native spiritualism had afforded a glimpse of man equating himself with nature. As a result, he was disposed to view the Indian as the superior of the white, dismissing conventional, historically-based notions of man's "progress" from "barbarism" to "civilization" precisely because they were premised on the necessity of destroying the equation.

A serial story entitled Two Little Savages appeared in Seton's column in the Ladies' Home Journal throughout 1903. Published in book form in time for Christmas of the same year,[80] the story is important for several reasons. In the first place it is set in Lindsay, Ontario, the source of Seton's childhood memory. Although a fictionalized rendering, it is an autobiographical account. It speaks to a conception of wilderness; it speaks to Indians; above

---

[80] Ernest Thompson Seton, "Two Little Savages: Part 1," Ladies Home Journal, 20 (Jan., 1903), 11-12; "Part 2," LHJ, 20 (Feb., 1903), 11-12; "Part 3," LHJ, 20 (March, 1903), 13-14; "Part 4," LHJ, 20 (April, 1903), 11-12; "Part 5," LHJ, 20 (May, 1903), 11-12; "Part 6," LHJ, 20 (June, 1903), 11-12; "Part 7," LHJ, 20 (July, 1903), 11-12; "Part 8," LHJ, 20 (Aug., 1903), 15-16, 32; Two Little Savages; Being the Adventures of Two Boys Who Lived As Indians Do and What They Learned (New York: Doubleday, Page, 1903).

all it speaks of and to children. The predictable Victorian piety is present, but it is brought to the service of an ecological end--that of reuniting man, or rather child, with his environment. Indeed, the narrative is so riddled with pantheistic heresy that we are forced to wonder how America's collective motherhood could have abided its implications. The protagonist, Yan, and his friend, Sam, take great pains to understand the rituals of various Indian tribes. The text is broken by diagrams and detailed explanations, showing the young reader how to build a teepee, how to arrange a council fire, how to make a bow and arrows. The emphasis in each case is on both the craftsmanship of the object and the diversity of its origins. We are not acquainted with <u>one</u> teepee or <u>one</u> bow, but with several, each representing the traditions of a different tribe, each reflecting different responses to different wilderness landscapes. In this way, although the story is located in Ojibway country, we learn about the Piegans, the Sioux, the Cheyenne, the Crow and the Blackfeet--and, necessarily, of their environments. While the tale itself remains important, it is adroitly juggled in a way that allows the author to teach and to philosophize. Caleb Clark, an old man who befriends Yan and Sam, instructs the boys in the making of an Indian drum. Once the reader has been taken through a step by step demonstration of its construction, Caleb begins to beat the drum and to chant. Both boys are so mesmerized by the sound that "their savage instincts seemed to revive." Soon they are both dancing round the singing drummer.

> But above all it worked on Yan. As he pranced around in step his whole nature seemed to respond; he felt himself a part of that dance. It was in himself; it thrilled him through and through and sent his blood exulting. He would gladly have given up all the White-man's "glorious gains" to live with the feeling called up by that Indian drum.

Seton continues to weave the Indian and nature study themes together, creating a sense of their mutual interdependence. By the end of the book Indian, animal, child and wilderness have been combined in purest innocence.[81]

By 1903, at latest, the groundwork had been laid for the reception of Hall's ideas. In fact, however, Seton had set organizational wheels in motion long before that, using the Wyndygoul estate as a kind of laboratory for assessing the appeal and workability of his ideas. The preparation of the manuscript for Two Little Savages was accompanied by his introduction of a regular series of weekend campouts for neighboring Greenwich and Cos Cob boys. These gatherings were so successful that word about them spread widely. As the number of boys attending continued to swell, the "Seton Indians" grew from one tribe into several, each with different leadership, all adhering to a flexible set of guidelines laid down in the first

---

[81] Seton, Two Little Savages, pp. 38-39, 62-70, 75-88, 163.

instance by the Chief himself.

As Seton tells it, the boys from his immediate area were incensed by his having fenced the farmland where, for generations, they had freely roamed. In an effort to regain lost prerogatives they entered upon a program of harrassment, shooting the animals and breaking down the fences. Seton responded by marching into the one-room Cos Cob schoolhouse early in the spring of 1902 to invite the vandals "to come up to the Indian Village on my place" to camp for the weekend.[82]

Seton had indeed built a modular village at Wyndygoul, including teepees of Sioux, Blackfoot and Cheyenne construction, "gorgeous with pictures of warriors and buffaloes."[83] Pulled up on the beach beside the lake were two Malecite birch-bark canoes. This juxtaposition of Plains and Algonkian wares prompted Charles G. D. Roberts to observe that "here at Wyndygoul, the East and West meet amicably."[84]

---

[82] Seton, Autobiography, pp. 377-78; T. Wyckoff, "Interview with Seton Indians in Greenwich, Conn., Sept. 22, 1966," MS Copy in SP. In Seton's published accounts there is some discrepancy about the exact date. It varies from Spring, 1901 to Fall, 1902; see, e.g., Ernest Thompson Seton, "Organized Boygood: The Boy Scout Movement, Its Purposes and Laws," Success Magazine, 13 (Dec., 1910), 804.

[83] Ernest Thompson Seton, "The Boy Scouts in America," Outlook, 95 (July 23, 1910), 631-32; "Ernest Thompson Seton's Boys, Part 5," p. 15; Lida Rose McCabe, "At Wyndygoul with Ernest Thompson Seton," Book Buyer, 25 (1903), 26.

[84] Charles G. D. Roberts, "The Home of a Naturalist," Country Life in America, 5 (Dec., 1903), 155.

A large council ring dominated the scene, surrounded by mammoth rock paintings executed by the master. In the middle of the ring was the council fire. Inspired by a Navajo sand painting, the fire itself was encircled by a necklace of stones to symbolize the Great Spirit. Equidistant around its circumference four arms extended to represent spirit, mind, body and service. At the end of each arm was attached a lamp, with three rays, each symbolizing the laws of woodcraft. The "fire centered circle," while obviously reflecting the cycle of nature, also "led men to think about the Great Mystery over all."[85] It was difficult to imagine a more picturesque setting. One observer marvelled that "so wild a scene could be found within forty odd minutes of New York. . . . Its wood, almost as wild as a forest in appearance; its lake with rocky shores and a series of cliffs breaking directly down to the water's edge, have been left untouched or else improved with an art which leaves them all perfectly natural."[86]

On the appointed Friday, several exuberant adolescents mustered at the gate, challenging Seton to make good on his invitation. Overwhelmed by the size of the group, and a bit intimidated by some of its older leaders, he gathered his composure, welcomed them, fed them all a good, hot meal

---

[85]Seton, "Spirit of the Woods," p. 219.

[86]"Are You a Seton Indian?" New York Herald, Oct. 11, 1903.

and then asked them to sit around the council fire while he spoke.

> I told them of Indians and Plains life, gauging my stories in a steady crescendo till I had renewed the Fenimore Cooper glamour of romance, and heightened it to a blaze of glory. . . . Meanwhile I was watching my material. I wanted them all interested and subdued, but not sleepy. Near eight o'clock the bickerings and rude practical jokes . . . had ceased. The group had tightened up about me. . . . I could feel the thrill of intense interest; I could feel . . . their regret that the noble red men were gone before their day.

With just the right proportion of theatre to reason, Seton seems to have created the desired mood instantaneously. And after three days with Black Wolf, as he asked to be known, these same boys had formed the first tribe of Seton Indians.[87]

Seton believed that the "boy is a cave man till he is sixteen or seventeen. . . . Loyalty to his gang is his real religion, far overpowering any loyalty he may feel to Church or State." Just as every gang had a leader, so must every tribe. The simplest thing to do was to manoeuvre the leader of the gang into the position of Head Chief, but under the guise of a direct democratic process. (The tribe itself would function internally as both Church and State.) Second and Third Chiefs were also elected, each responsible to fill the position above him during the absence of his superior. Next in line came the Wampum

---

[87]Seton, "Boy Scouts in America," p. 633; Autobiography, 380-81; Blazes on the Trail No. 2: The Rise of the Woodcraft Indians (Greenwich, Conn.: Indian Village, Little Peequo, 1928), pp. 7-8.

Chief who had charge of the money and public property of the tribe, except the records. The Chief of the Painted Robe kept the records, including "the Law Book, the Roster or Roll, the Winter Count, or Record of Camps and Seasons and the Feather-tally or record of honors and exploits. He enters nothing [in the records] except on instructions from the Council." Finally there was the Chief of the Council Fire who had the "exclusive privilege" of making fire. "He must do it without matches. He must also see that the camp and woods are kept clean." The Head Chief, the Second and Third Chiefs, the Wampum Chief, the Chief of the Painted Robe and the Chief of the Council Fire, together with the Medicine Man (Seton himself), and five others to be elected, constituted the Council of Twelve. "All are under the Chief. All disputes, etc. are settled by the Chief and Council. The Council makes the laws and fixes the dues. The Chief enforces the laws." The laws in turn were aimed squarely at the protection of and respect for the wilderness and its inhabitants, man or beast. Smoking and drinking were forbidden, but beyond that token gesture, the laws avoided questions of conventional social morality. It was clearly understood that rebellion against the elected Council was punishable by expulsion. Quite literally, the Laws of the Seton Indians were the laws of the land.[88]

---

[88] Ernest Thompson Seton, "The Laws of the Seton Indians," _Association Boys_, 4 (June, 1905), 100-102; _Autobiography_, pp. 378, 382.

By 1903 there were between 50 and 60 tribes in the United States modelled after Seton's example. Frequent guests at Wyndygoul were journalists who sent back inspiring accounts of their aspirations and activities.[89] The rapid growth stimulated by these reports eventually forced Black Wolf to introduce the Confederation, or Nation, concept, in an effort to provide a forum for other leaders who chose to adopt his methods. "The League of Seton Indians" would meet annually in a Grand Council and would be presided over by a Grand Head Chief, himself elected every three years by the Grand Council. Delegates to the Grand Council (one from each tribe) were to be democratically elected within the individual tribal Councils of Twelve. The Grand Council would also contain positions duplicating, on a larger scale, those of the tribal Councils--the Keeper of the Wampum, the Keeper of the Birch Bark Roll, the Keeper of the Tally, and so on. As the organization became more sophisticated, it adopted new titles and rituals reflecting the customs of linguistic groups native to the geographical regions represented in the Grand Council. All tribal members took Indian names of their own choosing.[90] Thus the

---

[89] "Indians at Wyndygoul," Greenwich News, Oct. 23, 1903; W. W. Storms, "The Woodcraft Indians," Holiday Magazine, 3 (Oct., 1904), 76-79.

[90] Ernest Thompson Seton, The League of the Seton Indians, MS notes, n.d., SP; Seton to Copper Eagle, July 28, 1908, SP.

structural framework for the separate tribes and for the League which united them promoted the ideal of pan-Indianism, while at the same time encouraging decentralization and a large measure of local autonomy. Adults, although present to offer direction, were not obviously in command. When setting up the original tribe, Seton himself had taken care to remain in the background of the Council. It was equally clear that within the Grand Council the young people could be in absolute control. All of these features combined to stimulate a democratic atmosphere where rights and responsibilities, both perceived as social instincts, could develop simultaneously.

Seton combined Hall's and William James' theories about the power of instincts (fear, love, curiosity, pride, pugnacity, ownership, ambition, constructiveness) with his own intuition that children were born innocent and then corrupted by external forces. The organic function of his Indian scheme defied what he called "the atrocity of the creed that the Church and Sunday school were . . . preaching." Committed to "the struggle for outdoor life," it fostered a pantheistic "love of all the wild things" and evoked "sorrow in their destruction." Childhood and adolescent instincts, extensions of their animal and Indian counterparts, were, because natural, inherently good, "power to be used--never to be thwarted or destroyed." They are, he maintained, "like the life forces of a tree, the power that is continually compelling and inspiring growth. . . . The exact shape and direction of that growth may be modified by externals. It is the province of the teacher to manipulate these externals and guide the final direction." Seton therefore viewed himself as an educator whose primary tools were a set of recognizable human instincts and a given environment. Existing social attitudes did not necessarily speak to the instinctual realities. And that these latter were frequently thwarted by the former was a crime against nature. The "direction" that Seton aspired to give these instincts required an environment in which they could first be recognized. Instinct found its home in the wilderness where it was born.[91]

Seton's first article of faith, that the "highest aim of education is not scholarship, but manhood," introduces his personalized conception of the theory of recapitulation. That is to say, "manhood" is the end of recapitulation. It does not mean bravery and courage, in the traditional sense, but maturity. Maturity, in turn, implies harmony with one's environment, social and physical. It is understood, however, that the social environment must be a direct product of the physical. Otherwise, harmony will forever remain an elusive objective. For the individual, "manhood" also means "character." Character was formed

---

[91] Seton, History of the Boy Scouts, pp. 6-7.

before the age of ten, in the animal stage of childhood. "After that, it will develop, but not change much." The fundamental social institutions governing character development are the church, the school, the home and the playground. On the playground, "The opinion of the gang is everything; gang instinct is the real religion of the normal boy." From Seton's perspective, "it was through woodcraft pursuits and gang government that our race attained manhood; and our children should recapitulate those experiences. It is the educator's business to pilot them through." Woodcraft, therefore, had two dimensions: instinct (the beginning) and manhood (the end). But manhood did not deny the continued existence of instinct. Believing that play was the unconscious vehicle for the outward expression of instincts, Seton argued that <u>directed</u> play could serve an educative role superior to that of the other three social institutions, all of which he felt were instinct-repressive. Evidence for the repression of instinct, and for the corresponding power of church, state and school, was to be found in the subdued condition of the land itself. The woodcraft way demanded a wilderness, that is to say, an ecologically balanced setting.[92]

In John Dewey's terms, Seton conceived education "retrospectively" rather than "prospectively," regarding

---

[92]Ibid., pp. 10-11.

it as "a process of accommodating the future to the past."[93] (Although it might be truer to say that he aspired to make the future accountable for the past). But Seton also conceived the past more in biological than in historical terms. At best, he regarded biology and history as mutually antagonistic explanations of the past. Biology implied a tangible continuum, a slow, organic change. Seton might easily be identified with the American Herbartians, especially given his interest in myth, folklore and music.[94]

One of Dewey's major objections to Herbartianism was that it set the influence of heredity above, or in opposition to, that of environment. Heredity, for Herbartians, predetermined the traits of the individual and "so fixed [them] that little serious change can be introduced into them."[95] Seton was not prepared to deny the necessity of change. But the quality and content of change were overriding concerns. Anthropocentric change had already demonstrated its power, but not necessarily its value, by the way in which it had affected environment. If that environment in turn was to influence man, it was also going to influence the animals, the trees and the rivers which had constituted

---

[93] John Dewey, Democracy and Education (New York: Macmillan, 1964), p. 79.

[94] Harold B. Dunkel, "Herbartianism Comes to America," History of Education Quarterly, 9 (Summer, 1969), 202-33.

[95] Dewey, Democracy and Education, p. 74.

an earlier environment which had also influenced man. History measured change in man and in man's society by virtue of the changes imposed by man upon his environment. Darwinian biology, on the other hand, measured natural change in all species, including man, and therefore constituted a history of the environment. As such, it measured the effects of anthropocentric change upon the environment, in the interest of the environment. An education which applauded anthropocentric change--social reform or social progress--at the expense of environment was unacceptable to Seton. Social progress and social reform, though eminently worthy objectives, could not be exalted to the exclusion of the environment which would be called upon, not only to absorb, but to sustain them. The material foundation of progressivism, he felt, imposed an impossible burden upon the land and its indigenous inhabitants. Fortunately, he argued, our instincts remain to connect us to both. Anything which purports to pass for reform, but denies those instincts, is in fact repressive. Nothing in this value system was contrived to negate man's humanity. Indeed, Seton's devotion to the Indian proved his overwhelming, and empirically based, social passion.

While acknowledging the Indian's intimacy with the land, Dewey had no desire to idealize the savage intelligence and volition. "The savage paid for highly specialized skill in all matters of personal adjustment, by incapacity in all that is impersonal, . . . remote, generalized, objectified, abstracted."[96]

---

[96] John Dewey, "Interpretation of Savage Mind," Psychological Review, 9 (May, 1902), 226.

Seton, who viewed the Indian as an exemplary citizen, did idealize him--and for the reason that he found the impersonal, the remote, the generalized and the abstracted reprehensible.

In his second criticism of Herbartianism, Dewey attacked "the theory that the proper subject matter of instruction is found in the culture products of past ages." He himself viewed education as a process of present (as opposed to a preparation for future) living. And the present did not result from past products but from "the life of which they were the products." For Dewey, history had meaning only insofar as it recorded man's social life and progress--from then, to now. The danger of "making the . . . remains of the past the main material of education is that it cuts the vital connection of present and past and tends to make the past a rival of the present and the present a more or less futile imitation of the past." Dewey's quest for unity in history even predisposed him to question the role of biology in the curriculum. He assailed nature study for the reason that "nature itself is not a unity; nature in itself is a number of diverse objects in space and time, and to attempt to make it the center of work by itself, is to introduce a principle of radiation rather than one of concentration."[97]

As far as Seton was concerned, Dewey's was an attempt

---

[97]John Dewey, "My Pedagogic Creed," in Dewey on Education, ed. Martin S. Dworkin (New York: Teachers' College Press, 1959), pp. 22-26; Democracy and Education, pp. 75-76.

to create unity. It undermined mystery, giving children a simplistic and patently false impression about unity. For him, nature was the sum of its diversities; it did have a definable order and unity which not only warranted the child's primary attention, but which, especially in play, spontaneously won it. The child deprived of wilderness, wanted wilderness. The proof of that could be found in the popularity of his own stories. In calling up the animal, the Indian and the wilderness, Seton was doing more than calling up the products of the past. He was calling up the past itself, in its entirety, arguing that nature demanded it. There was no question of selecting random bits of the past merely because they spoke to social progress. In its very diversity and mystery, nature gave to life dimensions which an anthropocentric history could not hope to capture. The past could not be a rival of the present because the past lived in the present. The animal, the Indian and the wilderness, he argued, are within us all and must be obeyed. In proportion as we subdue each, or all, of these, believing them to be external to ourselves, we contribute not to our social progress but to our extinction.

Seton was neither a Conservative Darwinist nor a Reform Darwinist.[98] Although a subscriber to Hall's theory

---

[98]For a concise analysis of the distinction, see Mathew Zachariah, "The Impact of Darwin's Theory of Evolution on Theories of Society," Social Studies for Teachers and Administration, 62 (Feb., 1971), 69-77.

of recapitulation, he was not a Herbartian. Insofar as he was conscious of being moulded by the ideas of others, it might be argued that he was a mixture of all. He abhorred competition, holding tenaciously to an ideal of mutual aid. He found concepts of progress, measured solely in technological and/or social terms, naïve. While he would have shared William Graham Sumner's concern about the man-land ratio, he would have done so for the sake of species other than man. In the same sense that Wayland Drew and George Grant can argue to the same ecological conclusions from polarized ideological traditions, Seton was a kind of Red Tory by inclination whose biologically induced ethics brought him to a Kropotkin-like social perspective. He loved the individual; but he loved the community more. And the community was not simply man, but man and all other species. An accurate record of the past activities of the community required biological as well as historical measurement. Seton persistently attacked the Reform Darwinist's dictum that nature was not always worthy of emulation. He denied the Pragmatists' view of man as the measure of all things but he endorsed their opinion that the organism and its environment are mutually interdependent. He shared the Herbartian love of ritual, folklore, music and myth. But again, he could not agree with their archaic perspective on the past. In Hayden White's very specific meaning of the word, Seton was a primitivist.

As an educator, Seton assumed that if man were reacquainted with the subtleties of woodcraft in childhood, he would carry an ecological perspective with him for the rest of his life. It would guide his attitude to change and would, in turn, be reflected in the quality and content of that change. That he did not see the curriculum, but rather play, as the vehicle for this process of reacquaintance, is another major characteristic that distinguishes him from a man like Dewey. At the core of woodcraft lay Seton's conviction that the primary prerogative of play was the enrichment of the imagination. He shared Dewey's belief that all of life's experiences constituted the education of the individual, but feared that in the public mind both "education" and "school" had been equated with "work." The true function of education should be

> the enlarging of the faculties which make for appreciation of play, and all play should be associated with a mental and spiritual development which must bring about a better capacity for work. . . . [E]verything that develops the youth is interrelated in life, and a work that is not joy and a play that is not productive are equally deleterious.[99]

He was much less optimistic than Dewey that school teachers would be able, "through the continual and sympathetic observation of childhood's interests . . . [to] enter into

---

[99]Ernest Thompson Seton, "The Twelve Secrets of the Woods," <u>Craftsman</u>, 30 (June, 1916), 330; "Nature and Human Nature," <u>Nature</u>, 3 (May, 1924), 318.

the child's life."[100] Even if they could, the prerogative of progressive education, the desire to "cast the school as a fundamental level of social and political regeneration"[101] would overcome the positive values of play. Although a propagandi: Seton was only indirectly a political being. Politics was relevant to the extent that it aided in effecting an ecological end. His social concerns were intimately tied to the same purpose. Woodcraft would, he maintained,

> bring together young people from various so-called stations, break down the barriers that society has foolishly placed between them, and establish in their minds while they are young a finer kind of humanity, a real understanding that the important thing is the association of the human spirit. . . . [A]nd although America is a democratic nation and the high school system is supposed to break down social barriers, . . . we do not accomplish much in the way of democratic education through the schools. . . . [W]ide human comradeship . . . can only exist . . . when based upon fundamental intimacy with Nature's ways--the only unchangeable, inevitable ways. And if the young people of this nation can be so trained that they will grow to look upon Nature with eager interest, if they become familiar with her traditions, her kindness, her discipline, her beauty, her tragedies, they will find themselves held together by a bond of sympathy that no superficial social structure can ever obliterate.[102]

These were hardly the warblings of utopian dreamer.

Education implied directed growth from one stage of being to another. While it might easily be separated from

---

[100] Dewey, "My Pedagogic Creed," p. 29.

[101] Lawrence A. Cremin, The Transformation of the School (New York: Knopf, 1961), p. 88.

[102] Seton, "Twelve Secrets," pp. 232-33.

the school, play definitely could not be separated from the growth of the individual. According to Seton, the play educator differed from his counterparts in other institutions by his willingness (and ability) to participate in the world of the child. He subscribed to the dictum that "when the right kind of teacher comes along, a little child shall lead him."[103] Johan Hiuzinga defined play as "a voluntary activity or occupation executed within certain fixed limits of time and place, according to rules freely accepted but absolutely binding, having its aim in itself and accompanied by a feeling of tension, joy and the consciousness that it is 'different' from 'ordinary life.'"[104] Woodcraft was all this and more. It was a play-oriented "plan to utilize and recapitulate the stages of our growth." By understanding the instincts of children through participation in their play, the educator could then employ their play, and consequently their instincts, to direct their growth through the various stages leading to manhood. Exactly what it was that should constitute the process of growth to manhood remained somewhat ambiguous. But the idea of achieving balance and harmony with nature, or a kind of ecological karma, was clearly dominant.[105] Dewey's concept

---

[103]Seton, History of the Boy Scouts, p. 10.

[104]Johan Huizinga, Homo Ludens: A Study of the Play Element in Culture (Boston: Beacon Press, 1955), p. 28.

[105]Seton, History of the Boy Scouts, p. 11; "Twelve Secrets," p. 234.

of culture as "the capacity for constantly expanding the range and accuracy of one's perception of meanings,"[106] found material realization for Seton in the wilderness. Manhood did not carry connotations of virility, strength and power. Indeed, growth to manhood meant the child's passage in spirit, mind and body, toward the pinnacle of human consciousness, the accumulated balances and harmonies with nature reflected in the best of all mankind. The surest measure of maturity was the shortness of the distance between man and pinnacle.

All those leaders choosing to unite with Seton were expected to share his belief in Rousseau's state of nature. "They [must] deal chiefly with the natural instincts of the growing child, offering a vast range of developmental activities, spiritual as well as physical, the choice of which is left to the child's free will."[107] Because woodcraft implied an organic interaction between man and the land, the Indian was a logical model for the child. The first contingent of boys to arrive at the gates of Wyndygoul had proved the worth of this insight. And the rapid expansion of the League of Seton Indians suggested that boys generally would find it appealing. A heroic figure, at once attuned to his environment and a believer in the necessity of ritual,

---

[106] Dewey, Democracy and Education, p. 123.

[107] Seton, History of the Boy Scouts, p. 13.

the Indian was "picturesque," possessed of noble instincts, an ideal. "By all the evidence at hand, his was a better system, a better thought. . . . He, more than any type I know, is the stuff that fires our highest dreams of manhood."[108]

Seton's long association with ethnologists had led him to conclude that the Indian, like the Wolf, was the victim of persecution based on ignorance. Martyrdom always had a distinct appeal. But when the martyrdom of an entire race could be juxtaposed with a set of values inimical to the survival of other species, the latent ecologist was quick to grasp the correlation. The theory of recapitulation which argued that the child must relive the history of the human race led here, however, to the dreadful inconsistency which was to plague Seton's later youth work. For as time passed, he no longer viewed the life of the Indian as a stage in the development of civilization. He had for some time been disposed to question many of the values of existing civilization. And considering his own past interests and experience, the rationalization of the "game" of woodcraft into the reality of life, where the "savage" state became the truly civilized state, where that which was to act merely as a preparatory stage itself became the end of moral striving, could not have been far

---

[108] Ibid., p. 20; Ernest Thompson Seton, "Woodcraft Extolled as the Science that Makes Men: A Gospel for Out-of-Doors," Current Opinion, 73 (Oct., 1922), 461-64.

off. The Indian was not only attuned to nature. The Indian was nature.

By choosing the Indian as a model Seton inherited a ready-made, if highly diversified set of rituals. It did not much matter that he mixed Siouan, Algonkian and Iroquoian traditions together. The important thing was to select those aspects of native life which would appeal to children and adolescents alike.[109] Despite his dalliance with socialism, Seton was fundamentally apolitical. But he never veered from an abiding faith in democracy which, he maintained, was of nature's own instigation. There was nothing inconsistent, therefore, in his introduction of democratic elections to positions on the Council, or in the democratic operation of both the Council and the League. That this internal democracy helped to make the scheme more comprehensible to boys born into the American tradition was an added advantage, not an explicit objective.[110] The democratic process would also discourage competition. None of the braves was able to inflict his views upon the others. "Rough and wild boys might defy the teacher, scoff at the opinions of their elders, but they cannot scoff at the public opinion of their playmates, nor defy [them]."[111]

---

[109] Seton to W. T. Talbot, April 2, 1905; Seton to E. M. Robinson, Feb. 21, 1928, SP.

[110] Ernest Thompson Seton, "Spiritual Thrift, Number 2: Helping to Develop Democracy with Nature's Aid Through the Woodcraft League," Touchstone, 1 (June, 1917), 119; "Twelve Secrets," p. 330.

[111] Seton, History of the Boy Scouts, p. 38.

If Seton's model was a socialist, democratic Indian who rejected competition and gloried in mythic ritual, his tribal activities followed suit. There were three classifications of things to do under the rubric of woodcraft. First came the category of athletic proficiency, stressing the achievement of individual standards and pointedly discouraging all unnecessary forms of competition. Seton viewed athletics as a stimulus rather than as an end. They functioned primarily as an incentive to entice boys into the outdoors and to involve them in their surroundings.[112] The second department was called woodlore, and its objective was to teach self-sufficiency in the wilderness. It was within this category that the skills of the Indian (not the pioneer) were given paramount attention. Craftsmanship and a familiarity with survival techniques were joined to the learning of Indian legends, dances, songs and religions. Broadly speaking, this section encouraged an ecological relationship between the child and his environment, demonstrating that man's survival in the future need not be predicated upon the extinction of other species. The use of firearms, for example, was altogether forbidden. "The gun stands for little skill, irresistible force supplied from an outside source, overwhelming unfair odds, and sure death to the victim." The bow and arrow, on the other hand, could

[112] Ibid., p. 37.

be manufactured from accessible natural materials. "The bow is silent and it sends the arrow with exactly the same power that the bowman's arm puts into it." It keeps the odds, as between animal and man, fair, and demands as much skill in the shooting as in the making. It "stands for all that is clever and fine in woodcraft."[113] The third area of concentration was nature study, aimed at a detailed understanding of environmental complexity. Writing in 1904, Seton explained that

> No one can comprehend the nature and purpose of a wheel in some complex machine by study of that wheel alone. He must consider the whole machine, and especially the parts nearest to that particular wheel. Man, therefore, seeking to know himself, must fail utterly, unless he remember that he is only a part of the great machine of the universe. He must therefore study the other wheels, that is, the life forms about him, which are parts of his environment and offspring of the same creative power as himself.

Above all, nature study would develop "powers of correct observation and clear reasoning from cause to effect."[114] All three departments bore the indelible stamp of Seton's personal experience in the wilderness. Their mutually supporting objects were to create a feeling of community, not only among the young people, but between them and their natural environment. As competition invariably excluded the physically weak, overexalting the achievements of the

---

[113] Seton, "Ernest Thompson Seton's Boys, Part 4," p. 16.

[114] Ernest Thompson Seton, "On Nature Study," in _The Library of Natural History_, ed. Richard Lydekker (New York: Saalfield, 1904), p. iv.

physically strong, he worked hard to replace glory in mere feats of strength with group participation in Indian games of skill. Throughout, the goal was to encourage individuals to achieve standards of excellence in each of the three categories. As he was as much concerned to foster good habits of mind, spirit and social behavior, there was nothing to be gained by dependence upon a strictly physical regimen.

The measure of excellence in the Seton tribes was the Siouan feather, or coup. Having reached a designated standard within a particular field--athletics, woodlore or nature study--the young brave would be awarded one or more of these by his peers in Council.[115] By establishing standards in the different areas of concentration, Seton attempted to encourage a frame of mind which recognized value in all forms of achievement, but particularly in those which reflected an intimacy with land. Furthermore, when one boy looked upon another, he was not inclined to draw comparisons. In the first place, there were no uniforms. Each individual designed his own regalia and had to be prepared to explain its significance. The coups themselves could be counted, but it was difficult to know, without close inspection, in which areas they had been won. In the end, their significance

---

[115] Seton, "Ernest Thompson Seton's Boys, Part 4," p. 17; Two Little Savages, pp. 146-49; American Boy Scout, pp. 10-11.

was subtly downplayed by their becoming simply another feature of a colorful wardrobe, distinctive in itself. The relatively wide availability of the coups encouraged all boys to better themselves, not at the expense of others, but in pursuit of standards of excellence measured individually.

The athletic coups were classified as "Red Honors" and were awarded to individuals who could meet standards of speed or endurance in swimming, running and canoeing. Although traditional sports were included, emphasis was placed upon those activities which had most relevance for the woodcraft ideal.[116] Seton disliked spectator sports, arguing that "it is ruinous to provoke a powerful emotional impulse and then let it dissipate as mere emotion without commendable action." Professionalism in sport contributed to the decadence of civilization by relegating the physically uncoordinated to the sidelines while the experts demonstrated their wizardry on the field. The spectators knew what the skills were, but they abandoned any hope of every mastering them. As far as Seton was concerned, mastery was not the issue. But participation was a law of biology. Sport could be traced to the hunt. "The hunting of an animal," he suggested, "could be set forth in ten chapters, of which the first nine, the pursuit, are wholly delightful." Anything contributing to the success of the chase was

---

[116] Seton, "Laws of the Seton Indians," pp. 103-05; American Boy Scout, pp. 12-25.

legitimate sport--but the tenth chapter, "the destruction of a harmless and beautiful animal," was to be left unwritten, except in emergencies. This gave to woodcraft athletics a certain raison d'être or context, not to mention a practical value.[117]

It is interesting to note in passing that the role of formal athletics declined in importance after Seton left the Boy Scouts in 1915, while the woodlore and nature study fields gained more than proportionately. This was a reflection of Seton's increased awareness, intensified by his own bitter experience in the Boy Scout movement and by the shadow of the war, that competition, associated in the public mind with harmless sport, was anathema. Hence, also, we find him gradually reducing the symbolic importance of the chase itself in favor of skills like animal photography which implicitly required considerable physical dexterity and endurance.[118]

The woodlore section was set under the heading, "White Honors," and covered everything from lighting a fire without matches to building a birch canoe. The number of coups awarded in this category increased substantially from 1905

---

[117] Ernest Thompson Seton, Sport, MS Notes, SP.

[118] Compare, e.g., Seton, American Boy Scout (1910) and Woodcraft Manual for Boys (1917). The first is 86 pages long, of which 14 are given to athletics. The second is 441 pages long, of which 11 are given to athletics. See also, Ernest Thompson Seton, "Hunting with the Camera," Recreation, 8 (April, 1898), 263-64.

to 1917.[119] But the actual demonstrations of skill, rewarded by coups, were further supplemented in the manual by a steadily increasing volume of explanatory material, the purpose of which was to link techniques to Indian customs and to substantiate their practical origins and values. There were subtler intentions as well, especially the desire to "justify the Indian as a model" and the hope "that this may lead to a measure of long-delayed justice being accorded him."[120] By 1912 readers of the eleventh edition of the "Birch Bark Roll" were being introduced to newly discovered aspects of Indian history in both the pre- and post-contact periods, complete with military records, ethnological data and biographical information gleaned from scholars like Mallery, Dorsey, Mooney, Morgan and others.[121] The historical features

---

[119]Seton, "Laws of the Seton Indians," pp. 105-07; American Boy Scout, pp. 26-32; Woodcraft Manual for Boys, pp. 356-62.

[120]Seton, Book of Woodcraft, pp. vi-vii.

[121]The "Birch Bark Roll" was the name Seton gave to various of his publications dealing with woodcraft. He identifies the 1st edition of the Roll as the first series of articles published in the Ladies' Home Journal, under the heading "Ernest Thompson Seton's Boys" (1902); the 2nd edition is called How to Play Indian (Philadelphia: Curtis, 1903); the 3rd edition is called The Red Book (New York: By the Author, 1904). Not until the 6th edition does it appear under its best known name, The Birch Bark Roll of the Woodcraft Indians (New York: Doubleday, Page, 1906), and by the 8th edition it has become The American Boy Scout (1910), by the 11th edition, The Book of Woodcraft and Indian Lore (1912), by the 15th edition, The Woodcraft Manual for Boys (1917). The last recorded Birch Bark Roll, the 29th edition, appears as The Birch Bark Roll of Woodcraft (New York: A. S. Barnes, 1931).

are supplemented by the inclusion of Indian songs, dances and ceremonies (including, interestingly, the Ghost Dance Song)[122] drawn for the most part from the research of Natalie Curtis, Frances Densmore and Alice Fletcher. Each ritual is identified with its tribe of origin and its significance described in minute detail. While most of the dances chosen related to animals--The Shoshoni Dog Dance, the Ojibway Snake Dance, and so on,--the songs have a more spiritual tenor. At the centre of Seton's ritual was the Omaha Tribal Prayer, "Wa-kon-da dhe-dhu Wa-pa dhin a-ton-he," and vigil.[123] In Omaha culture, the prayer was taught to the child entering adolescence, the time at which the mind "becomes white."

> This expression referred to the dawn, to the passing of night into day, and represented the coming of the child out of the period where nothing was clearly apprehended into a time when he could readily recall past events with their distinctness of detail. This seeming mastery of the minutiae of passing occurrences indicated that a stage of growth had been reached where the youth could be inducted into the religious mysteries through a distinct personal experience.

This distinct personal experience was acquired in the rite, "Non-zhin-zhon", observed during four nights and days of vigil and fasting in some isolated location. During this

---

[122] See, Dorsey, "The Ghost Dance Religion and the Sioux Outbreak of 1890;" Weston LaBarre, The Ghost Dance: Origins of Religion (New York: Delta, 1972).

[123] Seton, Book of Woodcraft, pp. 61-80. Freely translated, the prayer means, "Father, a needy one stands before thee; I who sing am he."

time the youth would repeat the prayer, weeping as he sang and wiping the tears with the palms of his hands, lifting them first toward heaven, then laying them on the ground.

> When at last he fell into a sleep or trance, and the vision came, of bird or beast or cloud, bringing with it a cadence, this song became ever after the medium of communication between the man and the mysterious power typified in his vision; and by it he summoned help and strength in the hour of his need.[124]

The prayer and the vigil clearly combined to create a Spartan imagery. But the entire ritual, and its explanation, reinforced Seton's idea of history.

Virtually all of the games centre upon Indian or nature lore.[125] They really serve to link all three aspects of woodcraft and they explain, at least partly, the gradual de-emphasis of formal athletics in the program. Most of the games demand group participation and each has a teaching function. For example, one of Seton's favorites was the "Deer Hunt," first introduced in 1902, but greatly embellished in Two Little Savages. It involved constructing a realistic dummy deer out of burlap, straw and wire mesh. Once the replica was erected, a young brave would take it into the woods and hide it in the undergrowth. Attached to the bottom of his boots, however, were protruding iron stencils in the shape of deer tracks. These left their mark on the

---

[124]Fletcher, Indian Story and Song, pp. 26-27.

[125]Seton, Book of Woodcraft, pp. 196-220.

soft ground but could be obscured by walking over rocks or through streams. Once the dummy had been concealed, the hunters, armed only with bows and arrows, went out in search of it. The first to find it had to call his fellows to the location where it stood. From a specified distance he had the first shot. If he missed, the next in line would advance five yards and shoot again--and so on, until someone had made a direct hit on the heart of the animal. The function of the game was intricate. In the first place it required familiarity with the anatomy of the deer itself. Its construction was a group affair, involving the skills of artist, biologist and engineer. The youth who hid the deer had to act like one. This meant not only moving quickly-as he had a limited time in which to perform his task--but also taking care to disguise the trail. The hunters on the other hand, had to learn to read the tracks and to follow them. They also had to demonstrate their proficiency with bows and arrows of their own creation.[126]

The woodlore category required a knowledge of Indian herbal remedies and tonics. These Seton described in detail, complete with illustrations of the wild plants from which they are made. He also discussed the nutritious emer-

---

[126]Seton, "Ernest Thompson Seton's Boys, Part 4," p. 16; <u>Two Little Savages</u>, pp. 140-46; "The Revival of the Bow and Arrow," <u>Country Life in America</u>, 7 (Jan., 1905), 273-75; "A New Deer Hunt with the Bow," <u>CLIA</u>, 7 (Feb., 1905), 370-71; <u>Book of Woodcraft</u>, pp. 199-202.

gency foods of the northern woods, teaching his readers how to make rabbit snares from natural materials and how to prepare palatable meals from ants, mice, bark, buds and lichens.[127] In describing the organization of a wilderness campsite, Seton urged the observation of Indian ways. The Seton Indians favored teepees over tents, moccasins over shoes, headbands over hats. Tribal rituals were to be observed in camp and elaborate instructions were itemized to ensure that each travelling band functioned democratically in the tradition of the League. To overland or canoe tripping parties was added the innovation of "Dog Soldiers," carefully selected by the Chief and pledged to respect an oath of obedience, whose role it was to maintain discipline.[128] The "Birch Bark Roll" detailed plans for building all the essential material accoutrements for a pan-Indian mobile encampment--everything from a Navajo mat loom to Siouan teepees to Arapaho beds.[129] Had he been writing for purists, Seton might legitimately have been criticized for indulging

---

[127] Seton, Book of Woodcraft, 221-35, 240-49; "Emergency Foods in the Northern Forest," Country Life in America, 6 (Sept., 1904), 438-40.

[128] Seton, Book of Woodcraft, pp. 172-95.

[129] Ibid., pp. 108-43, 468-508. Here again Seton took advantage of his friends at the Smithsonian. See, e.g., Mooney, "The Ghost Dance Religion and the Sioux Outbreak of 1890," p. 963; Washington Matthews, "Navajo Weavers," Third Annual Report of the Bureau of Ethnology to the Secretary of the Smithsonian Institution, 1881-82 (Washington: Government Printing Office, 1884), 375-91.

this inclination toward pan-Indianism and for obscuring tribal diversities. But it must be remembered that he was indirectly the product of an anthropological school which found in rituals like the Ghost Dance evidence for an underlying racial unity.[130] Besides, he was writing for children and adolescents. He did not deny them the diversities implicit in the whole; indeed, he took considerable pains to locate each dance, song, custom and artifact in both geographic and cultural contexts. He resisted efforts to lump him with popular nature faddists, and he hoped, by crediting the universally respected scholarship on which he based his thought, that he would not only protect himself against the charge of simplistic romanticism, but also give substance to his belief "that to forget the primitive nature of man is to court race ruin."[131]

The third area of concentration in woodcraft was nature study, classified under the heading, "Blue Honors."[132] There was a certain security in this field; like motherhood and conservation it implied respectability. On the surface, Seton linked it with sport and wildlife protection. "Nature study is so nearly in the same line as hunting that it is

---

[130]LaBarre, Ghost Dance, pp. 232, 249.

[131]Seton to W. T. Talbot, April 12, 1905, SP.

[132]Seton, "Laws of the Seton Indians," pp. 76-77; American Boy Scout, pp. 33-36.

easily substituted. It fosters all the fine art of Woodcraft without the brutalities." Underneath, however, he had more ideas in mind. Believing that the destruction of wilderness was a byproduct of the repressiveness of civilized society, he argued that children must obtain their ideas about nature from nature--not from books. Nature study meant intimate contact with and sympathy for the natural objects in the organic community.[133]

Formal nature study in the schools had had a relatively brief life by 1903. Until the middle of the nineteenth century enough children grew up in rural environments with ready access to the wild things that there was no apparent need to teach "about" nature. Besides, the school system was not interested in catering to the child's individual desires. Education was primarily intended to give him possession of the tools for acquiring knowledge. Spelling, reading, writing and arithmetic, with supplementary instruction in religion and morals were deemed adequate to the task. But as the balance began to tip in favor of urban living, and as more children found themselves in city schools where access to wild nature was denied them, the craving for it remained. What previously had been associated with play, now found a reason for inclusion in the curriculum.

---

[133]Ernest Thompson Seton, Introduction to the Beast and Bird Section of the Volume on Nature Study, MS notes, n.d., SP; "On Nature Study," p. v.

Initially it was the supplementary moral and religious instruction which afforded the best vehicle for nature study. "Nature stood as a source from which to draw great moral and religious truths, and recognizing . . . childhood as the time for impressing these truths upon the race, it was but a logical result of their reasoning that the study of nature should find favor in the elementary schools." By mid-century the best nature study instruction was to be had in the female seminaries where, it was understood, the teachers of the future were being prepared. But here the study was wholly confined to botany, "by far the most attractive, elegant, and precise study; so well adapted to the refinement of female education." While this pre-Darwinian period did recognize the instinctive interest of children in their environment, and while the school hoped to turn that interest to useful ends, the inevitable textbook eventually surfaced to keep nature in the classroom and the child locked out of his wilderness.[134]

Between 1860 and 1890, the textbook approach dominated what nature study children could muster the discipline to memorize. Botany still served as its focus until Darwin's controversial revelations began to filter through religious considerations in the form of zoology. But even though

---

[134] Ira Benton Meyers, "The Evolution of Aim and Method in the Teaching of Nature-Study in the Common Schools of the United States," Elementary School Teacher, 11 (Dec., 1910-Jan., 1911), 209-12.

zoology had made such a dramatic entrance, the textbook ensured that basic anatomy would provide its script, robbing it of any intrinsic interest.

During the late 1870's and early 1880's, Pestalozzi's conviction that elementary education should have more to do with things than with books captured the imagination of influential American pedagogues like Alpheus Hyatt, Lucretia Crocker, W. T. Harris, H. H. Straight and Wilbur S. Jackman, at least two of whom, Hyatt and Straight, had been students of the natural historian, Louis Agassiz. Through their combined efforts, nature study gradually made its way out of the classroom and back to the field. Liberty Hyde Bailey identified Jackman's scheme, fully elaborated in 1889, as the most influential. Its salient features included

> (1) . . . using all the material which the "Rolling Year", season by season, brought into the lives of the children; (2) . . . it rejected the idea of close and specialized study of inert or dead form and sought to place the children in the fields and woods that they might study all nature at work; and (3) instead of looking upon nature-study as being supplementary to reading, writing and other forms of expression, nature-study in itself became a demand that these subjects should be taught.[135]

Although the nature-study movement was born in the elementary schools, it did have its supporters in the university. Chief among them was Clifton F. Hodge, a biologist and

---

[135] Liberty Hyde Bailey, The Nature-Study Idea (New York: Doubleday, Page, 1903), pp. 10-11; Meyers, "Evolution of Aim and Method," pp. 241-244.

colleague of G. Stanley Hall at Clark, whose pathbreaking book, Nature Study and Life, urged careful consideration of feelings and sentiments in all elementary education. Where nature study in particular was concerned, Hodge challenged teachers to give free rein to their pupils' instinctive interest in nature. By January, 1905, when M. A. Bigelow founded his Nature-Study Review, the subject had become an important aspect of the curriculum in many of the urban schools.[136]

John Dewey, always a skeptic where nature study was concerned, felt that the subject should be united with geography. Rejecting Hall's ideas outright, he argued that nature must be "treated as a whole" so that "its phenomena fall into their natural relations of sympathy and association with human life."[137] Effectively this meant learning from the general to the particular, and, in a sense, was harmful to Dewey's own empirical theories. Certainly, if adopted, his anthropocentric bias guaranteed the return of the textbook and the end of nature study, so-called. Seton wanted nature study to be a part of play. Even if the schools did fall back into the old textbook pattern, he was prepared to see his work supplementing the curriculum. As a naturalist

---

[136] Ross, G. Stanley Hall, pp. 230, 234, 299; C. F. Hodge, Nature Study and Life (Boston: Ginn and Co., 1902). See also Hall, Adolescence, 2:144-231.

[137] Dewey, Democracy and Education, p. 212.

and an explorer, he felt that the inquiring mind of the child would be satisfied with nothing less than total participation. He would never have denied that one must view nature as a whole. But it was imperative that individuals assemble that whole from its diverse parts, avoiding simplistic generalizations. In the service of this perspective, he called up the memory of "Gilbert White, Catesby, Bartram, Huber [who] . . . went on piling up their little ascertained facts or fragments of facts, . . . clearing the way for Darwin and Pasteur and making them possible." The Seton Indians would be encouraged to pursue such exemplary conduct in their own investigations.[138]

Despite his own special interest in animals, Seton was careful to maintain balance in the nature study section of the "Birch Bark Roll," giving equal time to botany and zoology. He was consistent in suggesting that the study of nature demanded its own step by step analysis:

> First, take one thing at a time. Second, . . . [h]ave a simple but comprehensive guide book . . . that tells in simple, clear language the main facts. Later, you will want to go into more scientific study. Third, make a record in a notebook of what you see and either make drawings or preserve specimens. Fourth, if you have a friend "who knows" get information from him as to the specimen you have seen or have in your possession.

(In other words, he advocated the methods first taught him by Dr. Brodie in Toronto during the 1870's.) He introduced

---

[138]Seton, "On Nature Study," pp. v-vi.

the subject with a look at plants and flowers because they were "so easy to catch and preserve." John Macoun's techniques were employed in the description of flower, leaf and root curing. There followed a systematic list, arranged by dominant color, of many of the flowering plants indigenous to America, complete with scientific nomenclature, elaborate diagrams and descriptions of natural characteristics.[139] The same kind of treatment is afforded the trees. In 1910 Seton published The Forester's Manual as a separate volume, but by 1917 it had become a regular feature of the "Birch Bark Roll." Trees deserved special notice, he felt, because of the service they had rendered man throughout the ages. "A tree has always been a blessed and a glorious thing to me. Often I feel the axe chopping into my own soul when I see it laid to some splendid tree. . . ." As well as itemizing the coniferous and deciduous forms, again with scientific names, maps of geographical distribution, and diagrams, he offered sound advice about the control and prevention of forest fires. The fire of the Seton Indians would service two ends only: that of cooking and that of comradeship of the council ring. In neither case was a large one necessary. "No Woodcrafter ever builds a bonfire. It is wasteful, uncomfortable, unsocial, dangerous

---

[139] Seton, Woodcraft Manual for Boys, pp. 265-279.

and criminal."[140]

The botany section was followed by an introduction to the study of common birds and quadrupeds, involving descriptions of habits and range and again including the customary illustrations. Birds dominated the discussion, Seton assuming that his animal stories and treatises on mammalogy were household items and did not bear repetition. The reader was, however, treated to long digressions on tracking and taxidermy. Seton refused to sanction the childhood killing of any animal except for scientific purposes. He reasoned that a sound knowledge of specimen preparation would keep even this form of killing to a minimum.[141] The object of all serious nature study was the accumulation of knowledge. While this might lead to the accumulation of specimens, Seton preferred an accumulation of notebooks. Recognizing that many of his young followers were city boys, unaccustomed to a wilderness environment, he was constantly looking for ways to bring them close to those animals which, because they were never seen, were presumed not to exist.

---

[140] Ernest Thompson Seton, The Forester's Manual; or the Forest Trees of Eastern North America (New York: Doubleday, Page, 1910); Woodcraft Manual for Boys, pp. 280, 281, 285-313; Book of Woodcraft, pp. 327-467.

[141] Ernest Thompson Seton, "How to Stuff a Bird," Country Life in America, 6 (July, 1904), 267-69; "How to Study a Bird," Bird-Lore, 6 (Nov.-Dec., 1904), 181-84; Book of Woodcraft, pp. 268-78. Seton asked C. Hart Merriam to contribute a section on the mounting of small mammal skins; ibid., pp. 281-84.

He assured aspiring naturalists that within five miles of New York, or Chicago, or Boston they could be sure to find "twenty wild quadrupeds living their lives as they always did." Because they feared man, they did not show themselves until nightfall, a habit which placed the observer at a distinct disadvantage. Tracking skills, however, alleviated the problem, for "wherever a Four-foot [sic] goes it leaves behind a little written account of its visit, its name, the time, the place whence it came, what it did, whither it went away. We can find these accounts and read them and thus learn of the numbers and kinds of our wild neighbors." In a literal sense, the trail could be viewed as "an autobiographic chapter of the creature's life, written unwittingly . . . and in perfect sincerity." The tracks themselves could be sketched, or cast in plaster. The behavior of the animal, measured out in its trail, could be recorded in the notebook, along with similar observations about its predators, its prey and its congeners. In the long term, the conscientious observer could bring all his materials together and make his generalizations. To make the job easier, and to demonstrate the possibilities afforded by this method of investigation, especially in winter, Seton set up in trailing diagrams some hypothetical situations involving the interrelationships of various animals. Eventually the reader is qualified to determine and record their distinguishing characteristics and habits simply by analysing the marks they make.[142]

Seton's emphasis on trailing integrated the three major aspects of woodcraft--sport, woodlore and nature study.[143] Nevertheless, it was still understood that the mystery of nature was not something one solved. Woodcraft was intended to give the youth access to still more and greater mysteries, to make him reflect upon his own limitations within nature and to stress in his social relationships the humility born of that reflection. By learning respect for the Indian and the animal, Seton hoped that young people would be able to comprehend the largeness not only of their community, but of their responsibilities to it. Childhood contact with the wilderness, even if it could be gained only from play, was essential to the survival of diversity. For without this diversity the survival of man himself would be called into question. Wild nature was the womb of democracy. But not in Frederick Jackson

---

[142] Ernest Thompson Seton, "Tracks in the Snow," St. Nicholas, 15 (March, 1888), 338-41; "Stories on the Tree-Trunks," Country Life in America, 6 (May, 1904), 37-39, 90; "The Oldest of all Writing--Tracks," CLIA, 17 (Dec., 1909), 169-73; 242, 244, 246, 248, 250; "Ernest Thompson Seton's Boys, Part 1," p. 15; "Ernest Thompson Seton's Boys, Part 2," p. 15; Book of Woodcraft, p. 288; Woodcraft Manual for Boys, pp. 203-15, 267. See also Ernest Thompson Seton, Animal Tracks and Hunter Signs (Garden City, N.Y.: Doubleday, 1958).

[143] Ernest Thompson Seton, "What to Do When Lost in the Woods," Country Life in America, 6 (Aug., 1904), 359; "Blazes and Indian Signs," CLIA, 7 (April, 1905), 632-34; "The Woodcrafter and the Stars," CLIA, 7 (Nov., 1904), 61.

Turner's sense. The theory of community that inspired Seton's vision encompassed all species. Where Turner was a nationalist, Seton was a provincialist. Where Turner was anthropocentric, Seton was anti-anthropocentric. Nature was, and must remain, a local thing, altering and fluctuating in form and content through numberless geographical spaces. Indian cultures had recognized this diversity and had adapted accordingly their place. But the pioneer, at whose hand the Indian fell, sought to homogenize both the landscape and its inhabitants, to cancel diversity in the name of social progress. That his Cattle might live, the Wolves must die.

Woodcraft, as its name implied, meant craftsmanship. This, for Seton, was the essence of independence, individuality and resourcefulness. "Probably nothing is sadder," he wrote, "than to go into a home where everything is bought ready prepared; . . . [where] no brain or pride has gone into the making." No society could legitimately argue in defense of individualism if its individuals did not know "the pleasure of workmanship, the joy that comes from things made well by [one's] own hands." The wilderness demanded nothing less than absolute skill. It was the place where "life is made worth while, not by the few great moments, but by the making of the daily life . . . full of meaning."[144]

---

[144]Seton, Woodcraft Manual for Boys, pp. 216-17.

IV

The history of the Boy Scouts is a tangled web of conflicting interests and objectives. It has been recorded and re-recorded by many of its own leaders in England and in the United States. Until the arrival of David Macleod's excellent thesis in 1973, interested scholars were largely dependent, in the latter case, upon William Murray's official History of the Boy Scouts of America which, as Macleod notes, is frequently inaccurate.[145] As one pores over the documents touching on the birth and early development of the BSA, one is forced to ask whether Murray's history was not contrived to cast the organization in the best possible light. Ernest Thompson Seton clearly believed that he was not only inadequately, but unfairly, treated in the book and expended considerable rancor on former friends and enemies trying to get the record revised.[146]

---

[145] William D. Murray, The History of the Boy Scouts of America (New York: Boy Scouts of America, 1937); Macleod, "Good Boys Made Better," p. 484. See also, Will Oursler, The Boy Scout Story (Garden City, N.Y.: Doubleday, 1955); Harold P. Levy, Building a Popular Movement: A Case Study of the Public Relations of the Boy Scouts of America (New York: Russell Sage Foundation, 1944).

[146] Seton to James E. West, Oct. 11, 1937, Aug. 9, 1940, Sept. 7, 1940, Oct. 1, 1940, Oct. 22, 1940, Nov. 30, 1940; James E. West to Seton, Oct. 20, 1937, Oct. 14, 1940; E. M. Robinson to Seton, Feb. 20, 1939, March 12, 1938, Sept. 15, 1943; E. M. Robinson to Julia M. Seton, Feb. 13, 1941; Seton to W. D. Murray, Aug. 16, 1938; W. D. Murray to Seton, July 26, 1938, Aug. 26, 1938; E. S. Martin to Seton, Sept. 16, 1938; R. H. Vitalius to Seton, March 20, 1942, SP. Dan Beard also felt that Murray's History was inaccurate; Dan Beard to Seton, Nov. 4, 1938, Jan. 30, 1940, May 28, 1940; Seton to Dan Beard, May 23, 1940, SP. See also Dan Beard to Cyril Clemens, March 22, 1938, copy in SP.

Murray's book was actually the product of many minds, all of them attached to the Executive Board of the BSA. They had begun preparation of a definitive work as early as 1934, and by January, 1935, were informing Seton of its progress and soliciting his information.[147] But the apparent good will reflected in these gestures was misleading.

Seton had been Chief Scout of the BSA from 1910 to 1915. In the latter year he was expelled from the organization for reasons which had never been properly explained to the public. Seton had been threatening to set the record straight since April 9, 1926. On that date he wrote to Colin H. Livingstone, a member of the original National Council, first President and now honorary Vice-President, noting that he had just read the most recent edition of the Handbook for Scoutmasters wherein was recorded Baden-Powell's interpretation of the events leading up to the creation of the organization. "Baden-Powell's history of the Boy Scout Movement is so flatly contradicted in many particulars by documents, and so wholly unfair to everybody but himself, that I feel bound to write the facts." Livingstone acknowledged receipt of Seton's letter on April 17th.[148] Meanwhile, Livingstone had been in touch with James E. West, the Chief Scout Executive who, on April 20th, wrote to Seton informing him that the Boy Scouts had recently initiated

---

[147] James E. West to Seton, June 29, 1937; E. S. Martin to Seton, March 2, 1934; E. M. Robinson to Seton, Jan. 17, 1935; Seton to W. D. Murray, March 12, 1935; W. D. Murray to Seton, March 13, 1935, SP.

[148] Seton to Colin H. Livingston, April 9, 1926; Colin H. Livingstone to Seton, April 17, 1926, SP.

the Silver Buffalo Award for distinguished service to the organization and that "a report adopted by our Executive Board April 12th designating the persons to receive the first awards," included Seton's name.[149] This was no mere coincidence. As it happened, the awards were to be granted at the BSA Annual Meeting in Washington on the weekend of May 1st. Among the guests of honor were to be Robert Baden-Powell and the President of the United States. West having sent notice of the award, Livingstone followed up quickly with another letter.

> In further acknowledgement of your letter of the 9th instant may I ask that since we are to have Sir Robert Baden-Powell in this country at our great meeting in Washington . . . that the controversy which you indicate may arrise [sic] about the history of the Boy Scout movement be not developed until we have an opportunity of sitting down quietly and discussing the whole question as it appears to you and to ourselves.[150]

The award was intended to take the steam out of Seton's wrath and to delay potential embarrassment, at least until after Baden-Powell had come and gone. West waited until April 27th before inviting Seton to attend the meeting itself, apparently hoping that it would be too late for him to make the necessary arrangements.[151] Luckily for West,

---

[149] James E. West to Seton, April 20, 1926, SP.

[150] Colin H. Livingstone to Seton, April 23, 1926, SP.

[151] James E. West to Seton, April 27, 1926. This same tactic was employed for subsequent important events; see, e.g., James E. West to Seton, May 3, 1927; Seton to James E. West, May 12, 1927, SP.

Seton was on a lecture tour and could not have been present anyway.[152] Later in the same summer a circular letter emerged from Boy Scout headquarters in New York inviting correspondence concerning the history of the movement.[153]

From 1926 onward Seton's correspondence files are literally jammed with letters revolving around his early participation in scouting. These clearly reflect his bitterness, not only at being expelled from the organization in 1915, but at the bastardization of the principles which he had introduced to it before and during his tenure as Chief Scout. There were consistent counter efforts to disarm him peacefully. For example, Mortimer Schiff, Vice-President of the BSA, and James West both contributed funds to Seton's Woodcraft League in a patronizing attempt to buy him off. In 1927 Seton actually sent West a manuscript outlining his own interpretation of the origins of the BSA, and announcing his intention to publish it. West replied that his facts were wrong, and that Seton would be making a grave mistake if he did publish them.[154]

Seton always had the good sense to send drafts of his

---

[152]Seton to James E. West, April 28, 1926; Bolton Smith to Seton, May 4, 1926, SP.

[153]Seton to Walter W. Head, Aug. 16, 1926, SP.

[154]James E. West to Seton, Jan. 7, 1927, Oct. 27, 1927; Seton to James E. West, Oct. 25, 1927; Seton to Mortimer Schiff, June 17, 1927, March 1, 1928, March 11, 1928, SP.

historical critiques to Edgar M. Robinson before mailing them to West and Murray. A fellow Canadian, a YMCA secretary, and one of the most prepossessing instigators of the Boy Scouts of America,[155] Robinson invariably agreed with Seton's interpretation. He never hesitated to chastise him for sounding "as bad as a Methodist publishing an analysis of . . . some other denomination that is sincerely though possibly mistakenly working for the same general ends." But he also argued at national headquarters that "Seton has just cause for his dissatisfaction with the lack of expressed appreciation of his contributions both to Baden-Powell and to the Boy Scout movement in this country." Robinson doubted that Baden-Powell could have launched the organization successfully without Seton, and suggested that it was "a matter of opinion which [of the two] contributed the most to the Boy Scouts the world over."[156]

---

[155] Edgar M. Robinson, The Early Years: The Beginnings of Work with Boys in the Young Men's Christian Association (New York: Association Press, 1950); C. Howard Hopkins, A History of the YMCA in North America (New York: Association Press, 1951), pp. 462-63; Macleod, "Good Boys Made Better," pp. 237-41. One of Robinson's colleagues described him as "critical and somewhat radical. . . . He was a stimulant to all searchers of truth. To the contented and complacent he was a constant irritant, and a troublesome person to have around." A. G. Knebel, Four Decades with Men and Boys (New York: Association Press, 1936), p. 181.

[156] E. M. Robinson to Seton, June 24, 1927, Oct. 6, 1928, Nov. 6, 1934; Seton to E. M. Robinson, Oct. 11, 1928, SP; E. M. Robinson to W. D. Murray, copy in SP.

As animosity between Seton and Baden-Powell grew over the years, the latter became so concerned that his claims to authorship might be challenged that he prepared a "Note for . . . when I am dead,"[157] describing his own pre-eminent role. He also expressed great reluctance at having any of his correspondence with Seton published, and in 1940 asked West "to serve as his representative in considering any letters that [Seton] may present indicating that any injustice has been done to [Seton] in his writings." West informed Seton that he had the authority "to take any action that may be necessary . . . . in the interests of accuracy and fairness." Whether or not this was a veiled threat, Seton understood that "accuracy and fairness" at national headquarters usually attached to Baden-Powell's interpretation of events.[158]

W. D. Murray's History was written, at least partly, to anticipate Seton's prolonged threat to publish a personal vindication. Members of the BSA Executive Board despised Seton with a passion only vaguely obscured by attempts to humor his sense of dignity. He was never left in doubt, should he choose publicly to question the integrity of

---

[157]Robert S. S. Baden-Powell, "Note for office to keep in case of revival of arguments later when I am dead. R.B-P. 17.12.13," quoted in E. E. Reynolds, The Scout Movement (London: Oxford University Press, 1950), pp. 2-3.

[158]Robert S. S. Baden-Powell to Seton, Aug. 8, 1939; James E. West to Seton, Oct. 14, 1940, SP.

Baden-Powell, that he would be subjected to legal proceedings. Although Seton frequently allowed his persecution complex to color adversely his treatment by others, this fact alone is insufficient to explain the bitterness he felt in his old age. He obviously believed that the Seton Indians were the roots of the BSA tree and that their principles had been perverted by a league of Wall Street brokers and reactionary patriots who sacrificed the spontaneity of youth to the blandness of the status quo.[159]

V

We must now return to the origins of the Boy Scout movement to ascertain whether or not Seton's perspective was warranted. Here we shall be concerned not so much with the institutional as with the philosophical side of the controversy.

As a popular writer, Seton was much in demand on the lecture circuit. His agents in North America and Britain were able to fill huge auditoriums with adoring young people and their parents while he retold old stories and

---

[159]Ernest Thompson Seton, Scouting and Woodcraft; Submitted to C. J. Carlson, Jan. 16, 1930 and accepted by him with 2 emendations, MS notes, n.d., SP; For Union of Woodcraft and Boy Scouts, MS notes, n.d., SP; Memorandum Regarding Woodcraft and Boy Scouts, MS notes, n.d., SP.

created new ones. He also lectured on natural science topics before university audiences. By all accounts he was a performer of consummate skill, able, singlehandedly, to enchant a crowded hall for an entire evening.[160] On an early tour in Britain, Seton began to consider the possibility of using his lectures as a vehicle for spreading the "gospel of woodcraft" internationally. Taking as his subject, "The Red Indian as I Know Him," he launched a special series of informative and analytical talks intended to evoke a response by which he could measure the potential. His first recorded engagement was at the Congregational Church, Stoke Newington, London, on October 13, 1904. This was quickly followed by visits to schools throughout England where he also distributed copies of the "Birch Bark Roll," by then in its third edition. The earliest tangible results were reported in November, 1904, at Eccles, where a certain William Knight established the "Seton Nature and Athletic Association" based exclusively on woodcraft principles and incorporating intact all of the Indian lore.[161]

In March 1906, on a similar tour, he was introduced

---

[160]Elise Thompson, Taped Interviews, Toronto, Ont., Nov. 20, 1971, Jan. 25, 1972. Conger Roads to Seton, Nov. 15, 1908; Charles A. VanDeusen to Seton, Nov. 16, 1908; W. C. Glass to Seton, Nov. 23, 1908; C. Hart Merriam to Seton, Nov. 27, 1908; Howard Bradstreet to Seton, Dec. 7, 1908; J. G. Scorer to Seton, Dec. 9, 1908; Nicholas Murray Butler to Seton, Nov. 23, 1908, SP.

[161]Seton, History of the Boy Scouts, p. 40; William Y. Knight to Seton, n.d., SP.

by the Duke and Duchess of Bedford to Lord Roberts, who, since his retirement from the army in 1904, had been devoting his energies to the creation of a citizen army. Expressing sincere interest in the woodcraft program, Roberts promised to bring it to the attention of one of his former officers, Robert S. S. Baden-Powell, whom he knew to have related interests.[162] On May 6, Baden-Powell sent Lord Roberts a "paper on training boys in scouting," which prompted the latter to remark enthusiastically that "it would be better for boys to spend a day bicycling in the country near the large towns, and learning to scout and to ascertain the whereabouts of a skeleton enemy than to waste their time . . . in looking on at games."[163] Shortly afterwards, perhaps at Roberts' suggestion, Seton forwarded a copy of the "Birch Bark Roll," now in its 6th edition, to Baden-Powell. On August 1, Baden-Powell wrote to Seton, thanking him for the book.

> It may interest you to know that I had been drawing up a scheme, with a handbook to it for the education of boys as scouts--which curiously runs much on the lines of yours. So I need hardly say your work has a very special interest to me.

---

[162]Ernest Thompson Seton, History of the Scouting Movement, MS notes, n.d., SP; A History of the Woodcraft Movement: An Interview, MS copy, n.d. in SP; History of the Boy Scouts, p. 41.

[163]Lord Roberts to Robert S. S. Baden-Powell, May 18, 1906, quoted in E. K. Wade, Twenty-One Years of Scouting: The Official History of the Boy Scout Movement from Its Inception (London: C. Arthur Pearson, 1929), p. 41.

> I should very much like to meet you if you are . . . in England--and in any case if you would allow me to send you a copy of my scheme later on (it is not yet printed completely) and give me your criticism of it [sic] I should be very grateful indeed.[164]

In point of fact, if Baden-Powell's scheme involved digging out skeleton enemies, there was little reason to believe that it was remotely similar to Seton's.

Indeed, Baden-Powell's initial plans were a direct product of his military experience. As early as 1884 he had prepared a brief manual entitled Reconnaissance and Scouting, a Practical Course of Instruction . . . for Officers, Non-Commissioned Officers and Men.[165] When this proved inadequate, he rewrote it, "on improved lines, with a view to developing character--that is, manliness, self reliance and reliability--as well as field efficiency." The revised book, Aids to Scouting for N.C.O.'s and Men, was first published in 1899.[166] In 1903 Baden-Powell returned to England fresh from organizing the South African Constabulary, "on lines of my own, the principles of which were practically those of the Scout training applicable

---

[164] Robert S. S. Baden-Powell to Seton, August 1, 1906, SP. See also, William Hillcourt, Baden-Powell: The Two Lives of a Hero (London: Heinemann, 1964), pp. 256-57.

[165] Reynolds, Scout Movement, p. 4.

[166] Robert S. S. Baden-Powell, Scouting and Youth Movements (New York: Jonathon Cape, 1931), p. 22. The book was subsequently revised and enlarged; see, e.g., Aids to Scouting for NCO's and Men (London: Gale and Polden, 1909).

to men."[167] He was surprised to discover that Aids to Scouting had been adopted for use in the schools on the advice of Charlotte Mason, head of the Teachers' Training College at Ambleside.[168]

Also in 1903 Baden-Powell was invited by Sir William Smith to review the Boys' Brigade at Glasgow on the twentieth anniversay of its founding. Based upon church membership, with Bible study, drill and physical training as dominant features of its program, the Brigade was the embodiment of crusading Christian soldierdom. And it was Smith who recommended that Aids to Scouting be rewritten in a format acceptable to boys, a challenge that Baden-Powell accepted along with the Vice-Presidency of the Brigade itself. By 1906 he was consulting, not only with Roberts and Seton, but with Lord Grey, Lord Strathcona, officials in the YMCA, and sundry others, with a view to formalizing his theories still further. Until he read the "Birch Bark Roll," however, he had not seriously entertained the woodcraft idea. As he himself pointed out, during the siege of Mafeking, Lord Edward Cecil, his Chief

---

[167] Interrogatories . . . administered to Lieutenant General Sir Robert S. S. Baden-Powell, of London, England; Baden-Powell's testimony given May 24, 1918 at the U. S. Consulate in London in the case of Boy Scouts of America v. United States Boy Scouts, Supreme Court, New York County, copy in SP. Hereafter cited as Baden-Powell, Interrogatories.

[168] Baden-Powell, Scouting and Youth Movements, p. 24.

Staff Officer, had established a corps of local boy messengers mounted on bicycles whose primary duty it was to carry dispatches to and from the front. "The experience taught us that if their training was made to appeal to them boys would learn readily, and also that boys were capable of taking responsibility to a far greater degree than was generally believed, if only they were trusted."[169]

One is forced to ask how, in the exigencies of war, general principles about the behavior of boys could, with such assurance, have been made. Lord Cecil's messengers performed admirably without any previous scout training, a fact which begs the question. When his own survival is at stake, man, like all other species, seems to muddle through.

On October 30, 1906, the hero of Mafeking met with Seton over lunch to compare notes. Baden-Powell cited the decentralized, democratic nature of the woodcraft camps, their non-competitive, outdoor program, and the high quality of scouting practice as distinctive characteristics.[170] The next day he mailed Seton the same "preliminary notice which I sent out early this year regarding my scheme of 'Boy Scouts,'" and which, he suggested, was practically

---

[169] Ibid., p. 23; Baden-Powell, Interrogatories; Reynolds, Scout Movement, pp. 5-6.

[170] Robert S. S. Baden-Powell, Diary, Oct. 30, 1906, quoted in Reynolds, Scout Movement, pp. 7-7.

identical to Seton's, "except that mine does not necessarily make its own organization--it is applicable to existing ones. He concluded that it would be advisable for them "to work together in the same direction," and implied the "direction" he meant by forwarding a copy of Aids to Scouting.[171]

Seton's innate urge to rub shoulders with the mighty surfaced temporarily at this juncture, leaving Baden-Powell with the impression that woodcraft could easily be brought within the purview of military scouting. Reflecting back upon this initial, enthusiastic response to Aids to Scouting, Seton recalled, with remorse at his own stupidity, that it offered "no program of activities, no scheme of merit badges, no illustrations, no woodcraft, no definite attempt at character-building, no reference to youth in the formative stage, no attempt at an organization. It is all soldiering, by a soldier, and for soldiers."[172]

For the balance of the year 1906, the two men lectured extensively in England on their respective philosophies. Although they maintained a regular correspondence, they did not see each other again for some time.[173] In May, 1907,

---

[171]Robert S. S. Baden-Powell to Seton, Oct. 31, 1906, SP.

[172]Seton to Robert S. S. Baden-Powell, Nov. 3, 1906; Seton to Lord Roberts, Dec. 8, 1906, SP; Seton, History of the Woodcraft Movement: An Interview.

[173]Robert S. S. Baden-Powell to Seton, Nov. 3, 1906, Nov. 6, 1906, Nov. 7, 1906, Nov. 10, 1906, Nov. 23, 1906, Dec. 4, 1906, Dec. 11, 1906, SP.

Baden-Powell placed his first formal proposal before the British public in a small printed brochure. Opening with the premise that the same decadence which had precipitated the fall of the Roman Empire was at work in Britain, he pleaded with his countrymen to wrest themselves from their complacency and to "turn the rising generation on the right road for good citizenship." He hoped that his plan, which he saw as a vehicle for achieving this patriotic end, would appeal not only to boys and parents, but to all creeds and classes. He did not want his program to compete with those of existing organizations but hoped that it could be adopted as a supplement. In communities where Boys' Brigades, Church Lads or the YMCA had not been established, the Boy Scouts could supply an organization of its own. He also noted Seton's "phenomenal success" with a "similar idea" in the United States.[174]

In Part Two of the brochure he summarized the scheme itself. Men Scouts, he informed the reader, were of two kinds: War Scouts and Peace Scouts. The former was a soldier "selected for his reliability, courage, and intelligence to go ahead of a force to find out all about the enemy and the country." The Peace Scouts were epitomized by pioneers, trappers, prospectors, Mounties and, of course,

---

[174] Robert S. S. Baden-Powell, Boy Scouts: A Suggestion (London: By the Author, May, 1907), copy in SP.

the South African Constabulary, a copy of whose uniforms the Boy Scouts would later proudly wear. The qualities necessary to war and peace scouts alike included courage, endurance, chivalry, loyalty to duty, the power of observation--and woodcraft. Woodcraft was defined as their "ability to live in the open, to kill and cook for themselves, to look after their health, and to understand the ways of animals, and to have general resourcefulness and self-reliance." It was felt that the schools had abdicated their responsibility for inculcating good citizenship and that the Boy Scouts could fill the vacuum with the attractiveness of its novel features. Instruction would fall under the heading of "Scoutcraft" and would include, with all the qualities noted above, a special emphasis on discipline and patriotism. Patriotism implied an understanding of the history and geography of Britain and her colonies and "deeds that won the Empire." An adult Scoutmaster would organize a Troop by hand picking 6 or 8 "special boys" as Patrol Leaders, putting them through a course of instruction based on the handbook, and then permitting them to form and instruct a Patrol of 6 Scouts each on the same lines. In this way responsibility filtered down from Baden-Powell, through the adult Scoutmaster, to the boys themselves. The program was established by adults and directed by adults. There was no discussion of democracy, and the boys themselves had no part in the decision-making process affecting the internal organization of the Troop. "The principle is that each Scout is put on his honour to carry out orders whether his officer is present or not." Where Seton had intended woodcraft to acquaint the child with his own nature, in nature, Baden-Powell immediately brought it to the service of citizenship, assuming that nature was external to the child. In fact, Baden-Powell did not understand Seton's scheme at all. For him, character and good citizenship had a preordained mission: the maintenance of Empire. The child learned kindness for the animals in the same sense that he would learn paternalism for the Zulus. What Seton had hoped to do was to demonstrate that nature, in all its complexity, produced both the community and the wherewithal to live within it. The "Birch Bark Roll" contained no emphasis on patriotism and discipline because such concepts were alien to his notion of citizenship. Nature was an organic democracy requiring the observance of laws of limitation and mutual respect by all species, including man. Seton's citizens were the ecological products of nature, while Baden-Powell's would use nature as the field for the enactment of civilization.[175]

Early in the summer of 1907 the philanthropist-

---

[175]Robert S. S. Baden-Powell, Boy Scouts: Summary of Scheme (London: By the Author, May, 1907), quoted in Wade, Twenty-One Years of Scouting, pp. 35-40; "Note for . . . when I am dead," p. 3; Scouting and Youth Movements, p. 24.

publisher C. Arthur Pearson, founder of Pearson's Weekly and the Daily Express, entertained Baden-Powell at his country estate in Surrey to hear the plans in greater detail. Although almost blind, Pearson was nonetheless a vigorous, if poorly informed, promoter of what he perceived to be noble causes. He also had sound business instincts. Just as Bok had done with Seton at the Ladies' Home Journal in 1901, he agreed to publish Baden-Powell's program in serial installments guaranteed to win widespread public attention. The name of the Boer War hero was enough in itself to insure good returns on the investment. Accordingly, by December, 1907 Pearson had provided Baden-Powell with offices in London and with a promise not only to launch its official organ, The Scout, but also to underwrite the proposed organization for at least one year.[176]

Before committing the plan to print, Baden-Powell held an experimental camp at Brownsea Island, Dorset, in August, 1907. With the assistance of H. Robson, an officer in the Boys' Brigade, and Major Maclaren, an old fellow-soldier, he brought together a cross-section of boys, set them up in four Patrols, and embarked upon his training program.

> The training was based on that which I had employed with soldiers and the [South African] Constabulary, with considerable adaptation to make it suitable for

---

[176] Wade, Twenty-One Years of Scouting, pp. 43, 52-53; Robert S. S. Baden-Powell to Seton, June 17, 1907; Dec. 3, 1907, SP.

> boys. We followed some of the principles adopted by Zulus and other African tribes, which reflected some of the ideas of Epictetus and the methods of the Spartans. . . . I also looked into the Bushido of the Japanese, as well as many of the more modern methods, including that of John Pounds, that of Jahn for physical culture, and those put into practice by Sir William Smith, Ernest Thompson Seton, Dan Beard and others.

The style of the activities pursued in camp was similar to Seton's model. Only the underlying objectives were strictly Baden-Powell's invention. He did not adopt the ways and dress of North American Indians, but he told stories about and made object lessons of the Zulus, the Swazis and the Matabele "amongst whom ordeals are practised for testing their youths before they can rank as warriors." P. W. Everett, Pearson's literary manager and one of Baden-Powell's assistants at Brownsea, recalled the General standing in the light of the evening campfire imitating bird calls, demonstrating animal tracking, telling stories, leading in dances and songs. Among the songs he taught the boys was the Eengonyama chorus, of Zulu Impi derivation, which later became the Boy Scout "War Cry." A visiting reporter from the Daily Express also assured his readers that "some of the best tales about Mafeking ever told" made their way into the repertoire. The daily activities included cooking, sea and land scouting, nature study, pioneering and woodcraft, all "under orderly discipline."[177] Baden-Powell's version

---

[177] Baden-Powell, Scouting and Youth Movements, pp. 25-27; Wade, Twenty-One Years of Scouting, pp. 44-51.

of tracking involved "stalking," which taught the boy "to creep up unseen within 50 yards of the umpire in fair stalking country, and "spooring," which involved "follow[ing] a horse's trail for two miles over varying ground--reporting all his changes of pace."[178] The Spartans were present, the natives were present, many of the games from the "Birch Bark Roll" were present. But the South African Constabulary was also present. And that made all the difference.

Clearly, Baden-Powell did not appreciate this difference. The Brownsea camp having proved successful, he wrote to Seton that "we are going on with my scheme like your 'Woodcraft Indians'. . . . I hope you will allow me to make frequent mention of yourself and your tribes as examples to the Scouts." When the first sections of "Scouting for Boys" were almost ready to go to press he requested permission to quote from the "Birch Bark Roll," promising to acknowledge and to copyright in Seton's name "any wholesale extracts." He also asked permission to use the "Tracking Irons" employed in Seton's "Deer Hunt" game, again promising to protect their copyright. (After granting this request, Seton stopped by Boy Scout headquarters in London to see if the conditions were being observed and was surprised to be sold a pair of "B.P.

---

[178] Robert S. S. Baden-Powell to Seton, June 17, 1907, SP.

tracking irons."[179]

In the early half of 1908 Seton was again on a lecture tour of England. He joined Baden-Powell for dinner on March 1st and was presented with the first few installments of "Scouting for Boys" which Pearson had been publishing fortnightly since January.[180] At this point Seton began seriously to question Baden-Powell's integrity. Leafing quickly through the articles he "was astounded to find all my ideas taken, all my games appropriated, disguised with new names, the essentials of my plan utilized, and not a world of acknowledgment to me, or explanation why I should be left out of a movement that I began."[181] Not wanting to offend the General, but wishing to make a point of principle, Seton confronted him good naturedly with the plagiarism of the games, most of which had been renamed, reordered and altered in small details. Baden-Powell regretted that he had "omitted mentioning the source of several of the games" and promised later to include a notice to the effect that they had been lifted from the "Birch Bark Roll."[182] In fact, Seton was really reaping the

---

[179]Ibid.; Robert S. S. Baden-Powell to Seton, Jan. 24, 1908, Feb. 13, 1908, Feb. 24, 1908, SP; Ernest Thompson Seton, Note, MS note, 1940, in SP.

[180]Robert S. S. Baden-Powell to Seton, March 2, 1908, SP.

[181]Seton, History of the Boy Scouts, p. 44; Seton to Robert S. S. Baden-Powell, April 24, 1910, SP.

[182]Robert S. S. Baden-Powell to Seton, March 14, 1908, SP.

whirlwind for having tried to ingratiate himself with a military hero whom all the world respected. He had left Baden-Powell with the impression that their overall ends were identical, despite obvious evidence to the contrary in the "Birch Bark Roll." Seton was at such pains to be self-effacing before acknowledged greatness that he had forgotten to tend his own campfire which, at least in England, was already beginning to burn out.[183]

Seton had talked himself into believing that the Boy Scouts were simply Woodcraft Indians dressed up as soldiers. Yet Baden-Powell was clearly unwilling to adopt a model with relevance only for North Americans. In this sense it was interesting to note his use of Zulus, who figured, one assumes, as part of the Imperial tradition. One of his biographers made the absurd claim that Baden-Powell had "a more correct idea of the Red Indian" than Seton, and that "while he recognized their admirable qualities, he did not blind himself to their limitations."[184] One might have hoped for the same analytical perspective on soldiers. But given his experiences of life, and the inflated value attached to them by others,[185] it was not surprising that Baden-Powell adopted a Kiplingesque Tommy

---

[183]William Y. Knight to Seton, n.d., SP.

[184]Reynolds, Scout Movement, pp. 7-8.

[185]Brian Gardner, Mafeking: A Victorian Legend (London: Cassell, 1966), p. 229.

Atkins model, or that it was eagerly accepted. Seton understood this situation, and was probably not offended by it in principle. It was perfectly in keeping with his philosophy that woodcraft ideals could and should be adapted to local traditions and customs. And although that conviction still awaited articulation, it was implicit in his treatment of the Boy Scouts in England.[186] Still, there were probably snippets of self-delusion on both sides. Each man believed himself the founder of a new movement and each regarded the other as his collaborator, at least until 1908.

While Baden-Powell was not lacking in imagination, he did not have Seton's creative originality. If his own military experience had provided the model for his notions of character building, and if the Boys' Brigade had suggested structural possibilities, Seton's advice, without which the other two features would have floundered in a sea of patriotic mush, was central to his Boy Scout program. Furthermore, in placing scouting before the British public as he did, Arthur Pearson seems to have copied Bok's policy in the Ladies' Home Journal. Although there are differences in presentation, the serialized articles, "Ernest Thompson Seton's Boys" and "Two Little Savages," published between 1902 and 1903, deserve comparison with "Scouting for Boys," published in 1908. Curtis republished the first Seton series, and

---

[186]Seton, History of the Woodcraft Movement: An Interview.

Doubleday the second, in book form. It was not surprising, therefore, in May, 1908, to find Scouting for Boys selling in neighborhood bookstores.[187] In June, Arthur Pearson offered it to a New York publisher for reprinting in the United States.[188] The offer was accepted on condition that the book's "Briticisms" be removed. Baden-Powell wrote to Seton in July telling him that he had received a "request" for reprinting from an American house, but that he was "unwilling to go on with it for fear of interfering with your 'Woodcraft Indians' scheme. If however, you should like to take up the book, I should be very glad to co-operate with you and you could effect such alterations in it as you thought desirable for your country."[189] If Seton had been skeptical, he now felt threatened. It appeared to him that the Woodcraft Indian he had taken to England in 1904 had been given a British military education and was about to be sent back to America dressed in his new uniform. Tommy Atkins suddenly assumed new and horrifying

---

[187] Robert S. S. Baden-Powell, Scouting for Boys: A Handbook for Instruction in Good Citizenship (London: Horace Cox, 1908).

[188] International Press Service to W. T. Porter, June 22, 1908, copy in SP.

[189] Robert S. S. Baden-Powell to Seton, July 9, 1908, SP. Seton had already asked Baden-Powell to contribute a chapter on scouting to a revised issue of the "Birch Bark Roll." The chapter appears in both the 7th (1908) and 8th (1910) editions; Robert S. S. Baden-Powell to Seton, June 17, 1907, SP.

dimensions. What had begun as a shared enterprise had ended in two separate schemes. "My aim," said Seton, "was to make a man; Baden-Powell's to make a soldier."[190]

Insulted at being asked to re-introduce a bastardized version of his own program, Seton poured over Scouting for Boys looking for plagiarism on the one hand and aberrations on the other. He noted that the book was really Aids to Scouting modified mainly by the addition of his own woodcraft principles and games. The motto, "Be prepared," an egotistical play on Baden-Powell's initials, meant "be prepared (for war, understood)." The uniform reflected a desire to stamp out diversity and to impose homogeneity. The second and seventh scout laws, commanding loyalty to the King and to his officers and unquestioning obedience to the orders of parents, patrol leaders and scoutmasters, were singled out as tyrannical and autocratic. Seton made it clear that the military did not frighten him so long as its members had not become soldiers before their twenties, after a childhood and youth of freedom and spontaneity. By teaching children military methods and blind patriotism,

> You run the risk of making a lot of little prigs now and of disrupting their steady development into citizens of high and independent character. . . . It is like forcing an orchid to flower two seasons ahead, by using chemicals. . . . Yes, it blooms, but dies early, and . . . never seeds.[191]

---

[190] Seton, History of the Boy Scouts, p. 42

[191] Seton, History of the Woodcraft Movement: An Interview.

Worst of all, by reducing the importance of native lore, Baden-Powell had effectively eliminated not only the aesthetic and spiritual qualities of woodcraft, but the sense of biological continuity, as between child and environment, that went with them. Baden-Powell's Scouts were supposed to draw spiritual sustenance from woodland hikes where they could adopt the ways of those frontiersmen and explorers who epitomized manliness, efficiency and skill.[192] Evidence of plagiarism abounded. Bands, Tribes and Medicine Men had been duplicated, with altered functions, in Patrols, Troops and Scoutmasters. Braves, Sagamores and Sachems were replaced by Tenderfoot, Second and First Class Scouts. Seton's coups also suggested the merit badges which later became an integral characteristic of scouting. In correspondence Baden-Powell had requested him to forward articles and notes on aspects of woodlore such as tracking, measuring heights and distances without instruments, signal fires and so on. These, Seton claimed, had been incorporated almost verbatim. And the list went on.[193]

By 1909 Seton felt sufficiently abused to take the offensive, requesting his publisher to demand an accounting for copyright evasion. But by this time the second edition

---

[192] Baden-Powell, Scouting and Youth Movements, p. 30.

[193] Seton, History of the Boy Scouts, pp. 45-49; Robert S. S. Baden-Powell to Seton, Nov. 10, 1906, May 31, 1910; Seton to Robert S. S. Baden-Powell, April 24, 1910, June 24, 1910, SP; Hillcourt, Baden-Powell, pp. 270, 282.

of Scouting for Boys had appeared and Baden-Powell could enumerate specific references to Seton's work.[194] But Seton persisted, demanding nothing less than credit for establishing first principles. Although Baden-Powell went part of the way toward meeting this last ultimatum by consenting to publish in The Scout a statement "of my indebtedness . . . for several details," he was plainly losing patience with a man to whom he owed much more. Seton, of course, was being petulant as usual, harming rather than helping his case.[195]

VI

By 1908 the possibility of introducing the Boy Scouts to America was already being discussed. Seton's major publisher, Doubleday, Page, urged him to drop his Indian model, arguing that "There are too many Americans who think of Indians either as dirty and loafing degenerates, or as savages, to make the idea popular when they think of educating

---

[194] Robert S. S. Baden-Powell to H. C. Roberts, Sept. 3, 1909, copy in SP. See, Robert S. S. Baden-Powell, Scouting for Boys (London: C. Arthur Pearson, 1909), pp. 9, 18, 38, 70, 76, 91, 117, 139, 140, 171.

[195] Seton to Robert S. S. Baden-Powell, Sept. 30, 1909, Dec. 10, 1910; Seton to Edward Bok, Dec. 13, 1909; Robert S. S. Baden-Powell to Seton, Jan. 17, 1910.

their children."[196] But Seton kept doggedly to his chosen path, firmly believing that the expanding number of adherents represented a corresponding decline in widespread racism.[197] Outside the League of Woodcraft Indians, which had evolved into a forum for debate on principles, he did not yet have a national organization. Playgrounds, camp, church groups, YMCA's and even some independent private schools had picked up the woodcraft idea, forming their own tribes. In 1910 one journalist estimated that there were between one hundred and fifty and two hundred thousand Seton Indians in the United States, compared to one hundred thousand Boy Scouts in England.[198] But there is no way of gauging the accuracy of these figures. Because many of the tribes were allied with summer camps, there was good reason to suggest that Seton reached more boys superficially, while Baden-Powell reached fewer, indelibly. Baden-Powell shared Seton's reluctance to establish a centralized bureaucracy, creating a system of Local Associations which were encouraged, within the bounds of Scouting for Boys, to exercise and profit from their autonomy. In its multifarious

---

[196] H. W. Lanier to Seton, Aug. 11, 1908, Aug. 25, 1908, SP.

[197] Luther Gulick to Seton, Sept. 21, 1909; Harvey L. Smith to Seton, Sept. 22, 1909, SP.

[198] Seton to Sarah Splint, Dec. 20, 1909; Seton to George J. Fisher, Sept. 5, 1926, SP; M'Cready Sykes, "Let's Play Indian: Making a New American Boy Through Woodcraft," Everybody's Magazine, 23 (Oct., 1910), 481; Macleod, "Good Boys Made Better," p. 225.

editions, Seton's "Birch Bark Roll" outlined a basic structure but left the organizing entirely to local bodies. Scouting had an obvious edge in the mere fact that it was a year-round organization for all its members. Seton's strict emphasis on the outdoors, while it reflected, and was demanded by, his principles, had the adverse effect of making his program seasonal, thereby reducing its range. Nevertheless, the Woodcraft Indians went a long way to restoring Seton's credibility with people like John Burroughs and Theodore Roosevelt, both of whom forgot the trials of the nature faker controversy and eagerly endorsed the program. Burroughs aided the revision of the "Birch Bark Roll" and, as late as 1917, allowed his name to stand on the roster of the National Council of the Woodcraft League. Indeed, Burroughs' support meant much more to Seton than his relationship with Baden-Powell, a true indication of the direction he wanted the movement to take.[199]

But from Seton's viewpoint one of the most exhilarating aspects of the woodcraft work was the zeal with which its older graduates spread the gospel. One of the original Seton Indians, Miller Jordan from Rutherford, New Jersey,

---

[199] John Burroughs to Theodore Roosevelt, Aug. 27, 1906, quoted in Barrus, Life and Letters of John Burroughs, 2:116; Seton to Theodore Roosevelt, Sept. 11, 1906, LC TRP, Vol. 68 (Series 1), p. 150. Cf. Theodore Roosevelt to Seton, March 3, 1907, LC TRP, Vol. 71 (Series 2), p. 211; Seton to Theodore Roosevelt, March 8, 1907, SP. See also, Seton, American Boy Scout, p. 34; Woodcraft Manual for Boys, p. ix.

had achieved the status of Sagamore in 1908. Beginning in the autumn of that year, Jordan--or Silver Fox, as he was known within the League--launched his own tour of boys' clubs and playgrounds in the slums of East and West side New York. Supported by the Parks and Playgrounds Association of which both Jacob Riis and Luther Gulick were active directors, he demonstrated Indian dances and teepee construction to throngs of underprivileged children. By 1910 he himself had helped to organize several inner city tribes. His careful reports to Seton demonstrated the necessity of altering aspects of the "Birch Bark Roll" to accommodate the needs of children who could not leave the metropolitan area.[200] After several speaking engagements at the Henry Street Settlement and at Christ Church House, where he was able to observe at first hand the problems that Jordan was encountering, Seton introduced his first major innovation, "Roof Camping," which reflected the importance he attached to maintaining the integrity of the ideal. Roof camps could be established on the tops of buildings, "where there are thousands of acres of open sunlight, well-ventilated, unused space." Gardens with flowers, fruits and vegetables, vines and trailers could be grown here. Bird sanctuaries, complete with nesting boxes would eventually induce a wide

---

[200]Seton, Woodcraft Manual for Boys, p. 33; Miller Jordan to Seton, Jan. 4, 1910, July 21, 1910, Sept. 9, 1913; Howard Bradstreet to Seton, Sept. 6, 1908, SP.

range of birds to found colonies. Teepees could be set up beside a council fire, secured in a brazier over a stone hearth, so that "a hundred feet above the street, the boys . . . are in another world, and can dress and live much as in camp." Seton was genuine in his desire to reach all boys in all walks of life, never doubting that the inculcation of woodcraft principles would quickly stem the crime wave. The Miller Jordan experience was typical of many of the senior boys who maintained their connection with the woodcraft movement well into adulthood. Seton had begun with the intention of letting the children lead him. And lead him they did.[201]

Seton was never prepared to forsake the Woodcraft Indians, keeping them alive during his entire tenure as Chief Scout and continually trying better to adapt them to urban conditions. Early in January, 1910, Howard Bradstreet, a colleague of Miller Jordan and Chairman of the Neighborhood Worker's Association, called a public meeting for the purpose of investigating special woodcraft programs for downtown children. Among others present with Seton were Dr. and Mrs. Luther Gulick. President of the Playground Association of America and Chairman of the Playground Extension

---

[201]Howard Bradstreet to Seton, Dec. 7, 1908, Dec. 29, 1909, July 7, 1910, SP; Seton, Woodcraft Manual for Boys, pp. 163-64; Rev. Baxter Liebler to the author, July 5, 1973; Leonard Clark, Taped Interview, Greenwich, Conn., Dec. 17, 1971.

Committee of the Russell Sage Foundation, Gulick had a long and successful career of YMCA youth work behind him. (His wife, Charlotte, would be remembered for creating the Camp Fire Girls in 1912.) At this meeting Seton's program was endorsed, but with a change of name. Henceforth, the Woodcraft Indians would be known in Bradstreet's playgrounds and boys' clubs as "Indian Scouts," the object being to take advantage of Baden-Powell's more "advertisable name" without disrupting woodcraft principles. As the Boy Scout organization existed formally in almost every country except the United States by this time, it was not surprising that pressures of this nature arose. Then, too, there was a clear desire on the part of Gulick and others to retain continuity not only with existing organizations, but with the unique cultural features of the country. The decision having been taken to experiment in this way, Seton was called upon to lecture throughout the month of February, again at Christ Church House, to school teachers and other leaders who would be called upon to effect the revised scheme. In sanctioning this step Seton had manifested some doubts, at least about the effectiveness of woodcraft in the slums.[202]

In the meantime another colorful figure had entered

---

[202]Open letter from Howard Bradstreet, Jan. 11, 1910; Seton to E. M. Robinson, Oct. 13, 1934, SP; Ethel Josephine Dorgan, Luther Halsey Gulick, 1865-1918 (New York: Teachers' College, Columbia University, 1934), pp. 100, 112-25, 141-42; Hopkins, History of the YMCA, pp. 251-70; Seton, History of the Boy Scouts, p. 53.

the field with an organization which was a more obvious synthesis of Seton's Woodcraft Indians and Baden-Powell's Boy Scouts. This was Daniel Carter Beard whom Seton had originally met while studying at the Art Students' League in New York during the 1880's. Later, as fellow members of the Camp Fire Club of America they had become relatively close friends. Beard had established a reputation as an animal painter and illustrator in Scribner's, Harper's and St. Nicholas magazines and had gained a measure of notoriety for his controversial sketches in the first edition of Mark Twain's Connecticut Yankee in King Arthur's Court. In 1882 Charles Scribner published his American Boys' Handy Book which for two generations commanded the attention of all good, red-blooded boys in the United States. Camping skills and nature lore played a part in, but did not dominate, this work, and certainly gave no grounds for Beard's later claims that Seton had plagiarized him. Beard was no scientist and he harbored no love for the American Indian. He was, however, a staunch conservationist and for two years, 1905-1906, was editor of Recreation magazine. Here he instituted a Boys' Department, written by himself, in an effort to capitalize on Seton's successes at the Ladies' Home Journal.

Beard's column yielded another parallel for the Woodcraft Indians, the "Sons of Daniel Boone," structured not around tribes or leagues or patrols or troops, but around stockades and forts. The President was called Daniel Boone and the

Secretary, Davy Crockett--a subtle ploy to link the organization with the prestigious New York Boone and Crockett Club of which many prominent public figures were members. When Recreation was sold, Beard transferred his column and the home base of his operations to the Woman's Home Companion where, until he quit in 1908, he tried to compete with Seton's work at the rival magazine. The Companion retained copyright to the name of the organization so that when Beard left, the column remained, now with reprinted and retitled extracts from Baden-Powell's Scouting for Boys. Beard moved briefly to the Pictorial Review, where he changed the name of his organization to the "Boy Pioneers," and, like Seton and Baden-Powell before him, condensed his magazine articles into a manual.[203]

These facts of life never discouraged Beard from believing that he had originated the scouting idea and that both Seton and Baden-Powell were nothing but junior partners.[204] In fact he was little more than a fellow-traveller, distinctive for his personal bravado and pompous patriotism. The only characteristics he shared with Seton were a zeal for youth

---

[203]Daniel Carter Beard, Hardly a Man is Now Alive: The Autobiography of Dan Beard (New York: Doubleday, Doran, 1939); The Buckskin Book of the Boy Pioneers of America (n.p., 1911); Macleod, "Good Boys Made Better," pp. 313-15.

[204]World Book, 1918 ed., s.v. "The Boy Scouts," by Daniel Carter Beard; Seton to Edmund Seymour, Jan. 11, 1919; Caspar Hodgson to Seton, Feb. 3, 1920; M. V. O'Shea to Seton, May 2, 1920, May 7, 1920, June 18, 1920; Seton to Colin H. Livingston, June 7, 1920, SP.

work, a considerable artistic ability, and petulance. His patriotic turn of mind and his sense of history, however, made him a kind of soul brother to Baden-Powell, who had recognized frontiersmen and explorers as stellar examples of peace scouts. With Baden-Powell he agreed that the object of scouting was "to turn out useful, self-reliant, alert, honest citizens," imbued with strong character.[205] "Character is to be found among those men who still have to fight their way with nature."[206] By focussing his own program on the pioneers, "those stalwart Americans, the men of the wilderness," and by dressing his boys (and himself) in buckskin clothes and coonskin caps, Beard painted "citizens" with his own Turnerian conception of democracy. For him,

> Scouting is typically and intensely American. It is safe to say that no full-grown man can appreciate the real meaning to the youth of the United States of the word Scout, unless that person is in full sympathy with American institutions, tradition and history, and familiar with the potential power, manly self-respect, personal integrity and personal dignity only to be realized under a Republican form of government, the only form of government that has no tendencies to make menials of its citizens.
>
> It is the American spirit of conscious individuality and initiative in the weave and woof of the Boy Scout idea, which gives to it its vim, life and vitality.[207]

---

[205] Daniel Carter Beard, "The Boy Scouts of America," American Review of Reviews, 44 (Oct. 1911), 432.

[206] Robert S. S. Baden-Powell, "Educational Possibilities of the Boy Scouts' Training," Nineteenth Century, 70 (Aug., 1911), 30.

[207] Beard, Hardly a Man is Now Alive, pp. 353-54; "Boy Scouts of America," p. 437.

This kind of rhetorical bombast was the very thing that Seton found most objectionable in Beard and Baden-Powell alike. As far as he was concerned, once their boys were dressed in uniforms, the threat of militarism would be realized. And despite the patriots' disclaimers, there was ample reason to be afraid.[208]

VII

The Boy Scouts of America were incorporated in the District of Columbia on February 8, 1910, by William D. Boyce, a Chicago publisher who had reportedly been rescued in a London fog by an English Scout who refused to accept a tip. Boyce did not know Baden-Powell, Seton or Beard before this momentous occasion, and although there were already in existence several scout troops of varying degrees of authenticity--including Seton's Indian Scouts at Christ Church House and at the West Side Branch of the New York YMCA--he has

[208]See, e.g., Ralph D. Blumenfeld, "The Boy Scouts," Outlook, 95 (July 23, 1910), 617-29. Blumenfeld was the editor of Arthur Pearson's Daily Express. See also, Theodore Roosevelt to Robert S. S. Baden-Powell, 1909, quoted in Wade, Twenty-One Years of Scouting, pp. 126-27; Theodore Roosevelt to Seton, Sept. 23, 1910; Seton to Robert S. S. Baden-Powell, June 24, 1910; William B. Kelsey to Seton, Sept. 18, 1910, SP; E. M. Robinson to F. P. Gignilliat Nov. 2, 1910, copy in SP; W. T. Hornaday to Dan Beard, July 10, 1910, Hornaday Letterbook No. 8, pp. 109-111, LC WTHP. Cf. Macleod, "Good Boys Made Better," pp. 229-32.

generally been credited with founding the American movement.[209] Working through the YMCA, Seton had already come to know Edgar M. Robinson, Chairman of the Boys' Work Subcommittee of the International Committee, and W. D. Murray, also a member of the Subcommittee. Because Robinson had been most vocal about his interest in the Boy Scouts, his office had become a kind of unofficial headquarters for discussions about them. On May 9, 1910, Robinson, Murray and Seton met with W. B. Wakefield and Charles E. Heald, representing the British Boy Scouts and YMCA respectively, at the Union League Club in New York, to compare their experiences in youth work, and to explore possibilities for the formal introduction of Scouting to the United States through the medium of the YMCA. Following the meeting, Seton, Heald, Wakefield and Robinson took the night train to Washington where, on the 10th, they joined W. D. Boyce to apply (unsuccessfully) before the Congressional Committee on Education for a federal charter. Although Robinson had met him earlier in the month, the Washington trip marked Seton's first introduction to Boyce. In any case, failure to acquire the national charter effectively terminated Boyce's active participation, leaving the reins in Robinson's hands.[210]

---

[209] E. K. Wade, Twenty-One Years in Scouting, p. 127; W. D. Murray, History of the Boy Scouts of America, pp. 1-21; James E. West to Seton, July 29, 1941, SP.

[210] Seton to Crump, July 22, 1941; William D. Murray, As He Journeyed: The Autobiography of William D. Murray

In 1910 militaristic Boy Scout organizations began to pop up spontaneously all over the country. The most devious example was William Randolph Hearst's American Boy Scouts who appeared on the scene early in May, bearing arms and drilling in public. Later in the month Robinson and Seton visited Colonel Peter S. Bomus, who had organized the Boy Scouts of the United States, and Brigadier-General William Verbeck, principal of a military school in Manlius, New York, who had announced plans to launch the National Scouts of America. The potential dangers posed by these men had the effect of casting Baden-Powell's version as the least of all evils--even for Seton. Although his hostility for Baden-Powell never abated, he was quite comfortable in his friendship with Edgar Robinson. Robinson had always remained loyal to Seton's woodcraft principles and there was every reason to believe that their combined efforts would yield a unique brand of scouting premised not on the values of the pioneer, but on those of the Indian. With this hope in mind, Seton published The American Boy Scout in the spring of 1910. Subtitled the Eighth Edition of 'The Birch Bark Roll,' it preserved intact the traditional Indian motif.[211]

On June 2, 1910, Seton, Gulick, Wakefield, Murray and

(New York: Association Press, 1929), pp. 346-47; History of the Boy Scouts of America, pp. 23-27.

[211]Seton to E. M. Robinson, Oct. 13, 1934, SP.

John L. Alexander, another YMCA Boys' Work Secretary, met in Robinson's office to plan a conference of representatives from thirty-seven youth work agencies for the purpose of organizing the Boy Scouts of America in opposition to Hearst's American Boy Scouts. The conference was held at the YMCA on June 15 and was attended by, among others, Murray, Seton, Robinson, Luther Gulick and Lee Hanmer, both representing the Russell Sage Foundation, Jacob Riis, Lincoln Steffens, George D. Pratt, a YMCA layman, prominent New York Republican and member of the Camp Fire Club of America, and Colin H. Livingstone, Vice-President of the American National Bank in Washington, D.C. Although educated at McGill, Livingstone, Like Robinson, was a native of New Brunswick. He was called upon to chair the meeting as he was the only person present who had been named in W. D. Boyce's proposed national charter. Dan Beard was not present although he had probably been invited. The composition of this group reflected Seton's and Robinson's primary concern that the poor be reached and that the money be made available with which to reach them. On June 21, the Boy Scouts of America was officially launched.[212]

The New York conference elected a temporary Executive Committee with Seton as Chairman, Lee Hanmer, Secretary, and

---

[212] W. D. Murray, As He Journeyed, p. 347; History of the Boy Scouts of America, pp. 27-32; E. M. Robinson to Seton, Oct. 4, 1934; Seton to E. M. Robinson, Oct. 13, 1934, SP.

George Pratt, Treasurer. They in turn created an Organization Committee comprised of Robinson, who was given a year's leave of absence from the YMCA, Gulick, Riis and Livingstone. Dan Beard and Frank Presbrey joined later. John L. Alexander was appointed Managing Secretary. Plans were also made to duplicate Baden-Powell's Brownsea Island camping experiment at Silver Bay, New York, Seton promising to do the administrative work and to absorb the expenses. Seton also recommended the preparation of a new manual, amalgamating reworked portions of Baden-Powell's Scouting for Boys with the "Birch Bark Roll." The former was to be included because it provided a sounder structural framework and method of presentation, items Seton himself thought necessary if the organization were to win its federal charter.[213] Eventually Seton was given the job of compiling the manual, a project which involved the agonizing task of re-reading Scouting for Boys to determine what was to be kept, what altered, and what eliminated. He became so harrassed in the process that he could not repress the urge to write to Baden-Powell:

> There is not an important idea in "Scouting for Boys" that I did not publish years ago in "Two Little Savages," . . . and my Woodcraft and Scouting articles of which I furnished you with copies. The only important

---

[213]Seton to Robert S. S. Baden-Powell, May 20, 1910, June 13, 1910, Sept. 17, 1910; Robert S. S. Baden-Powell to Seton, Sept. 24, 1910; Seton to F. M. Barton, June 30, 1910; William Byron Forbush to Seton, June 22, 1910,SP; Robert S. S. Baden-Powell to Doubleday, Page, July 5, 1910, July 31, 1910, Sept. 12, 1910, copies in SP.

> change you make is to add a little military drill, give things new names and assume their authorship for yourself.[214]

Not surprisingly, therefore, the first official BSA Handbook appeared later in the year under Seton's name.[215]

The partial inclusion, by agreement, of Baden-Powell's Scouting for Boys in the new Handbook might have carried with it the tacit assumption that Seton had adopted the General's philosophical perspective. But this was not true. Indeed, it was a well-established fact in most New York circles that Seton was the moving force behind the movement. This was particularly grating for Dan Beard who was now eagerly looking for responsibilities within the organization for which he proclaimed his Sons of Daniel Boone and Boy Pioneers to be the major inspiration. However, there was within the existing heirarchy a group of men, notably Livingstone, Pratt and Hanmer who did not share Seton's enthusiasm for the Indian and who viewed the incorporation of Baden-Powell's Scouting for Boys, revised or not, as a tangible victory. Seton, who was well aware of their animosity, had the allegiance of Robinson and Gulick, both of whom were devotees of Hall's theory of recapitulation, Jacob Riis, who looked to the Scouts for leadership in the slums, and John

---

[214]Seton to Robert S. S. Baden-Powell, June 24, 1910, SP; Seton's italics.

[215]Ernest Thompson Seton, Boy Scouts of America: A Handbook of Woodcraft, Scouting and Life-Craft, with Which is Incorporated by Special Arrangement, General Sir Robert Baden-Powell's Scouting for Boys (New York: Doubleday, Page, 1910).

Alexander. W. D. Murray seems to have been a fence sitter, waiting to see where the cards would fall. Seton had every reason to be concerned about the presence of Dan Beard whose particular views of youth work would inevitably ally him with the pro-Baden-Powell faction. As things stood, Seton's reputation, combined with his charismatic style and stage manner, made him the obvious mouthpiece for the movement. He gave to it both credibility and a colorful dynamism which was useful in attracting adherents. Beard had the same exterior qualities as Seton, but his ideas were decidedly different. He was waiting in the wings and would be called upon later to fill the shoes of the master, against the master.[216]

After a heady round of lectures to YMCA's, Chautauquas and charitable organizations on his own interpretation of scouting, Seton landed at Silver Bay on August 16, 1910, to conduct the Brownsea equivalent. The titular leaders at the camp included Robinson, Murray, Alexander and Seton, although Seton ran the entire program, with the help of Miller Jordan who was also there. Altogether there were present approximately 150 people divided into groups of six boys, each under an

---

[216]Sarah Splint to Seton, June 25, 1910; Dan Beard to Seton, Sept. 19, 1910; Seton to Edmund Seymour, Dec. 2, 1918; John L. Alexander to Seton, Aug. 10, 1910; George Dugan to Seton, Sept. 16, 1910, W. T. Hornaday to Seton, June 26, 1910, June 31, [1910], Nov. 10, 1910, SP; W. T. Hornaday to Dan Beard, June 28, 1910, July 21, 1910, Hornaday Letterbook No. 8, pp. 37, 109-11; W. T. Hornaday to Seton, Dec. 5, 1910, Hornaday Letterbook No. 8, p. 365, LC WTHP; Daniel Carter Beard, "The Boy Scouts," Outlook, 95 (July 23, 1910), 696-97.

adult, drawn from centres like Pittsburgh, Newark, New York, Jacksonville, and Boston. There was also a delegation from Toronto. The boys were housed in teepees, did their own cooking, and in the evenings sat around the council fire learning all the rituals to which the Woodcraft Indians had long since become accustomed. While the boys stayed with Seton, however, their leaders took instruction in Scoutmastership from W. B. Wakefield who had once more come from England to prevent this heresy from tainting the orthodox faith. Dan Beard arrived in camp on August 28, just in time to be photographed with the leaders. On the following day he instructed the campers in the art of throwing the hatchet. "As a consequence," wrote Seton, "every big tree in camp was disfigured and I got the blame." The Silver Bay camp was not a howling success. Outside of Robinson, none of the other leaders was committed to Seton's idea of scouting, and Wakefield was working directly at cross purposes. As it was the leadership which was to decide the verdict, we know nothing about the response of the boys. On September 5, however, at a meeting in New York, it was decided that Seton had not demonstrated the superiority of the Indian mode and that Baden-Powell's Scout provided the best alternative.[217]

---

[217] Seton to E. M. Robinson, Oct. 13, 1934; E. M. Robinson to Seton, July 26, 1910, Oct. 4, 1934; Miller Jordan to Seton, Sept. 19, 1910, SP; E. M. Robinson to E. M. Willis, copy in SP; W. D. Murray, <u>As He Journeyed</u>, p. 347; <u>History of the Boy Scouts of America</u>, p. 33.

While the Silver Bay camp was in progress, Baden-Powell had been on a lecture tour through Canada. But he had agreed, before returning to England, to address a dinner in his honor at the Waldorf-Astoria in New York on September 23. Because he was the only member of the executive who actually knew Baden-Powell, Seton was delegated the responsibilities of serving as his host and of presiding over the dinner. The irony was not lost on Baden-Powell who, in a mood of deference and gracious conviviality, allowed before the audience that Seton and Beard might be called the fathers of scouting, while he was merely one of the uncles.[218] Below this surface show of camaraderie, which was designed mainly for public consumption in the interests of deflating Hearst's organization, there lurked considerable dissent. Indeed, Baden-Powell was approached separately by the Livingstone faction who asked whether a complete edition of Scouting for Boys could not be brought out in the United States, "since it gives a more definite line of action for a Scoutmaster to work upon in organizing and training than is shown in the present handbook." Seton's manual, while "full of interesting instruction in details," was still inadequate to meet the structural

---

[218]Seton to Robert S. S. Baden-Powell, Sept. 17, 1919, SP; H. W. Gibson, "The History of Organized Camping," in Outdoor Education, ed. Donald R. Hammerman and William M. Hammerman (Minneapolis, Minn.: Burgess, 1968), p. 97.

Plate 12. Ernest Thompson Seton, Robert S.S. Baden-Powell and Daniel Carter Beard in New York (1910)

problems. Scoutmasters required "a progressive course of instruction laid down for them" so that the movement would be left open to a minimum of individual interpretation.[219] The drift to Baden-Powell's views on this matter was already in motion. The Executive Committee and the Organization Subcommittee had functionally amalgamated. William Verbeck had since joined it, and the voices of Colin Livingstone and Dan Beard were becoming more pronounced. Although Seton was still chairman, his own authority and control were gradually disintegrating. Despite the success of the Handbook, the inefficiency of the organizers prevented the rationalization of the program which, on its own, looked rather ambiguous and created confusion in the field. Sporadic financial contributions came from people like Boyce, but these were inadequate to meet the demands being placed on headquarters by inquisitive boys and aspiring scoutmasters. By the middle of October Seton realized that his own organizational abilities were legitimately suspect and that the cause he represented had no chance of surviving within the hierarchy. Even Robinson, who liked Seton's theory, was forced to admit that functionally it was too decentralized and uncoordinated to win national approval, something he deemed essential to success. This was a particularly bitter pill for Seton to have to swallow,

---

[219] Robert S. S. Baden-Powell to Seton, Sept. 24, 1910, SP.

especially in the face of Beard's corresponding ascendancy. His antipathy to Beard's ideas about youth work required that they not be confused. Consequently the barrier between them, although disguised from public view, was reinforced internally. By this time Murray and other fence-sitters had drifted into the Livingstone camp where plans were afoot to alter the entire structure of the organization. Gulick and Riis, regardless of their previous leanings, could see the writing on the wall and, with Robinson, agreed that the objective must continue to be the creation of a truly national organization. If a federal charter were to be sought after and obtained, evidence of greater cohesion and efficiency was essential.[220]

This meant returning to the incorporators of the BSA in the District of Columbia who, in law, were the only officials with the requisite authority to direct operations in the first place. W. D. Boyce had effectively delegated this authority to the temporary Committee on Organization of which he was nominally a member, but which Seton chaired. It was the measure of Seton's lack of control over that Committee which allowed it to go back to Boyce, and to his

---

[220]John L. Alexander to Seton, Sept. 29, 1910; Edward L. Wertheim to Seton, Oct. 11, 1910; Caspar Whitney to Seton, Oct. 13, 1910; Seton to Caspar Whitney, Oct. 17, 1910; S. S. Aplin to Seton, Oct. 14, 1910, SP; Macleod, "Good Boys Made Better," p. 242; Murray, History of the Boy Scouts of America, p. 36.

colleagues named in the original articles of incorporation, to request a new and permanent executive.

The incorporators, acting under directions from Boyce's friend, Colin Livingstone, elected an Executive Board composed of Boyce, Livingstone, Pratt, Murray, Presbrey, Hanmer and Robinson, all of whom had sat on the temporary Committee on Organization, as well as several new individuals. Conspicuous by their absence were Riis, Gulick and Seton. The former two were included in the newly constituted National Council which was otherwise dominated by prominent military men, bankers, businessmen, social workers, educators and Republican politicians. Seton was awarded the title "Chief Scout," the glamour of which disguised its <u>ex officio</u> impotence. Colonel Peter Bomus, Brigadier-General William Verbeck and Dan Beard were also neutralised as "National Scout Commissioners." The creation of these offices had the effect on paper of uniting the Woodcraft Indians, the Boy Pioneers, the National Scouts of America and the Boy Scouts of the United States with the BSA. It also pandered to the egos of Seton and Beard. Seton obviously had to be allowed the senior position for his role prior to 1910. But Beard was the more valued friend of the leadership. Neither man had any effective power, but both were colorful advertisements. The purposes of the BSA were spelled out in the preamble to a set of By-Laws, identified by Murray as the first constitution:

> The object of the Boy Scouts of America is to organize the boys of the District of Columbia and elsewhere in the United States, into units, and to teach them or cause them to be taught through duly designated leaders, discipline, patriotism, courage, habits of preservation and self-control and ability to care for themselves in all exigencies of life.[221]

Few of these objectives sat well with Seton, but Beard could be counted on to embrace them all wholeheartedly.

The new Executive Board assumed its powers at the end of October, 1910. Its strongly Republican composition, linked to its avowed purpose of making the organization efficient, would probably have been enough to guarantee bureaucratic centralization. But the appointment of James E. West to the position of Executive Secretary in January, 1911, eliminated any doubt. West was a thirty-four year old Washington lawyer with a long career of social work to his credit. An orphan from the age of six, he had always demonstrated a particular interest in the problems of young people. Like Edgar Robinson, the man whom he now replaced, he had spent many years in the employ of the YMCA. He was a close friend of Theodore Roosevelt and had served his administration in several capacities. Through the offices of Luther Gulick and Lee Hanmer, the Russell Sage Foundation was induced to provide the funds to support his full time appointment. The standardized program and structure of the BSA today bears testimony to his control

---

[221]Murray, History of the Boy Scouts of America, p. 41.

over the organization from 1911 to 1943.

Where Baden-Powell was concerned, West counted himself among the believers, although he was adamant about moulding a revised program and administration to uniquely American conditions. Although discipline was de-emphasized, formalism in the ranking of Scouts was augmented. Evidence of class distinctions was eliminated, while greater stress was placed upon morality and the civic virtues. The addition of three Laws touching on the values of bravery, cleanliness and reverence expanded Baden-Powell's originals but did not otherwise alter them. With regard to patriotism, West's ideas were probably stronger than those of the General. On the advice of bureaucrats in the YMCA, West directed the BSA administration away from their decentralized model. Henceforth national headquarters chartered the local councils primarily on the basis of their ability to fulfill the objectives of the BSA.[222]

In keeping with the desire for independence, the head office was moved out of the YMCA to new quarters on Fifth Avenue. At the first annual meeting in 1911 the National Council elected Colin Livingstone, President, Lee Hanmer, Secretary, and George D. Pratt, Treasurer. Although Robinson remained on both the Executive Board and the National Council, his leave of absence from the YMCA was over and he was

---

[222] Ibid., pp. 54-64; Macleod, "Good Boys Made Better," pp. 243-50, 253-55, 261-66.

compelled to return to official duties. With the addition of West, therefore, the organization reflected the absolute triumph of the Livingstone faction. Firmly entrenched at the top of the hierarchy was the financial wizardry to oil the administrative machinery engineered by West. Indeed, "the movement tended to become a businessman's organization locally as well as nationally." Although a paid employee of the BSA, West was the dominant power behind its operations and its functional, if not its titular, head.

> In dealings with the Executive Board he was the man in the middle and used that position to lobby against proposals by individual board members. His agenda and his report laid out the business to follow. Friction occasionally resulted, but West's policies usually won out; his colleagues respected his formidable dedication to the expansion of Scouting.

In the jargon of the progressive era, West was an efficiency expert of the highest order. And steadily increasing enrollment figures were deemed adequate evidence of his success.[223]

Once West had secured control, Seton was used, literally, as a figurehead whose name lent prestige and, for many, credibility, to the BSA package. While the movement was growing into an organization, prestige was an essential advertising tool. In addition, Seton's presence guaranteed the emasculation of potential competition

---

[223]Macleod, "Good Boys Made Better," p. 267.

by leaving him very little time to carry on the Woodcraft Indians. For his part, Seton enjoyed his new title but was aware that he had been sidestepped and effectively lobbied out. For a time he was confused about the role of the newly organized body. His instincts refused to let go of the woodcraft ideal, yet he recognized the need for a sounder structure if a large audience were to be reached As a result he vacillated back and forth between the virtues of one idea over the other and frequently achieved a synthesis of the two.[224] This became particularly apparent when Seton began preparing a revised manual early in 1911. He visited Baden-Powell in England to discuss the incorporation of more material on organization and then put together several of his old magazine articles, along with much new and more interesting detail, and submitted the lot for publication under the auspices of the BSA. The Editorial Board, from which Seton had also been excluded, was directed By Murray, Pratt and A. A. Jameson, all of whom felt that the new book "reflected too much the adult point of view." They went through the manuscript, dropping bits and pieces, maintaining other sections intact, and adding essays by different authors, creating a compendium of information which was the root for all future manuals until 1927.[225] Seton rescued all the extracted

---

[224]Seton to Arthur B. Reeve, Oct. 28, 1910; H. Chessman Kittredge to Seton, June 26, 1911, SP.

[225]Murray, As He Journeyed, p. 348; Robert S. S.

items, embellished them still more, and included them in his own Book of Woodcraft, published the following year against the wishes of the Editorial Board who argued that it was modelled after the Boy Scout Handbook. The irony did not escape Seton who replied immediately, reminding his colleagues that the Handbook had been modelled on his work.[226]

In other words, Seton was down but not out. He was not prepared to allow the impression that the Woodcraft Indians were dead and buried and took pains to keep past and present members of the BSA Executive Board informed of their activities, even going so far as to invite some of them to policy discussions. Most of his bitterness was indirectly aimed at West and Livingstone whom he called Philistines for having removed from Scouting "spirituality, the power of ceremony, the charm of romance and the importance of the beautiful." Seton's relationship with Dan Beard, in most matters of principle a partisan of the

---

Baden-Powell to Seton, Jan. 31, 1911, Feb. 3, 1911, SP; Boy Scouts of America, Boy Scouts of America: The Official Handbook for Boys (Garden City, N.Y.: Doubleday, Page, 1911).

[226] W. D. Murray, George D. Pratt and A. A. Jameson to Seton, Aug. 15, 1912; Seton to the Editorial Board, Aug. 25, 1912; W. D. Murray to Seton, Dec. 21, 1912, SP. In 1911 Seton also published (what he called) a sequel to Two Little Savages, in which he inveighed against the horrors of war, while at the same time praising the moral superiority of the North American Indian. This he dedicated to the Boy Scouts of America. See Ernest Thompson Seton, Rolf in the Woods (New York: Doubleday, Page, 1911).

Executive Board, had completely disintegrated by the end of his tenure as Chief Scout. On the other side of the Atlantic West's ascendancy was also viewed with alarm, especially by those who distrusted the centralized bureaucracy he had created. Aware that these feelings existed, West was likely worried by Seton's frequent trips to England where he was known to meet Baden-Powell and with whom he would obviously discuss the situation in America. Again the ironies abounded.[227]

The working atmosphere was severely tainted by Seton's intransigence on all matters of principle. The militarist issue in particular was flogged, though not without considerable justification. As an official spokesman for the BSA, he was constantly on tour throughout the country lecturing to audiences of all ages. "All the opposition we have had," he wrote, "is based on these totally unnecessary military methods." He maintained that in many cases the first in line to become scoutmasters were aspiring drill sergeants whose "entire notion of Scout

---

[227]Seton to John L. Alexander, Jan. 13, 1913; John L. Alexander to Seton, Oct. 27, 1914; Lee Hanmer to Seton, Dec. 5, 1913, Jan. 8, 1914; Philip D. Fagans to Seton, Dec. 9, 1913, Jan. 3, 1914; E. C. Bishop to Seton, Dec. 30, 1913; Seton to Dan Beard, Nov. 26, 1915; Dan Beard to Seton, Nov. 30, 1915; Robert S. S. Baden-Powell to Seton, March 3, 1913, March 10, 1913, March 12, 1913, April 17, 1913, April 30, 1913, May 9, 1913, SP; Macleod, "Good Boys Made Better," p. 267.

activity is military evolution." And these charges could be supported with evidence.[228] Nevertheless, Seton's behavior was clearly verging on the paranoid. He was deeply hurt at the way he had been treated. He had given his time, his energy, his ideas, and large amounts of his money to build an organization which would meet the needs of boys as boys perceived them. He felt that those needs had been misunderstood and that large enrollment figures disguised the fact that parents were <u>sending</u> their sons to the local Troop, some for disciplinary reasons, others because they liked the utilitarian emphasis. While woodcraft was still a part of the scheme, it was no longer the most important part. West, Livingstone, Pratt and others on the Board were so preoccupied with administration that Seton wondered whether they really cared at all about the substance of program activities of whether they considered themselves custodians of a mould that stamped uniformity on American boys. Under the pomposity and the petulance and the genuine anger, Seton was an extremely sad and disheartened man.

---

[228] Seton to the Editorial Board of the BSA, Aug. 25, 1912; Seton to the President and Executive Board of the BSA, Nov. 21, 1912, SP.

VIII

The end of Seton's association with the Boy Scouts of America came about through a set of circumstances which focussed attention on what had become for him the two most important issues: the necessity of woodcraft and the dangers of militarism. By sheer quirk of fate the controversy converged with the outbreak of the First World War.

In July, 1911, Seton engaged to publish for a popular magazine, The American Boy, a series of ten articles on "Scouting." In keeping with his own particular interpretation of that word, however, the articles were strongly reminiscent of those he had produced for Edward Bok. There were obvious sacrifices to orthodoxy under such headings as "How to Begin as a Scout," "Courtesy and Scouting in the Street," "What is a Gentleman?," and so on, but by far the greater part of the series was given over to what the Indian had to teach. The text is suffused again with detailed diagrams of totems and pictographs, hand signs for use in the sign language. With these are included instructions on how to make a Plains Indian willow bed or on how to decipher animal trails or on the proper way of conducting a camp. There is a section on Indian dances complete with music, another on how to light a fire with the bow drill. New touches are added, and old ones retained, but most descriptive articles are accompanied by a short story to sustain interest.[229]

---

[229]Ernest Thompson Seton, "Scouting: Number 1,"

By the time Seton had completed this series in 1912, BSA headquarters had purchased a magazine of its own called Boys' Life. Apparently no one had been unduly alarmed by Seton's emphasis on woodcraft in his other articles (probably because no one in authority had bothered to read them) for his pen was enlisted by the new periodical almost immediately. His first article was entitled "Smoke Signals, Sign Talk and Totems" and appeared in the December, 1912 number. It was continued in February, 1913[230] but was divided by his New Year's message in January. As Seton always regarded the primary value of the sign language to be its universal application and contribution to inter-tribal understanding, it was appropriate to find sandwiched between the two articles describing it a greeting which centered upon the approaching one hundredth anniversay of the War of 1812. The intervening years of peace between

---

American Boy, 12 (July, 1911), 5, 21; "Number 2," AB, 12 (Aug., 1911), 4-5; "Number 3," AB, 12 (Sept.,1911), 3-4; "Number 4," AB, 12 (Oct., 1911), 3-4; "Number 5," AB, 12 (Nov., 1911), 6, 29; "Number 6," AB, 12 (Dec., 1911), 3, 31; "Number 7," AB, 13 (Jan., 1912), 3, 31; "Number 8," AB, 13 (Feb., 1912), 3, 26; "Number 9," AB, 13 (March, 1912), 3, 27; "Number 10," AB, 13 (April, 1912), 7, 27. These articles had originally been requested by Theodore Dreiser for publication in The Delineator; Theodore Dreiser to Seton, Sept. 12, 1910; Seton to Theodore Dreiser, Sept. 27, 1910; H. W. Lanier to Seton, Oct. 12, 1909.

[230] Murray, As He Journeyed, p. 349; Macleod, "Good Boys Made Better," p. 252. Ernest Thompson Seton, "Smoke Signals, Sign Talk and Totems," Boy's Life, 3 (Dec., 1912), 24; "The Indian Sign Language for Boy Scouts," BL, 4 (Feb., 1913), 15.

Canada and the United States, marked by "three thousand miles of common frontier without a gun or a sentry on guard," was a clear reflection of the "spirit of brotherhood [which] has hovered over the councils of the two nations." The Boy Scout movement, as he viewed it, should have as its major object this same desire to foster the brotherhood of man. "Other men are striving to impress the thought on grown-ups. We seek the more plastic material and work with the youth, the boys who will be voters a few years from now." These thoughts were followed in the same year by the story of "The Badger Who Was Brother to a Boy," bringing attention to the links between man and animal.[231] In other words, Seton made a concerted effort, through his writing, to keep alive the very things he felt were being neglected by the movement as a whole. He must have been fairly successful at winning an audience for the fledgling magazine. By October, 1913, he had been awarded a monthly column, which he titled "Around the Camp Fire." Members of the editorial board may have felt guilty for the way they had insulted him in the matter of the Book of Woodcraft, seeing this as a means of making amends. And Frank Presbrey, a member of the board and a prominent

---

[231] Ernest Thompson Seton, "Peace and Patrol Names," Boy's Life, 3 (Jan., 1913), 21; "The Badger Who Was Brother to a Boy: A Record of a Lad's Remarkable Experience with a Kindly Wild Animal on the Canadian Plains," BL, 3 (Nov., 1913), 12-13, 34.

New York advertising executive, could not have missed the opportunity to use Seton's name as a gimmick to stimulate circulation.

Seton exploited his brief columns wisely and well, acquainting Boy Scouts with woodcraft ideals at every turn, stressing Indian ways and nature study in particular. In one issue he recounted a recent visit to Camp Dudley where he introduced boys to the building of a sweat lodge. Never content with simple "how to" explanations, Seton always gave these objects a context which expanded the reader's awareness of their significance. After suggesting that every Scout camp should erect a similar structure, he went on to point out that several tribes made regular use of them. "With most it was religion as well as hygenic performance, and was accompanied by special prayers, as well as the taking of purgatives. And yet some of the worthless Indian agents our Government allows to tyrannize over the Indians have been allowed to suppress the sweat lodge." The June, 1914, number details the construction of a Council Ring, "a fixed, prepared and consecrated place . . . with all the embellishments necessary for a true Council." The necessary embellishments included a decorative hand-made chair for the Chief, a separate place for the Tally Keeper and a large totem. The council fire was to be ringed by seats constructed by individual patrols from natural materials, and so on. No one could afford to be "blind to the foundations of our civilization and religion, all of which found in the camp-fire the focus at the beginning." And the Wolf was back as well. In asking all patrols to choose a totem, preferably an indigenous animal from their immediate region, Seton was quick to point out that the Gray Wolf was "the peerless Scout" and represented perfection. This meant that "any patrols adopting the name 'Gray Wolf' are committing themselves to the highest standard of exploit, are indeed announcing themselves to be in the very first rank in all things."[232]

But the regularity of the Chief Scout's articles on subjects not noticeably scoutlike had been allowed to pass for reasons which would not have been obvious to most people close to the organization. Early in 1914 Seton registered a complaint with headquarters about the revision of the 1911 manual, a large part of which had been drawn from his own writing. The revised edition, which was to be issued in 1912 by a new publisher, had been altered by the editorial board in such a way that material appearing under Seton's name did not represent his ideas. On principle the

---

[232] Ernest Thompson Seton, "Around the Camp Fire: 1," Boy's Life, 3 (Oct., 1913), 13; "2," BL, 3 (Dec., 1913), 28; "3," BL, 4 (Jan., 1914), 20; "4," BL, 4 (March, 1914), 20; "5," BL, 4 (April, 1914), 23; "6," BL, 4 (May, 1914), 24; "7," BL, 4 (June, 1914), 26; "8," BL, 4 (Sept., 1914), 26-27; "9," BL, 4 (Oct., 1914), 16, 25; "10," BL, 4 (Dec., 1914), 24; "A Greeting from the Chief Scout," BL, 4 (Jan., 1914), 2.

editorial board did have a responsibility, at the very least, to ask his permission to make changes in the text. But there was more to the story. When Seton had initially submitted his manuscript draft for the 1911 edition, it was on the understanding that he would be chairman of the editorial board itself. But he had not even been invited to attend its meetings, let alone guide its policies, for fear that the Indian and nature study themes would continue to dominate official publications of the organization. James West, who lived in mortal fear of any public airing of dirty laundry, attempted to employ diplomacy in an effort to disarm his only serious antagonist. Accordingly, in May, 1914, the editorial board extended the hand of friendship by appointing Seton Associate Editor of _Boy's Life_. But even this could not be done gracefully. Seton was expected to share the position with Dan Beard.[233]

Frank Doubleday, who had produced the 1911 _Handbook_, was extremely disturbed at losing the substantial profits which he knew would accrue to Grosset and Dunlap, proposed publishers of the revised version. It soon became clear that it was Doubleday's personal friendship with Seton which had much to do with the editorial board's decision to change

[233] W. D. Murray, _As He Journeyed_, p. 348; Macleod, "Good Boys Made Better," pp. 268-69; Seton to the Executive Council [sic], Boy Scouts of America, June 21, 1914, SP; Frank Presbrey, "The Associate Editors of Boy's Life," _Boy's Life_, 4 (May, 1914), 6.

publishers. Consequently, after negotiating with representatives of the Boy Scouts, Doubleday could write to Seton on August 19, 1914, that George Pratt thought Seton's resignation "would straighten out many complications. I think it would be unfortunate for everybody if it got around that there was a row; it would do neither you nor the Boy Scouts any good." He then frankly advised Seton to send his resignation to Colin Livingstone along with a note to the effect that "as some of your material which has been used in the Manual has been changed so that it does not represent your ideas, you would like to have your name and your material omitted from the next edition." (It is worth noting here that while Doubleday lost the 1914 edition, he managed to regain the Handbook rights once Seton had left the BSA.)[234]

By this time Seton had already begun to entertain thoughts of resigning anyway. But he was not prepared to go quietly as there were matters of principle at stake. James West, of course, realized that his labored diplomatic efforts had failed to placate the Chief Scout, but he was still eager to avoid a public showdown. It would have been decidedly impolitic to allow the impression that the BSA had been unjust to Seton who had clearly been a major force

---

[234] F. N. Doubleday to Seton, Aug. 19, 1914; Seton to James E. West, July 18, 1914, SP; Boy Scouts of America, The Official Handbook for Boys. Rev. ed. (New York: Grosset and Dunlap, 1914).

in getting the movement off the ground and in winning and holding, so many boys. On the other hand, West did not want it to appear that Seton had a legitimate grievance. Above all, he did not want the woodcraft issue to enter the debate as he was clearly trying to steer the organization away from it and onto another course. The only pretext left with which to exorcise Seton from BSA officialdom was the question of his citizenship. Seton had never become an American.

By September 1914 Seton was threatening to resign unless he were given a larger voice in policy decisions. West refused to admit either possibility while he canvassed opinion on the citizenship question. There was no doubt in his mind but that Seton had to go. The problem lay in making his departure as quiet and painless as possible, allowing it to reflect nothing sinister on the hierarchy. One of those whose opinion he solicited was John L. Alexander, the original Managing Secretary of the BSA. Alexander also took the trouble to get Seton's view.

> [Seton] tells me . . . that your Committee will not accept his resignation and are very much disappointed that he even offers it. While on the other hand, you will not allow him to take any definite part in the moulding of the policies of the organization. He even says that you do not wish anything else but the use of his name in an official, ornamental way. I think this pretty nearly sums up the way he put the matter to me, and I have an idea, judging from my past contact with the inner councils of the organization, that this pretty nearly sums up the whole matter from your end as well. . . . I readily agreed . . . that the Boy Scouts of America should

> not expect him to hold the Chief Scoutship against his wishes and furthermore . . . advised him to insist on his resignation and to go his own way and develop his own organization.

Alexander described West's citizenship ploy as "flimsy," arguing that if he were in Seton's place "and anyone was to club me into being an American citizen by threatening to withdraw from me an honorary or honorable position, I should not hesitate to let the position go."[235] West, of course, knew that in his heart of hearts Seton did not want to become an American citizen. Otherwise he would not have felt so safe in using it as a vehicle for effecting his ouster. Once Seton had discerned West's petty intrigue he decided to pursue the matter to its conclusion. Aboard the Lusitania on his way to England late in January, 1915, Seton wrote to his wife that before debarking he had communicated to the Executive Board "my intention to take out my papers as soon as the European situation cleared."[236] This seemed effectively to remove the last egg from the executive basket.

The Chief Scout was bound by the constitution to stand for annual re-election. But as Seton was conveniently absent when the annual meeting was called on February 11,

---

[235] John L. Alexander to James E. West, Sept. 22, 1914, Oct. 27, 1914, copies in SP; John L. Alexander to Seton, Oct. 27, 1914, Dec. 7, 1914, SP.

[236] Seton to Grace Gallatin Seton, Jan. 31, 1915, Grace Gallatin Seton Papers, Old Greenwich, Conn.; Seton to E. S. Martin, March 6, 1934, SP

1915, and as no one else was nominated for the position, the office was merely left vacant.[237] When Seton returned in the spring he had discussions with members of the Board including Livingstone, Pratt, West and Schiff by whom he was told, presumably, that he was ineligible to hold any office in the BSA unless he was then, at that very moment, an American citizen. In May, Seton sent a formal letter of resignation to the Board, refusing to acknowledge that "the question of my British birth has [any] bearing on the case whatever."[238] Boy's Life, of which he himself had been made Associate Editor only two years earlier, reported vindictively that Seton had been "dropped" by the National Council at its February meeting and "did not resign from the position of Chief Scout as he alleged." The anonymous author of this notice reassured the readership that

> The Scout Movement is in no wise affected by the dropping of Mr. Seton. The program, followed since 1910, is not dependent upon any individual. Mr. Seton was not a member of the Editorial Board which compiled the original handbook. . . . The comparatively small number of pages of material written by Mr. Seton are interesting, but nothing essential to the program of Scouting. They can be easily replaced in future editions by eminent American citizens.[239]

---

[237] James E. West to Seton, Aug. 27, 1937; "Ernest Thompson Seton's Statement Answered by Executive Board of the National Council," New York Times, Dec. 11, 1915.

[238] Seton to Charles C. Jackson, May 12, 1915; Seton to James E. West, Aug. 14, 1937; James E. West to Seton, Aug. 27, 1937, SP.

[239] "Why Mr. Seton is Not Chief Scout," Boy's Life, 6 (Jan., 1916), 28.

This calculated attack on Seton, published in January, 1916, was for the information of young boys. It was clearly slanted to create the impression that Seton had done nothing of value for the organization, thereby attempting to eliminate earlier fears that the Executive Board would be accused of abusing the senior partner of scouting. But what was it that created the necessity to say so? And why did the magazine wait until January, 1916, to take its stand?

On November 30, 1915, Theodore Roosevelt wrote to James E. West noting that General Leonard Wood had informed him that the BSA had been harboring in its ranks "certain leaders . . . [who] have used the Boy Scouts organization as a medium for the dissemination of pacifist literature and . . . as a propaganda for interfering with the training of our boys to a standard of military efficiency." Threatening to withdraw his support from the movement if this intolerable situation were not soon remedied he argued that a "Boy Scout who is not trained actively and affirmatively that it is his duty to bear arms for the country in time of need is at least negatively trained to be a sissy."[240] Roosevelt's rebuke was probably sufficient to make several of the Republican Board members squirm. And Seton himself added ammunition when, in an interview for the New York

---

[240]Theodore Roosevelt to James E. West, Nov. 30, 1915, in Theodore Roosevelt, The Letters of Theodore Roosevelt, ed. Elting Morison (8 vols., Cambridge, Mass.: Harvard University Press, 1951), 8:992-93.

Times, he pointed out that he had resigned from the BSA because of personal disagreements with West whom he styled as "a man of great executive ability but without knowledge of the activities of boys, who has no point of contact with boys, and who, I might almost say, has never seen the blue sky in his life."[241] On the following day, West responded with the charge that "Mr. Seton was in harmony with the views of anarchists and radical socialists on the question as to whether the Boy Scouts of America should stand for patriotism and good citizenship." He had, claimed West, persistently objected to the chapter on patriotism in the Handbook, opposing in particular the pledge of allegiance and arguing that boys should be left "free to support our country when they thought our country was right and to damn it when they thought it was wrong." As Seton himself had demonstrated a predilection for damning, it had been felt necessary to expel him. "This is the real and only reason that Mr. Seton is not now the Chief Scout of the Boy Scouts of America."[242] Seton was rather proud of being the object of this attack and made it abundantly clear that he was not in sympathy with the organization or its present trends and vowed publicly to give the rest of his life to the

---

[241]"Seton Still Insists on Quitting Scouts," New York Times, Dec. 6, 1915.

[242]"West Says Seton is Not a Patriot," New York Times, Dec. 7, 1915.

Woodcraft Indians. Which he did.[243] On June 15, 1916, the Boy Scouts of America were granted a federal charter by Congress. This charter limits membership to American citizens.

---

[243]Ibid., Seton to Dan Beard, Dec. 22, 1938, SP.

Chapter 5

CONCLUSION

Because I have known the torment of thirst I would dig a well where others may drink.

Ernest Thompson Seton

In Mutual Aid, Peter Kropotkin maintained that human ethics should abide by the complex laws of nature. He also suggested that those laws were themselves ethical in substance because of their very naturalness. The concept of mutual aid, a repudiation of the ideas of Herbert Spencer, was carefully deduced from personal observations in the field by an astute naturalist who did not have access to a huge corpus of integrated ethological and ecological theory. The same options that were open to Kropotkin were open to others like Seton, who possessed similar tools for the investigation. Indeed, it is safe to argue with William Morton Wheeler that the inquiries and subsequent hypotheses of those naturalists who followed Kropotkin did much to assist and to influence the research of biologists. A classic example is afforded by Aldo Leopold, one of the greatest ecologists of our time, who devised what he called a "land ethic." This he defined as

> a mode of guidance for meeting ecological situations so new or intricate, or involving such deferred reactions that the path of social expediency is not discernible to the average individual. Animal instincts are modes of guidance for the individual in meeting such situations. Ethics are possibly a kind of community instinct in-the-making.[1]

Leopold's idea of community encompassed all living things. Man was simply one of many species of citizens. He condemned

---

[1]Leopold, Sand County Almanac, p. 239. For a close look at Leopold, see Susan L. Flader, "Thinking Like a Mountain: A Biographical Study of Aldo Leopold," Forest History, 17 (April, 1973), 14-28.

the conservation movement for its anthropocentric goals and for its assumption "that the economic parts of the biotic clock will function without the uneconomic parts."[2] The acceptance of a universal land ethic required a re-interpretation of history in ecological terms, as well as the development of an "ecological conscience" wherein "a thing is right when it tends to preserve the integrity, stability and beauty of the biotic community," and "wrong when it tends otherwise."[3] Like Kropotkin, Leopold believed that man could learn most from the animals, through diligent observation and measured deduction. For this and for many other reasons, wilderness was essential:

> Ability to see the cultural value of wilderness boils down . . . to a question of intellectual humility. The shallow-minded modern who has lost his rootage in the land assumes that he has already discovered what is important; it is such who prate of empires, political or economic, that will last a thousand years. It is only the scholar who appreciates that all history consists of successive excursions from a single starting point, to which man returns again and again to organize yet another durable search for a durable scale of values. It is only the scholar who understands why the raw wilderness gives definition and meaning to the human enterprise.[4]

The real ethical question, then, was not whether nature is moral on man's terms, but whether man was moral on nature's

---

[2]Leopold, Sand County Almanac, p. 251.

[3]Ibid., p. 262.

[4]Ibid., p. 279. For extensions of this theme, see Frank Fraser Darling, Wilderness and Plenty (New York: Ballantine Books, 1971); Paul Shepard, Man in the Landscape: A Historic View of the Aesthetics of Nature (New York: Ballantine Books, 1972).

terms. Leopold and Kropotkin did not propose that we sit in judgment on the promiscuity of Monkeys or applaud the virtues of the monogamous Wolf. They did, however, seek to investigate those aspects of animal behavior which had counterparts in abstract human morality. In this respect especially they have had countless disciples, many of them biologists. Konrad Lorenz, for example, was fascinated by the multitude of ways in which the instincts of animals operated to prevent useless killing within species. In this light human athletic activities assumed for him a new significance as ritualized belligerence, fulfilling all the demands of aggressive drives but eliminating the ominous side-effects of war.[5] Ethologists continue to recognize in animals an instinctive morality which evolves constantly to promote the survival of the species; they perceive in man an abstract morality, born of reason, which seems to conspire against it.[6] Yet they acknowledge that man still possesses, and frequently expresses, his own instinctive behavior. Thus W. H. Thorpe postulates that some of man's artistic activities, "notably his dances and displays involving head dresses and other ornaments, may have been stimulated by watching the ceremonial displays of birds." This raises the question as to why behavior patterns produced in evolution as signals for sexual activity

---

[5]Lorenz, On Aggression, p. 241.

[6]Thorpe, Biology and the Nature of Man, p. 91.

should strike us as beautiful. "Does it," asks Thorpe, "imply some fundamental unity between the mind and perceptual systems of groups as far apart as the Aves and mankind?"[7] Most ethologists are compelled by their research to respond in the affirmative.

If animals can survive happily on the basis of instinctive morality, and if the environment which sustains them is not altered negatively by that morality but rather flourishes because of it, is it possible that man's ability to reason--which not only distinguishes him from the animals but also creates the abstract morality to generate ideology, to generate war--is responsible for repressing instinctive morality, pejoratively labelled "the law of the jungle?" Has the reasoned act of repression severed forever the connection between instinct and morality? Or is there hope that the two may be reconciled? For Kropotkin and Leopold it was imperative for man to employ his reason in the service, not only of possibility, but also of limitation and restraint. Far from denying change, nature's laws worked through evolution to produce it. It fell to man to keep his own possibilities within the boundaries allowed by ecology. Science, evidence of reason, was the understanding of nature's laws. Technology, also evidence of reason, was a modified application of nature's laws. But if modified applications spawned unmeasureable alterations of those laws, they ceased to be

---

[7] Thorpe, Science, Man and Morals, p. 92. For an anarchist's extension of this theme, see Herbert Read, The Origins of Form in Art (London: Thames and Hudson, 1965).

reasonable. The problem of abstract morality was inextricably linked with the problem of technology and its products. Indeed, abstract morality was an expedient accommodation of the homogenizing machine and a denial of nature's diversity--a glorification of rational, efficient use for the user, a rejection of the living beings used.

Ernest Thompson Seton's ideas about man's place in nature paralleled those of Kropotkin and anticipated those of Leopold. Born in England, raised in Canada, trained in France and resident in the United States, Seton was truly a cosmopolitan figure. He was also a kind of renaissance man. With his variety of vocations and multiplicity of pseudonymns, he was a living monument to the diversity he worshipped. Although fate deposited him squarely in the middle of the Progressive Era, he never embraced its technocratic, liberal faith. In his art, in his animal stories and in his early scientific writing, there were two major factors at work, pressing him toward an unpremeditated conclusion. On the one hand there was a concerted effort to escape from the pious Victorian Calvinism and unrelenting persecution of his father. This was attended by a desire to recreate a childhood filled with play and fantasy, never sanctioned by his family and only glimpsed by him in fleeting moments of rebellious exuberance. With each real and imagined persecution, this rebelliousness became gradually more pronounced. Associated with his profound love for the wilderness and its inhabitants, it led Seton to a kind of pantheism, an intensely personal religious feeling that accounts, at least in part, for his ecological intuition and for his avowed desire to force his way inside the skins of the animals, to look out upon the world of men through their eyes. On the other hand there was Seton the naturalist, eager to contribute significantly to the science to which this pantheism inclined him. In most aspects of his life his pomposity and arrogance mounted in direct proportion to his prestige. But in the scientific realm, where he was aided by spiritual cleansings at the hands of Robert Ridgway and John Burroughs, Seton learned an uncharacteristic humility.

Seton's political perspective, such as it was, grew naturally out of his biological insights. Despite a continuing fascination for Britain, and particularly for titled nobility, his heart was attached to the wilderness of North America. Although most of his original work prior to 1915 was based upon his Canadian experience, he always spoke in continental terms--not because he was a Goldwin Smith liberal, but because Merriam's Life Zone Theory acknowledged no man-made boundaries. From his own detailed observations in the field, Seton concluded that instinct and learning in animals deserved the careful attention of scientists, philosophers and politicians. While adapting G. Stanley Hall's theory of recapitulation to his youth work, he was also blending the thought of Ernst Haeckel, Lewis Henry Morgan and William Morton Wheeler in a practical attempt to demonstrate the essential unity of man and nature. Committed to the principle that competition was evil, he maintained that reason deprived of instinct militated against the survival of the species. He viewed the Indian as the first American, an exemplary socialist whose greatest virtues emerged spontaneously from a spiritual union with the land. Blasphemed, persecuted and all but exterminated by Turner's avaricious pioneer, the Indian, like the Wolf, symbolized nobility, strength and intelligence. The Indian also knew the celebration of diversity to be the true source of community, the sine qua non of peace among men and the guarantor of harmony with nature. Seton felt comfortably virtuous in his alliance with the virgin against the dynamo. Unwittingly he transferred his Presbyterian piety to his pantheism, alienating many of his adult friends and contemporaries in the process. Nevertheless, through his charismatic presence and devotion to principle he was still able to win the loyal allegiance of thousands of young people. He regarded himself as the radical antithesis of an age where technology, viewed by many as a panacea, was quickly reducing humanity to the slavery of Taylorism. Although he participated actively in many of the events with which historians have come to equate the Progressive Era, his hope was to spark a reversal of many of their trends. In the affairs of men he abhorred bureaucracy just as in science he feared generalization. Anything that was not particularist defied the laws of nature.

There is a certain poignant irony in all of this. Seton was never able to shed his Victorian cocoon. Growing up in Lindsay and Toronto he was exposed to all the excesses of parochial prudery and Imperialist rhetoric endemic to the new Canadian nation. Within this garrison culture he studied art under British teachers whose painful academic style remained riveted to a subject matter uniquely his own. At Charlotte Schreiber's bidding he joined the Canadian exodus to France, studying under the "best" painters at the "best" atelier. Under the influence of Bouguereau and Gérôme he sank to the depths of the genre tradition with his painting, Awaited in Vain. Because another piece of his work had been accepted by the Salon (then acknowledged in Europe to be entering an unprecedented period of decline) he assumed that his grotesque bit of "realism" demanded the attention of his compatriots. He was encouraged in this view, not by Canadian artists, but by friendly Canadian politicians, who forced acceptance of the work at the Columbian Exposition on the basis of the painter's European record. Seton never hesitated to exploit the connections he had taken pains to cultivate wherever he went. His personal old boys' network extended deeply into every sphere where he had demonstrated a serious interest. The artistic career furthered by the intervention of political figures and civil servants first encountered in Manitoba quickly became a vehicle for the establishment of new contacts in the world of science.

Spencer F. Baird, J. A. Allen, Elliott Coues, Robert Ridgway, C. Hart Merriam, William Hornaday, Frank M. Chapman, David Starr Jordan--the leading naturalists in America--all were called upon to drop a kind word here or to feather a nest there. Beginning as an illustrator of their monographs, he was well placed to absorb, to weigh and ultimately to synthesize their science. With their assistance he was introduced to the élite of New York and Washington society and enabled to join some of the most exclusive clubs. Among his closest friends were his own publishers, Charles Scribner, Frank Doubleday and Edward Bok, who went beyond the traditional demands of their profession in promoting his books and in forwarding generous advances. At his huge estate in Connecticut, Seton was able to project two separate images: one embodied in the rugged, if somewhat contrived, wilderness surrounding it, the other symbolized by the grandiose mansion itself. Here, Theodore Roosevelt, Irving Bacheller, Hamlin Garland, Charles Lummis, Mark Twain, Charles, Will and Mary Fanton Roberts, Lincoln Steffens and countless other illustrious guests could be seen coming and going at their leisure. Seton calculated his personal associations carefully, not always because he liked the people, but because he aspired to push himself forward into the limelight of universal respectability. While there was undeniably a Machiavellian element in these methods, Seton never lost sight of the ends they serviced.

With tangible triumphs like Salon hangings and best-selling books to his credit, Seton's ego swelled and flourished. However, while he himself felt that recognition by the mighty was essential to these successes, the implications of his most original hypotheses remained antagonistic, not only to his Victorian heritage, but also to the values of his contemporaries. Beginning with the anti-anthropocentric subject matter of his paintings and continuing through the sympathetic biographies of his animals, the themes of Seton's work reverberate more emphatically than the artistry which contains them. Although he never used the words "ecology" or "ethology," he had obviously reached a primitive understanding of their meaning. Through his youth work he had also begun to extend them into the social sphere, expanding Kropotkin's theory of mutual aid and anticipating Leopold's ecological conscience. Enduring the rebukes of those, like John Burroughs, who refused to accept his speculations on behavior, he rewrote his animal stories in a scientific context, embellishing his interpretations with detailed evidence of reading and personal observation. Employing the accepted (now obsolete) technical perspective of C. Hart Merriam, he converted The Life-Histories of Northern Animals into a definitive descriptive work, re-establishing his credibility with scientists and challenging the competence of those who had attacked him during the Nature Faker debacle. A revised and enlarged version of the Life-Histories, published in four volumes between 1925 and 1928 under the title, Lives of Game Animals,[8] earned for its author not only the coveted John Burroughs Award, but also the Daniel Giraud Elliot Gold Medal of the National Academy of Sciences. Exalting in this nemesis, Seton himself failed to notice that the accolades were directed, not at his original behavioral analysis, but at his comprehensive synthesis and description.

Ernest Thompson Seton was a pivotal figure in the founding of the Boy Scout movement. The Woodcraft Indians, which he originated in Connecticut in 1902 and which he introduced in England on his lecture tours, was observed, revised and later packaged by Robert S. S. Baden-Powell. In the original scheme, Seton combined the results of his own biological investigations with the child psychology of G. Stanley Hall, creating a program which focussed attention on innate and learned behavior. Convinced that progressive technological achievements were too readily interpreted as the victory of man over nature, he sought to reacquaint children with their place in evolution. Deploring the conventional wisdom that man's superiority was measured by the reasoned emasculation of "primitive" instincts, Seton believed that the improvement of the human condition was wholly dependent upon man's ability to follow examples set

---

[8]Ernest Thompson Seton, Lives of Game Animals (4 vols.; New York: Doubleday, Page, 1925-28).

by nature in the animal kingdom. In choosing the Indian as a model he was not advocating a return to the wilderness condition or an abandonment of civilization. He was, however, attempting to demonstrate the value of those human cultures which acknowledged their limitations within nature. He was terrified by the prospect that these cultures might be annihilated in the name of progress. The Indian was unique in North America for his ability to see himself as part of a biotic community. If the white man could emulate this simple principle, Seton believed, he might slow the social and ecological damage wrought by modern industry, accept the diversity of all living beings and eliminate war.

In Seton's view the educational process went far beyond the development of human reason. It also required the acceptance of animal instinct. Instincts, like reason, continued to evolve. It was fruitless to regard the object of education as the repression of instinct. The Woodcraft Indians constituted Seton's effort to balance the formal school curriculum. Within his organization he aspired to amalgamate all his primary interests, conveying, in an artistic way, a scientific comprehension of man's place in nature. For him, education demanded an understanding of the human need for a true community, of evolution as the major source of change, of the persistence of instinct despite reason, and of the superficiality of technological progress. He tended to view the school as a socializing agency committed to a number of biologically untenable propositions. The Woodcraft Indians were supposed to become a body of critical youth, challenging accepted values and suggesting constructive alternatives. With these thoughts in mind, Seton emphasized the importance of individual craftsmanship, openly defended pacifism and fostered the spirit and letter of socialist democracy by allowing boys to govern themselves in small groups. He discouraged centralism in the organization and helped to cultivate a love of ritual and diversity by introducing young people to the laws, beliefs, dances, songs and customs of a multitude of North American Indian tribes. He refused to sanction a single mode of dress for his boys, trusting rather in their ingenuity to create, from their own imaginations, distinctive, individual costumes. Although he believed that the development of skills was essential to individual growth, he did not regard competition as a suitable vehicle for that purpose. Not surprisingly, the study of natural history was central to the entire program.

Baden-Powell, the last of the mighty with whom Seton chose to rub shoulders, duplicated many of his ideas, games and local forms of organization. He also revised the basic principles of the Woodcraft Indians, placing the boys under adult leadership, dressing them in paramilitary uniforms and extracting from them oaths of allegiance to God, to country and to flag. For Seton's Indian rituals were substituted others of African origin as well as modified forms of drill. Where Seton's scheme was devised expressly for young people, Baden-Powell's was designed to appeal also to parents. Where Seton's required children to see themselves as citizens of a biotic community, Baden-Powell's stressed socialization within and acceptance of the existing class structure. The latter also reinforced prevailing notions of the environment as object. On the other hand, parallels for much of what is considered "good" and "constructive" in scouting can be found in the early "Laws of the Seton Indians." Coups are replaced by badges. Tribes are replaced by troops. "Nation" as conceived by Indians is replaced by "State" as conceived by Englishmen. Baden-Powell copied Seton's entire method of presentation, publishing his scheme in magazine installments, later compiling them in book form.

When it was clear to Seton that Baden-Powell's Boy Scout plan would displace his own body at home--largely through its superior organization and obvious ability to attract funding from those whose ideological perspective it buttressed--he tried to win the leadership for himself. Aided by members of the YMCA who shared his own views, Seton attempted to weld the principles of the Woodcraft Indians onto the Boy Scouts of America in an effort to rescue the fledgling organization from its militarist leanings. Eventually, however, he lost what small measure of control he had gained to a faction of Republican businessmen and politicians who joined Daniel Beard in identifying the pioneer as the first

American and proper mythic figurehead of the body. The acknowledged hero of thousands of American boys, Seton was retained as Chief Scout of the BSA for five years. But once he had outlived his usefulness he was eased out of the organization on the pretext that he was not an American citizen. In fact, however, the real reason for his expulsion had more to do with his pacifism than with his citizenship. As the clouds of war drifted over the European continent and Britain, patriotism assumed greater proportions as the root of all virtue in America. Under the leadership of James E. West, the BSA adopted and expanded Baden-Powell's scheme. In keeping with the traditions of Progressivism, the organization was centralized, given a nationalist slant and primed for the impending conflict. The role of spiritual guide devolved upon Daniel Beard, while Theodore Roosevelt, a man whose admiration Seton had coveted since the 1880's, railed from the sidelines against anarchists and pacifists who would prevent the BSA from fulfilling its true mission.

By 1915 most of the mighty social elite had, like Roosevelt, abandoned Seton to his eccentricities. Hurt by the loss of so many friends, but wholly unrepentant, he reconstituted the Woodcraft Indians, this time admitting girls as well as boys, and embarked on a fruitless quest to displace the BSA. In an age where technology, homogenization and efficiency were equated with integrity, there was little hope that Seton would be considered a social asset, let alone a realist. The prerogatives of world war demanded reason, not instinct.

Bibliography

This bibliography separates all of Ernest Thompson Seton's published and unpublished work from the broad range of material investigated. However, with the exception of a section identifying the major secondary studies on Seton, the remainder follows closely the chapters in the dissertation itself. In rare cases where the same material was used in two or more chapters, I have placed the reference under the most obvious heading. In other words, none of the items appears more than once.

(1) Unpublished Material

Gagen, Robert F. Ontario Art Chronicle, Unpublished MS, Art Gallery of Ontario Library, Toronto, Ontario.

Hornaday, William T. Papers, Library of Congress, Manuscript Division, Washington D.C.

Macoun, John and Macoun, James. Papers, Public Archives of Canada, Ottawa, Ontario.

Roosevelt, Theodore. Papers, Library of Congress, Manuscript Division, Washington, D.C.

Schreiber, Howard G. A Brief History of Charlotte Schreiber (1834-1922) and Her Husband, Weymouth George Schreiber (1828-1898), Toronto, 24 December 1971, Unpublished MS in the author's possession.

Seton, Ernest Thompson. Journals, 1879-1946 (36 MS Volumes), American Museum of Natural History, New York.

________. Paintings, Drawings and Etchings (About 2000 Items), Seton Castle, Santa Fe, New Mexico and Seton Museum, Cimarron, New Mexico.

________. Papers, Seton Museum, Cimarron, New Mexico.

Seton, Grace Gallatin. Papers, Old Greenwich, Connecticut.

Seton, Julia Moss. Papers, Seton Castle, Santa Fe, New Mexico.

(2) Interviews

Chase, Anya Seton. Old Greenwich, Connecticut. Taped Interview, 15 December 1971.

Clark, Leonard. Greenwich, Connecticut. Taped Interview, 17 December 1971.

Ketchum, Mrs. Phillip. Port Hope, Ontario. Taped Interview, 13 January 1972.

Thompson, Elise. Toronto, Ontario. Taped Interviews, 20 November 1971; 25 January 1972.

Thompson, Jean. Toronto, Ontario. Taped Interview, 30 November 1971.

(3) The Published Work of Ernest Thompson Seton

This list includes most of Seton's published articles, notes, short stories, poems, books and illustrations. Only the first editions of books have been cited, except where there has been a change in title or in content. This list corrects many errors which appeared in a little-known bibliography produced in 1925 by Seton's publishers (Ernest Thompson Seton: A Biographical Sketch Done By Various Hands to Which is Attached a Complete Bibliography of the Works of This Author. Garden City, N.Y.: Doubleday, Page and Co., 1925), and makes substantial additions. It does not include material published by Seton in the Totem Board, the organ of the Woodcraft League of America, Inc., which was under Seton's direction from its inception in 1917 to its demise in 1934. Several Seton articles which appeared elsewhere, reappeared in the Totem Board. Other new editorials, notes and opinions also appear in this journal. As many of these are unsigned, however, it is difficult to determine what is and what is not attributable to him. The Totem Board had an extremely shaky career. Extant copies of back issues are difficult to locate, but the most comprehensive collection may be found at the Library of Congress. Altogether there are 13½ volumes, from 1917 to 1934. Volumes 4 to 9, No. 1 (May, 1920 to May, 1926) are titled the Woodcraft Totem Board. Publication was suspended from July, 1926 to April, 1927 and Volume 5, No. 1 is omitted from the numbering. The periodical ceased publication in May, 1934. All Seton items are listed in alphabetical order by title. I have also included the name under which Seton was writing in each instance.

Bailey, Florence Augusta. Birds of Village and Field. Boston: Houghton, Mifflin, 1898. [Illustrations by Ernest Thompson Seton.]

Century Company. The Century Dictionary; an Encyclopedic Lexicon of the English Language. Compiled by William Dwight Whitney. 6 vols. New York: Century, 1889-1891. [Illustrations by Ernest Thompson Seton.]

Chapman, Frank M. Bird Life. New York: D. Appleton, 1900. [Illustrations by Ernest Thompson Seton.]

________. Handbook of Birds of Eastern North America. New York: D. Appleton, 1895. [Illustrations by Ernest Thompson Seton.]

Charles Scribner's Sons. *Pictures by Popular American Artists Selected from "Scribner's Magazine."* New York: Scribner's, 1900. [Illustrations by Ernest Thompson Seton.]

Grinnell, George Bird. "Climbing for White Goats." *Scribner's Magazine*, 15 (May, 1894), 644-648. [Illustrations by Ernest Thompson Seton.]

________. "Wolves and Wolf Nature." *Trail and Camp-Fire.* Edited by George Bird Grinnell and Theodore Roosevelt. New York: Forest and Stream, 1897, pp. 152-203. [Illustrations by Ernest Thompson Seton.]

Hoffman, Ralph. *Bird Portraits.* Boston: Ginn, 1901. [Illustrations by Ernest Thompson Seton.]

Hornaday, William T. "The Extermination of the American Bison, with a Sketch of Its Discovery and Life History." *Annual Report of the Board of Regents of the Smithsonian Institution, 1887: Report of the United States National Museum, Part II.* Washington: Government Printing Office, 1889, pp. 367-548. [Illustrations by Ernest Thompson Seton.]

Johnson, E. Pauline. *The Shagganappi.* Introduction by Ernest Thompson Seton. Toronto: Ryerson Press, 1913.

Macoun, John. *The Autobiography of John Macoun, M.A.* Introduction by Ernest Thompson Seton. Ottawa: Ottawa Field Naturalists' Club, 1922.

Medsger, Oliver Perry. *Edible Wild Plants.* Introduction by Ernest Thompson Seton. New York: Macmillan, 1939.

Nelson, E. W. "Larger North American Mammals." *National Geographic Magazine*, 30 (November, 1916), 385-472. [Illustrations by Ernest Thompson Seton.]

________. "Smaller Mammals of North America." *National Geographic Magazine*, 33 (May, 1918), 371-493. [Illustrations by Ernest Thompson Seton.]

________. *Wild Animals of North America.* Washington, D.C.: National Geographic Society, 1918. [Illustrations by Ernest Thompson Seton.]

Pratt, Alice Day. *Animals of a Sagebrush Ranch.* Foreword by Ernest Thompson Seton. New York: Rand McNally, 1938.

Stickney, J. H. Bird World. Boston: Ginn, 1898. [Illustrations by Ernest Thompson Seton.]

Wright, Mabel Osgood. Four-Footed Americans and Their Kin. Edited by Frank M. Chapman. New York: Macmillan, 1898. [Illustrations by Ernest Thompson Seton.]

Seton, Ernest Thompson. "Achieving the Picturesque in Building." Country Life in America, 35 (October, 1918), 44-47.

________. "Acrobatic Skunks." Journal of Mammalogy, 1 (May, 1920), 140.

Thompson, Ernest E. "Additions to the List of Manitoba Birds." Auk, 10 (January, 1893), 49-50.

Seton, Ernest Thompson. "An Amateur Circus I Once Gave." Ladies' Home Journal, 33 (September, 1916), 57.

________. "The American Bison or Buffalo." Scribner's Magazine, 40 (October, 1906), 385-405.

________. The American Boy Scout; the Official Handbook of Woodcraft for the Boy Scouts of America. New York: Doubleday, Page, 1910.

________. Animal Heroes. New York: Scribner's, 1905.

________. Animal Tracks and Hunter Signs. Garden City, N.Y.: Doubleday, 1958.

________. Animals. Garden City, N.Y.: Doubleday, Page, 1926.

________. Animals Worth Knowing. Garden City, N.Y.: Doubleday, Doran, 1934.

________. "The Annals of a Fur Farm, Chapter 1." Country Life in America, 23 (November, 1912), 38-40

________. "The Annals of a Fur Farm, Chapter 2." Country Life in America, 23 (December, 1912), 61-62.

________. "The Annals of a Fur Farm, Chapter 3." Country Life in America, 23 (January, 1913), 30-32.

________. The Arctic Prairies. New York: Scribner's, 1911.

________. "The Arctic Prairies: Part 1, The Land of the Buffalo." Scribner's Magazine, 48 (November, 1910), 513-532.

________. "The Arctic Prairies: Part 2, The Land of the Caribou." *Scribner's Magazine*, 48 (December, 1910), 725-734.

________. "The Arctic Prairies: Part 3, The Land of the Caribou (Cont'd.)." *Scribner's Magazine*, 49 (January, 1911), 61-72.

________. "The Arctic Prairies: Part 4, The Land of the Musk-Ox." *Scribner's Magazine*, 49 (February, 1911), 207-223.

________. "Arnaux, The Homing Pigeon." *Ladies' Home Journal*, 22 (April, 1905), 13-14.

________. "Around the Camp Fire: 1." *Boy's Life*, 3 (October, 1913), 13.

________. "Around the Camp Fire: 2." *Boy's Life*, 3 (December, 1913), 28.

________. "Around the Camp Fire: 3." *Boy's Life*, 4 (January, 1914), 20.

________. "Around the Camp Fire: 4." *Boy's Life*, 4 (March, 1914), 20.

________. "Around the Camp Fire: 5." *Boy's Life*, 4 (April, 1914), 23.

________. "Around the Camp Fire: 6." *Boy's Life*, 4 (May, 1914), 24.

________. "Around the Camp Fire: 7." *Boy's Life*, 4 (June, 1914), 26

________. "Around the Camp Fire: 8." *Boy's Life*, 4 (September, 1914), 26-27.

________. "Around the Camp Fire: 9." *Boy's Life*, 4 (October, 1914), 16, 25.

________. "Around the Camp Fire: 10." *Boy's Life*, 4 (December, 1914), 24.

________. "The Badger Who Was Brother to a Boy." *Boy's Life*, 3 (November, 1913), 12-13, 34.

________. *Bannertail, the Story of a Gray Squirrel*. New York: Scribner's, 1922.

________. "Bannertail, the Story of a Gray Squirrel: Part 1." Ladies' Home Journal, 39 (February, 1922), 157-158.

________. "Bannertail, the Story of a Gray Squirrel: Part 2." Ladies' Home Journal, 39 (March, 1922), 141-142.

________. "Bannertail, the Story of a Gray Squirrel: Part 3." Ladies' Home Journal, 39 (April, 1922), 177-178.

________. "Bannertail, the Story of a Gray Squirrel: Part 4." Ladies' Home Journal, 39 (May, 1922), 149-150.

Thompson, Ernest Seton. "The Baron and the Wolves." Forest and Stream, 47 (December 26, 1896), 504-506.

Seton, Ernest E. T. "Bear and Moose Queries." Forest and Stream, 26 (June 24, 1886), 427.

Seton, Ernest Thompson. The Best of Ernest Thompson Seton. Edited by W. Kay Robinson. London: Hodder and Stoughton, 1949.

Thompson, Ernest E. "The Big Buck We Didn't Shoot." Forest and Stream, 37 (September 10, 1891), 143.

Seton, Ernest Thompson. "Billy, the Big Wolf: Part 1." Ladies' Home Journal, 22 (August, 1905), 5-6.

________. "Billy, the Big Wolf: Part 2." Ladies' Home Journal, 22 (September, 1905), 10, 49.

________. The Biography of an Arctic Fox. New York: Appleton-Century, 1937.

Thompson, Ernest Seton. Biography of a Grizzly. New York: Century, 1900.

Seton-Thompson, Ernest. "The Biography of a Grizzly: Part 1, The Cubhood of Wahb." Century, 59 (November, 1899), 25-40

________. "The Biography of a Grizzly: Part 2, The Days of His Strength." Century, 59 (December, 1899), 200-212.

________. "The Biography of a Grizzly: Part 3, The Waning of Wahb." Century, 59 (January, 1900), 351-362.

Seton, Ernest Thompson. The Biography of a Silver Fox; or Domino Reynard of Goldur Town. New York: Century, 1909.

________. The Birch Bark Roll of the Woodcraft Indians. New York: Doubleday, Page, 1906. Published in several editions from 1906 to 1931. Each edition is distinctive.

________. "Bird Records from Great Slave Lake Region: A Preliminary List of Birds Observed by My 1907 Expedition into the Arctic Barren Grounds of Canada." Auk, 25 (January, 1908), 68-74.

Thompson, Ernest E. "The Birds of Manitoba." Proceedings of the United States National Museum, 13 (May 29, 1891), 457-643.

Seton, Ernest E. T. "The Birds of Western Manitoba." Auk, 3 (April, 1886), 145-156.

________. "The Birds of Western Manitoba: Conclusion." Auk, 3 (July, 1886), 320-329.

Thompson, Ernest E. "The Birds of Western Manitoba: Addenda." Auk, 3 (October, 1886), 453.

________. "The Birds That We See." Scribner's Magazine, 13 (June, 1893), 759-776.

Seton, Ernest Thompson. "Blazes and Indian Signs." Country Life in America, 7 (April, 1905), 632-634.

________. "Bobcats and Wild Turkeys." Journal of Mammalogy, 1 (May, 1920), 140.

________. The Book of Woodcraft and Indian Lore. New York: Doubleday, Page, 1912. [Published, with alterations, in several subsequent editions.]

________. "'Bow Skirmish' or 'Arrow Fight'." Forest and Stream, 79 (November 30, 1912), 693.

________. "The Boy and the Lynx." Ladies' Home Journal, 20 (November, 1903), 13-14

________. Boy Scouts of America. New York: Doubleday, Page, 1910.

________. "The Boy Scouts in America." Outlook, 95 (July 23, 1910), 630-635.

________. "Boy Scout Initiations." Boy's Life, 3 (January, 1913), 19.

______. The Brownie Wigwam. New York: Woodcraft League of America Inc., 1921.

______. The Buffalo Wind. Santa Fe: Seton Village Press, 1938.

______. "Building a Log Cabin." Country Life in America, 8 (May, 1905), 79-80

Thompson, Ernest E. "Canadian Game and Fish Resorts." Forest and Stream, 28 (February 10, 1887), 42-43.

Seton, Ernest E. T. "A Carberry Deer Hunt." Forest and Stream, 26 (June 3, 1886), 366-368.

Seton, Ernest Thompson. "The Caribou and His Kindred." Scribner's Magazine, 39 (April, 1906), 426-443.

Thompson, Ernest Seton. "Chanticleer vs. Egret." Forest and Stream, 49 (October 9, 1897), 283.

Seton, Ernest Thompson. "Chicaree." World Outlook, 5 (December, 1919), 38-39.

Seton-Thompson, Ernest. "Chink: The Development of a Pup." Youth's Companion, 75 (January 17, 1901), 28-29.

Thompson, Ernest Seton. "Comments on the Art Anatomy of Animals." Quartier Latin, 2 (1897), 177.

______. "Communication Regarding the Needs of Artists in the Zoological Park." First Annual Report of the New York Zoological Society. New York: New York Zoological Society, 1897.

Thompson, Ernest E. "Critical Note on Mr. J. B. Tyrrell's Paper, Entitled 'Catalogue of the Mammalia of Canada Exclusive of the Cetacea'." Proceedings of the Canadian Institute, 7 (October, 1889), 178-180.

Seton, Ernest Thompson. "Dipo: Sprite of the Desert." Century, 105 (November, 1922), 106-115.

Thompson, Ernest Seton. "Directive Coloration of Birds." Auk, 14 (October, 1897), 395-396.

Thompson, Ernest E. "Disappearance of Forest Birds." Transactions of the Canadian Institute, 1 (1889-90), 47.

______. "Do Squirrels Hibernate?" Forest and Stream, 28 (February, 1887), 65.

Seton, Ernest Thompson. "Does the *Cuterebra* Ever Emasculate Its Host?" *Journal of Mammalogy*, 1 (February, 1920), 94-95.

________. "Dogs of Song." *Success Magazine*, 10 (August, 1907), 537-539, 562-563.

Seton, Ernest E. T. "Domestication of the Buffalo." *Forest and Stream*, 26 (July 8, 1886), 467.

Seton, Ernest Thompson. "Domino Reynard of Goldur Town: Part 1." *Century*, 77 (December, 1908), 208-219.

________. "Domino Reynard of Goldur Town: Part 2." *Century*, 77 (January, 1909), 374-389.

________. "Domino Reynard of Goldur Town: Part 3." *Century*, 77 (Feburary, 1909), 545-554.

Thompson, Ernest E. "The Drummer on Snowshoes." *St. Nicholas*, 14 (April, 1887), 414-417.

Seton, Ernest Thompson. "Early Bird Banding." *Auk*, 38 (October, 1921), 611.

________. "Elephant Hunting in New York." *Four Track News*, 9 (December, 1905), 469-470.

Thompson, Ernest Seton. "Elkland, 1." *Recreation*, 7 (September, 1897), 199-201.

________. "Elkland, 2: The Beaver Pond." *Recreation*, 7 (October, 1897), 286-290.

________. "Elkland, 3: Old-Timers." *Recreation*, 7 (November, 1897), 369-372.

________. "Elkland, 4: Flies and Weather." *Recreation*, 7 (December, 1897), 456-457.

________. "Elkland, 5: Puss and the Bear." *Recreation*, 8 (January, 1898), 33-34.

________. "Elkland, 6: Duels." *Recreation*, 8 (February, 1898), 117-119.

Seton, Ernest Thompson. "Emergency Foods in the Northern Forest." *Country Life in America*, 6 (September, 1904), 438-440.

________. "English Names of Mammals." Journal of Mammalogy, 1 (February, 1920), 104-105.

________. Ernest Thompson Seton's America. Edited by Farida A. Wiley. New York: Devin-Adair, 1954.

________. "Ernest Thompson Seton's Boys: Part 1, Trailing." Ladies' Home Journal, 19 (May, 1902), 15, 41.

________. "Ernest Thompson Seton's Boys: Part 2, The Second Chapter on Tracks." Ladies' Home Journal, 19 (June, 1902), 15.

________. "Ernest Thompson Seton's Boys: Part 3, Playing 'Injun'." Ladies' Home Journal, 19 (July, 1902), 17.

________. "Ernest Thompson Seton's Boys: Part 4, Archery." Ladies' Home Journal, 19 (August, 1902), 16.

________. "Ernest Thompson Seton's Boys: Part 5, Teepees." Ladies' Home Journal, 19 (September, 1902), 15.

________. "Ernest Thompson Seton's Boys: Part 6, Woodcraft Indians and Getting Lost." Ladies' Home Journal, 19 (October, 1902), 14.

________. "Ernest Thompson Seton's Boys: Part 7, Freezing." Ladies' Home Journal, 19 (November, 1902), 15.

Thompson, Ernest E. "Evening and Pine Grosbeaks in Ontario." Auk, 7 (April, 1890), 211.

Seton, Ernest Thompson. "The Evolutionary Force of a Wide Range." Journal of Mammalogy, 3 (August, 1922), 167-169.

Thompson, Ernest E. "An Exhibition of Snakes." Forest and Stream, 28 (June 16, 1887), 451.

Seton, Ernest Thompson. "Fable and Woodmyth: Part 1." Century, 67 (November, 1903), 35-39

________. "Fable and Woodmyth: Part 2." Century, 67 December, 1903), 276-279.

________. "Fable and Woodmyth: Part 3." Century, 67 (January, 1904), 346-351.

________. "Fable and Woodmyth: Part 4." Century, 67 (February, 1904), 496-500.

________. "Fable and Woodmyth: Part 5." Century, 67 (March, 1904), 750-756.

________, ed. Famous Animal Stories. New York: Brentanos, 1932.

Thompson, Ernest Seton. "The Fantail, Flagtail, or Gazelle Deer, Odocoileus texanus. Mearns." Recreation, 10 (January, 1899), 59-60.

Seton, Ernest Thompson. "Fauna of Manitoba." In the British Association Handbook. Winnipeg: British Association, 1909.

Seton-Thompson, Ernest. "A Fifth Avenue Troubadour." Ladies' Home Journal, 18 (October, 1901), 13-14.

Thompson, Ernest E. "First Sialia sialis at Lorne Park." Transactions of the Canadian Institute, 1 (1889-90), 56.

Seton, Ernest Thompson. "Food of the Red Fox." Journal of Mammalogy, 1 (May, 1920), 140.

________. "For a Methodic Study of Life Histories of Mammals." Journal of Mammalogy, 1 (February, 1920), 67-69.

________. "Forest Secrets--The Soul Song of Baba-Moss-Anib." Country Life in America, 20 (July, 1911), 39-42, 66, 68.

________. The Forester's Manual; or The Forest Trees of Eastern North America. Garden City, N.Y.: Doubleday, Page, 1910.

________. "Further Annals of a Hollow Tree." Country Life in America, 22 (October, 1912), 64, 66.

Thompson, Ernest E. "The Goldenrod." Our Animal Friends, 21 (August, 1894), 280-281.

Seton, Ernest Thompson, and Seton, Julia M., eds. The Gospel of the Red Man: An Indian Bible. Garden City, N.Y.: Doubleday, Doran, 1936.

Seton, Ernest Thompson. "Gray Squirrels and Nuts." Journal of Mammalogy, 2 (November, 1921), 41.

________. Great Historic Animals. New York: Scribner's, 1937.

Seton, Ernest E. T. "The Great North West." Herts and Essex Observer, March 24, 1883.

Seton, Ernest Thompson. "A Greeting From the Chief Scout." Boy's Life, 4 (January, 1914), 2.

________. "The Habits of Wolves: Including Many Facts about Animal Marriage." American Magazine, 64 (October, 1907), 636-645.

Thompson, Ernest E. "Hibernation of the Hare." Forest and Stream, 28 (April 7, 1887), 226.

Seton, Ernest Thompson. "The Hollow Tree." Bird-Lore, 11 (January, 1909), 1-3.

________. "The Hollow Tree." Country Life in America, 20 (September, 1911), 88.

________. "Holy Smoke!" Rotarian, 45 (August, 1934), 9-11, 57.

Thompson, Ernest Seton. "A Horned Cow Elk." Forest and Stream, 48 (February 20, 1897), 145.

Seton, Ernest Thompson. "The House That Is Mine." House and Garden, 44 (October, 1923), 62, 63, 112, 114, 116.

Thompson, Ernest E. Seton. "How Bull-Fighting Was Suppressed in France." Our Animal Friends, 24 (February, 1897), 128-129.

Seton, Ernest Thompson. "How Long Do Animals Live?" Ladies' Home Journal, 27 (August, 1910), 6.

________. "How to Get Cool Water." Boy's Life, 3 (May, 1913).

________. "How to Make a Fire By Rubbing Sticks." Country Life in America, 6 (June, 1904), 145-146, 204.

Thompson, Ernest Seton. "How to Measure an Animal." Recreation, 8 (April, 1898), 270-272.

Seton, Ernest Thompson. How to Play Indian. Philadelphia: Curtis Publishing Co., 1903.

________. "How to Study a Bird." Bird-Lore, 6 (November-December, 1904), 181-184.

________. "How to Stuff a Bird." Country Life in America, 6 (July, 1904), 267-269.

Thompson, Ernest Seton. "Hunting with the Camera." Recreation, 8 (April, 1898), 263-264.

Thompson, Ernest E. "Hybrid Pinicola enucleator + Carpodacus purpureus." Auk, 11 (January, 1894), 1-3.

Seton, Ernest Thompson. "If Da Vinci Came to Town." American City, 5 (November, 1911), 252-254.

________. "The Indian Sign Language for Boy Scouts." Boy's Life, 3 (February, 1913), 15.

________. "Indian Words in Common Use." Forest and Stream, 79 (July 8, 1911), 55.

Thompson, Ernest Seton. "The Injun's Christmas Spree." International Magazine, January, 1899, pp. 35-36.

________. "Intercommunication of Wolves." Forest and Stream, 48 (January 23, 1897), 64-65.

Seton, Ernest E. T. "Interesting Records from Toronto, Canada." Auk, 2 (October, 1885), 334-337.

Thompson, Ernest Seton. "An Interview With Whitman's Spirit." Recreation, 6 (June, 1897), 480.

Seton, Ernest Thompson. "Is Our Fur Supply in Danger?" World's Work, 47 (March, 1924), 491-502.

________. "Is the World's Fur Supply in Danger?" World Today, 45 (February, 1925), 200-203.

________. "The Jaguar in Colorado." Journal of Mammalogy, 1 (November, 1920), 241.

Seton-Thompson, Ernest. "Johnny Bear." Scribner's Magazine, 28 (December, 1900), 658-671.

Seton, Ernest Thompson. Johnny Bear, Lobo and Other Stories. New York: Scribner's, 1935.

________. "Joy Comes With a Job Well Done." New York American, November 8, 1922, p. 10.

Seton-Thompson, Ernest. "The Kangaroo Rat." Scribner's Magazine, 27 (April, 1900), 418-427.

Thompson, Ernest E. "The King of Currumpaw: A Wolf Story." Scribner's Magazine, 16 (November 7, 1894), 618-628.

Seton, Ernest Thompson. Krag and Johnny Bear. New York: Scribner's, 1902.

Seton-Thompson, Ernest. "Krag, The Kootenay Ram: Part 1." Scribner's Magazine, 29 (June, 1901), 693-707.

_______. "Krag, The Kootenay Ram: Part 2." Scribner's Magazine, 30 (July, 1901), 43-51.

Seton, Ernest Thompson. "Krag, The Kootenay Ram." Reader's Digest, 79 (July, 1961), 162-165.

_______. "Laws of the Seton Indians." Association Boys, 4 (June, 1905), 99-108.

Seton-Thompson, Ernest. "A Legend of the Bloodroot." Outing Magazine, 36 (May, 1900), 177.

_______. "The Legend of the White Reindeer." Century, 63 (November, 1901), 79-89.

Seton, Ernest Thompson. Life Histories of Northern Animals. 2 vols. New York: Scribner's, 1909.

Thompson, Ernest E. "Linota cannabrua at Toronto." Transactions of the Canadian Institute, 1 (1889-90), 54-55.

Thompson, Ernest Seton. "A List of the Big Game in North America." Forest and Stream, 51 (October 8, 1898), 285-286.

Seton-Thompson, Ernest. "A List of Fishes Known to Occur in Manitoba." Forest and Stream, 51 (September 10, 1898), 214.

Thompson, Ernest E. "A List of the Mammals of Manitoba." Transactions of the Manitoba Historical and Scientific Society, 23 (May, 1886), 1-26.

Seton, Ernest Thompson. "A List of the Turtles, Snakes and Batrachians of Manitoba." Ottawa Naturalist, 32 (November, 1918), 79-83.

_______. "Little Burnt All-Over." American Girl, 8 (April, 1924), 5, 6, 7, 26 and 27.

_______. "Little Warhorse: The Story of a Jack Rabbit." Ladies' Home Journal, 21 (June, 1904), 13-14.

_______. Lives of Game Animals. 4 vols. Garden City, N.Y.: Doubleday, Page, 1925-1928.

Seton-Thompson, Ernest. Lives of the Hunted. New York: Scribner's, 1901.

Seton, Ernest Thompson. "Lobo." Golden Book, 12 (July, 1930), 58-65.

________. Lobo. Philadelphia: Pennsylvania Institution for the Instruction of the Blind, 1900.

________. "Lobo, King of the Currumpaw." Reader's Digest, 41 (November, 1942), 103-106.

Thompson, Ernest Seton. Lobo, Rag and Vixen, New York: Scribner's, 1899.

Seton, Ernest Thompson. "Locality, Date and Name with Observations." Journal of Mammalogy, 1 (February, 1920), 107.

________. "The Lovers and the Shining One: A Prose Poem." Scribner's Magazine, 61 (June, 1917), 751-753.

________. "Lure of the Wild: The Story of a Dog." Coronet, 36 (September, 1954), 123-128.

________. "The Mackenzie River Ghost." Boy's Life, 28 (December, 1938), 16-17.

________. "Magpie as Sentinel for Rabbits." Journal of Mammalogy, 4 (May, 1922), 119.

________. "Making a Hollow Tree and What Came Into It: Bulletin 1." Country Life in America, 15 (November, 1908), 47, 84.

________. "Making a Hollow Tree and What Came Into It: Bulletin 2." Country Life in America, 15 (December, 1908), 226.

________. "Making a Hollow Tree and What Came Into It: Bulletin 3." Country Life in America, 15 (January, 1909), 310.

________. "Making a Hollow Tree and What Came Into It: Bulletin 4." Country Life in America, 15 (February, 1909), 414.

________. "Making a Hollow Tree and What Came Into It: Bulletin 5." Country Life in America, 15 (March, 1909), 542.

______. "Making a Hollow Tree and What Came Into It: Bulletin 6." Country Life in America, 15 (April, 1909), 658, 660, 662, 664.

______. "Making a Hollow Tree and What Came Into It: Bulletin 7." Country Life in America, 16 (July, 1909), 354, 356.

______. "Making a Hollow Tree and What Came Into It: Bulletin 8." Country Life in America, 16 (August, 1909), 455.

______. "Making a Hollow Tree and What Came Into It: Bulletin 9." Country Life in America, 16 (September, 1909), 552.

______. "Making a Hollow Tree and What Came Into It: Bulletin 10." Country Life in America, 16 (October, 1909), 662.

______. "Making Permanent Records of Animal Tracks." Country Life in America, 8 (June, 1905), 228-229.

______. "The Making of Silly Billy." National Sunday Magazine, January 23, 1916, p. 620.

______. "Mammals of the Yellowstone Park." Recreation, 8 (May, 1898), 365-371.

Seton, Ernest Thompson. "The Mane on the Tail of the Gray Fox." Journal of Mammalogy. 5 (August, 1923), 180-182.

Seton, Ernest E. T. "A Manitoban Blizzard." In Robert Miller Christy. Manitoba Described. London: Wyman and Sons, 1885, pp. 57-58.

______. "Manitoban Notes." Auk, 2 (January, 1885), 21-24.

Seton, Ernest Thompson. Manual of the Brownies, the Little Lodge of the Woodcraft League of America. New York: Woodcraft League of America, Inc., 1922.

______. The Manual of the Woodcraft Indians. Garden City, N.Y.: Doubleday, Page, 1915.

______. "The Master Plowman of the West." Century, 68 (June, 1904), 300-307.

______. "Meddy She Was Sorry: A Poem." Bookman, 29 (November, 1905), 93.

________. "The Merry Chipmunk." Success Magazine, 10 (May, 1907), 328-331, 368-370.

________. "Migrations of the Gray Squirrel (Sciurus carolinensis). Journal of Mammalogy, 1 (February, 1920), 53-58.

________. "The Mole-Mouse, Potato-Mouse, or Pine-Mouse." Journal of Mammalogy, 1 (August, 1920), 185.

________. Monarch, the Big Bear of Tallac. New York: Scribner's, 1904.

________. "Monarch the Grizzly: Part 1." Ladies' Home Journal, 21 (February, 1904), 5-6, 47.

________. "Monarch the Grizzly: Part 2." Ladies Home Journal, 21 (March, 1904), 13-14.

________. "Monarch the Grizzly: Part 3." Ladies' Home Journal, 21 (April, 1904), 15-16, 60.

Seton, Ernest E. T. "Moose and Bear Queries." Forest and Stream, 26 (June 24, 1886), 427.

Seton, Ernest Thompson. "The Moose and His Antlers." Scribner's Magazine, 39 (February, 1906), 157-178.

Thompson, Ernest E. "More About the Scream (of the Panther)." Forest and Stream, 39 (November 17, 1892), 421.

Seton, Ernest Thompson. "More About the Yak." Country Life in America, 21 (December, 1911), 44.

________. "More Acrobatic Skunks." Journal of Mammalogy, 3 (February, 1922), 53.

________. "More Wild Animals I Have Known: The Boy and the Lynx." Ladies' Home Journal, 20 (November, 1903), 13-14.

Seton-Thompson, Ernest. "Mother Love in the Cows of the Western Range." Breeder's Gazette, 38 (December 19, 1900), 947-948.

________. "The Mother Teal and the Overland Route." Ladies' Home Journal, 18 (July, 1901), 5-6.

Thompson, Ernest Seton. "The Myth of the Song Sparrow." Bird-Lore, 1 (April, 1899), 59.

Seton, Ernest Thompson. "The Narrowest Escape I Ever Had From a Wild Beast." Farm and Fireside, 46 (January, 1922), 13, 34, 35, 37.

________. "The Nation Awaits a Song." New York Tribune, December 1, 1918.

Thompson, Ernest Seton. "The National Zoo at Washington: Part 1." Century, 59 (March, 1900), 649-660.

________. "The National Zoo at Washington: Part 2." Century, 60 (May, 1900), 110.

Seton, Ernest Thompson. "The National Zoo at Washington, a Study of Its Animals in Relation to Their Natural Environment." Annual Report of the Board of Regents of the Smithsonian Institution, 1901. Washington: Government Printing Office, 1902, pp. 697-716.

________. "The Natural History of the Ten Commandments." Century, 75 (November, 1907), 24-33.

________. The Natural History of the Ten Commandments. New York: Scribner's, 1907.

________. "Nature and Human Nature." Nature, 3 (May, 1924), 279-80, 318.

Seton, Ernest E. T. "Nest and Eggs of the Philadelphia Vireo." Auk, 2 (July, 1885), 305-306.

________. "Nest and Habits of the Connecticut Warbler (Oporornis agilis)." Auk, 1 (April, 1884), 192-193.

Seton, Ernest Thompson. "A New Deer Hunt With the Bow." Country Life in America, 7 (February, 1905), 370-371.

Seton-Thompson, Ernest. "New Music From the Old Harp." Century, 60 (August, 1900), 639.

Thompson, Ernest Seton. "The New Sportsman." Recreation, 10 (January, 1899), 39-40.

Thompson, Ernest E. "Nights With the Coons." Forest and Stream, 30 (July 19, 1888), 518-520.

________. "Not Caught Yet." Our Animal Friends, 21 (December, 1893), 89.

Thompson, Ernest Seton. "Notes for Observation of Habits of Birds." Osprey, 4 (September, 1899), 6-8.

Thompson, Ernest E. "Notes of a Trip to Manitoba: Part 1." Forest and Stream, 41 (November 18, 1893), 424-425.

________. "Notes of a Trip to Manitoba: Part 2." Forest and Stream, 41 (November 25, 1893), 446-448.

Seton, Ernest Thompson. "Notes on the Breeding Habits of Captive Deermice." Journal of Mammalogy, 1 (May, 1920), 134-138.

Thompson, Ernest E. "Notes on the English Sparrow." Forest and Stream, 30 (April 5, 1888), 204-205.

Seton, Ernest Thompson. "Notes on Snapper." Forest and Stream, 90 (September, 1920), 499.

Thompson, Ernest E. "Occurrence of the Evening Grosbeak (Coccothraustes vespertina) at Toronto, Canada." Auk, 4 (July, 1887), 256-257.

Seton, Ernest Thompson. "Occurrence of the Least Weasel (Mustela rixora) near Denver." Journal of Mammalogy, 14 (1933), 70.

________. "The Oldest of All Writing." Country Life in America, 17 (December, 1909), 169-173, 242, 244, 246, 248, 250.

Seton, Ernest E. T. "On Architecture." Toronto Truth, April 4, 1885.

Seton, Ernest Thompson. "On Journal Keeping." Bird-Lore, 4 (November-December, 1902), 175-176.

Seton-Thompson, Ernest. "On Nature Study." The Library of Natural History. Edited by Richard Lydekker. New York: Saalfield Publishing Co., 1904, pp. iii-viii.

Seton, Ernest Thompson. "On the Popular Names of Birds." Auk, 26 (April, 1919), 229-235.

________. "On the Study of Scatology." Journal of Mammalogy, 6 (February, 1925), 47-49.

Thompson, Ernest E. "On the Use of Faunal Lists." Proceedings of the Canadian Institute, 7 (1888-1889), 275-280.

Seton-Ernest Thompson. "One Egg, Separated." Ladies' Home Journal, 37 (March, 1920), 184.

Thompson, Ernest E. "Ontario Game Laws." Forest and Stream, 30 (March 29, 1888), 185.

Seton-Ernest Thompson. "Organized Boyhood: The Boy Scout Movement, Its Purposes and Its Laws." Success Magazine, 13 (December, 1910), 804, 805, 843, 849.

Seton-Thompson, Ernest. "The Origin of Dick Cissel." Bird-Lore, 2 (June, 1900), 88.

Seton, Ernest E. T. "Our Canadian Birds: Paper 1." Canadian Science Monthly, 1 (June, 1884), 55-57.

________. "Our Canadian Birds: Paper 2." Canadian Science Monthly, 1 (July and August, 1884), 75-78.

________. "Our Canadian Birds: Paper 3." Canadian Science Monthly, 1 (September, 1884), 108-112.

________. "Our Canadian Birds: Paper 4." Canadian Science Monthly, 1 (October, 1884), 130-135.

________. "Our Canadian Birds: Paper 5." Canadian Science Monthly, 2 (February, 1885), 19-21.

________. "Outlines of Ornithology." Proceedings of the Canadian Institute, 3 (June, 1886), 180.

Thompson, Ernest E. "The Ovenbird." St. Nicholas, 17 (April, 1890), 520-521.

Thompson, Ernest E. S. "The Pack-Rat." Quartier Latin, 1 (1896), 25.

Seton, Ernest Thompson. "Peace and Patrol Names." Boy's Life, 3 (January, 1913), 21.

Seton-Thompson, Ernest. Pictures of Wild Animals. New York: Scribner's, 1901.

Thompson, Ernest E. "The Pintail." St. Nicholas, 15 (September, 1888), 826-827.

________. "Playing Pretend." Our Animal Friends, 23 (November, 1895), 65-66.

Thompson, Ernest Seton. "Poem Addressed to a Brass Paper Weight in the Form of a Mouse." Recreation, 8 (February, 1898), 153.

________. "A Poetic Exchange of Compliments." Recreation, 10 (February, 1899), 106.

Seton, Ernest Thompson. "Poison Ivy." Boy's Life, 3 (July, 1913), 31.

________. "The Porcupine at Home." Country Life in America, 17 (February, 1910), 415-416.

________. "Practical Fur Farming: Part 1." Field and Stream, 18 (March, 1914), 1146-1151.

________. "Practical Fur Farming: Part 2, The Care and Feeding of Skunks." Field and Stream, 18 (April, 1914), 1299-1302.

________. "Practical Fur Farming: Part 3, The Diseases and Breeding of Skunks." Field and Stream, 19 (May, 1914), 19-24.

________. "Practical Fur Farming: Part 4, Marking and Disarming." Field and Stream, 19 (July, 1914), 296-299.

________. "Practical Fur Farming: Part 5, Marketing Skins." Field and Stream, 19 (August, 1914), 391-394.

________. "Practical Fur Farming: Part 6, Mink Farming, With a Foot Note on Skunk Raising." Field and Stream, 19 (November, 1914), 746-751.

________. "Practical Fur Farming: Part 7, Marten Farming." Field and Stream, 19 (January, 1915), 923-925.

Seton, Ernest E. T. "The Prairie Chicken." Transactions of the Manitoba Historical and Scientific Society, 14 (May, 1884), 13-18.

________. "The Prairie Chicken or Sharptailed Grouse, Pedioecetes phasianellus (Baird)." Proceedings of the Canadian Institute, 1 (1879-1883), 405-412.

Seton, Ernest Thompson. "The Prairie Dogs (Cynomys ludovicianus) at Washington Zoo." Journal of Mammalogy, 7 (1926), 229-230.

Seton, Ernest E. T. "Prairie Fires." Transactions of the Manitoba Historical and Scientific Society, 16 (1884), 13-14.

________. "Prairie Fires." Annual Report of the Manitoba Department of Agriculture and Statistics, 1883. Winnipeg: Queen's Printer, 1883, pp. 491-492.

Seton, Ernest Thompson. The Preacher of Cedar Mountain. New York: Doubleday, Page, 1917.

Seton-Thompson, Ernest. "Preliminary Description of a New Caribou. Rangifer montanus." Ottawa Naturalist, 13 (August, 1899), 129-130.

Seton, Ernest Thompson. "The Prong-Horned Antelope." Scribner's Magazine, 40 (July, 1906), 33-49.

Thompson, Ernest E. Seton. "The Protection of Birds in France." Our Animal Friends, 26 (November, 1898), 56-58.

Seton-Thompson, Ernest. Raggylug. Philadelphia: Pennsylvania Institution for Instruction of the Blind, 1900.

Seton, Ernest Thompson. "Raising Fur-Bearing Animals for Profit." Country Life in America, 9 (January, 1906), 294-297.

Seton-Thompson, Ernest. "Rangifer dawsoni, Preliminary Description of a New Caribou from Queen Charlotte's Islands." Ottawa Naturalist, 13 (February, 1900), 257-261.

Thompson, Ernest E. "Rare Birds in Toronto University Museum." Transactions of the Canadian Institute, 1 (1889-90), 55-56.

Seton, Ernest Thompson. "Recent Bird Records for Manitoba." Auk, 25 (October, 1908), 450-454.

Seton-Thompson, Ernest. "Recognition Marks of Birds." Bird-Lore, 3 (November-December, 1901), 187-189.

Thompson, Ernest E. "A Record of Failures: Part 1." Forest and Stream, 28 (March 24, 1887), 178-179.

________. "A Record of Failures: Part 2." Forest and Stream, 28 (March 31, 1887), 198.

Seton, Ernest Thompson. The Red Book. New York: By the Author, 1904.

________. The Red Lodge. New York: By the Author, 1912.

Seton-Thompson, Ernest. Redruff. Philadelphia: The Pennsylvania Institution for the Instruction of the Blind, 1900.

Thompson, Ernest E. "Reply to Mr. Tyrrell's Note." Proceedings of Canadian Institute, 7 (1888-89), 285-286.

Thompson, Ernest E. S. "Reply to a Puritan's Denunciation of The Decameron." Quartier Latin, 2 (1897), 210.

Thompson, Ernest E. "Report of the Occurrence of the Evening Grosbeak (Coccothraustes vespertina) in Ontario During the Winter of 1889-90." Transactions of the Canadian Institute, 3 (December, 1892), 111.

Seton, Ernest Thompson. "The Revival of the Bow and Arrow." Country Life in America, 7 (January, 1905), 273-275.

________. "The Road to Fairyland." St. Nicholas, 31 (December, 1903), 103.

________. "The Rocky Mountain Goat." Records of North American Big Game. Edited by Prentiss N. Gray. New York: Derrydale Press, 1932.

________. Rolf in the Woods. New York: Doubleday, Page, 1911.

________. "A Roving Band of Say's Bats." Journal of Mammalogy, 3 (February, 1922), 52.

Seton, Ernest E. T. "The Ruminants of the North-West." Proceedings of the Canadian Institute, 3 (February, 1886), 113-117.

Seton, Ernest Thompson. Santana: The Hero Dog of France. Los Angeles: Phoenix Press, 1945.

Thompson, Ernest Seton. "A School of Animal Painting and Sculpture in the New York Zoological Park." Second Annual Report of the New York Zoological Society. New York: New York Zoological Society, 1898, pp. 69-75.

Seton, Ernest Thompson. "Scouting: Number 1." American Boy, 12 (July, 1911), 5, 21.

———. "Scouting: Number 2." American Boy, 12 (August, 1911), 4-5.

———. "Scouting: Number 3." American Boy, 12 (September, 1911), 3-4.

———. "Scouting: Number 4." American Boy, 12 (October, 1911), 3-4.

———. "Scouting: Number 5." American Boy, 12 (November, 1911), 6, 29.

———. "Scouting: Number 6." American Boy, 12 (December, 1911), 3, 31.

———. "Scouting: Number 7." American Boy, 13 (January, 1912), 3, 31.

———. "Scouting: Number 8." American Boy, 13 (February, 1912), 3, 26.

———. "Scouting: Number 9." American Boy, 13 (March, 1912), 3, 27.

———. "Scouting: Number 10." American Boy, 13 (April, 1912), 7, 27.

———. "Scouting on the Train." Boy's Life, 3 (May, 1913), 22.

Thompson, Ernest E. "The Screech-Owl." St. Nicholas, 17 (March, 1890), 432-3.

Seton, Ernest Thompson. "The Sea-Mink, Mustela macrodon (Prentiss)." Journal of Mammalogy, 2 (August, 1921), 168.

———. "The Sea Otter." The Book of Naturalists: An Anthrology of the Best Natural History. Edited by William Beebe. New York: Alfred A. Knopf, 1945.

———. "The Secrets of the Trail." Country Life in America, 8 (June, 1905), 202-205.

———. "Shishoka." Boy's Life, 35 (February, 1945), 8-9, 40.

———. Sign Talk. Garden City, N.Y.: Doubleday, Page, 1918.

Thompson, Ernest E. "Silverspot." Scribner's Magazine, 23 (February, 1898), 212-218.

Seton, Ernest Thompson. "The Slum Cat." Ladies' Home Journal, 21 (August, 1904), 9-10.

_______. "Smoke Signals, Sign Talk and Totems." Boy's Life, 2 (December, 1912), 24.

_______. "Snap, the Bull-Terrier: The Story of a Christmas Dog." Ladies' Home Journal, 21 (December, 1903), 10, 49.

_______. "The Snow-Show Rabbit." Everybody's Magazine, 16 (May, 1907), 599-608.

Seton-Thompson, Ernest. "Some Famous Wolves of France." Collier's, 26 (October, 1900), 6.

Thompson, Ernest Seton. "Some More About Wolves." Forest and Stream, 48 (March 6, 1897), 183-184.

Seton, Ernest Thompson. "Some Prints of Leaves." Nature, 2 (September, 1923), 142-143, 191.

Thompson, Ernest E. "The Song of the Prairie Lark." American Magazine, 7 (April, 1888), 717-720.

Seton, Ernest Thompson. "The Song of the Porcupine (Erethizon epixanthum)." Journal of Mammalogy, 13 (May, 1932), 168-169.

Thompson, Ernest E. "The Song Sparrow, An Investigation of His Life History." Forest and Stream, 30 (April 19, 1888), 244.

Seton, Ernest Thompson. "The Spirit of the Woods." Century, 103 (December, 1921), 213-224.

_______. "Spiritual Thrift, Number 2: Helping to Develop Democracy with Nature's Aid Through the Woodcraft League." Touchstone, 1 (June, 1917), 119-122.

Thompson, Ernest E. "'Sport' in France, 1." Forest and Stream, 38 (June 16, 1892), 560.

_______. "'Sport' in France, 2." Forest and Stream, 38 (June 23, 1892), 586.

_______. "'Sport' in France, 3." Forest and Stream, 38 (June 30, 1892), 611.

______. "Spring Notes." Transactions of the Canadian Institute, 1 (1889-90), 60.

Seton, Ernest Thompson. "Stories on the Tree Trunks." Country Life in America, 6 (May, 1904), 37-39, 90.

______. "The Story of Atalapha, a Winged Brownie." Scribner's Magazine, 59 (April, 1916), 441-458.

______. "The Story of Carrots." Farm and Fireside, 47 (June, 1923), 12, 13, 30, 31, 33.

______. "The Story of Coaly-Bay." May Court Magazine, February 1909, pp. 5-11.

______. "The Story of Wyndygoul." Country Life in America, 16 (August, 1909), 399-404, 446, 448, 450.

______. "The Story of Wyndygoul." Country Life in America, 16 (September, 1909), 505-508, 540, 542, 544.

Thompson, Ernest E. "The Strange Animals of Thibet." Forest and Stream, 42 (April 14, 1894), 311-312.

______. "Strawberry Finch." Transactions of the Canadian Institute, 1 (1889-90), 41-42.

Seton, Ernest E. T. "The Striped Gopher, Spermophilus tredecemlineatus (Mitchell)." Annual Report of the Manitoba Department of Agriculture and Statistics, 1882. Winnipeg: Queen's Printer, 1882, pp. 169-172.

Thompson, Ernest E. Studies in the Art Anatomy of Animals. London: Macmillan, 1896.

______. "Sturnella magna Wintering Near Toronto." Transactions of the Canadian Institute, 1 (1889-90), 42.

Seton, Ernest E. T. "The Swallow-tailed Flycatcher in Manitoba and at York Factory." Auk, 2 (April, 1885), 218.

Seton, Ernest Thompson. "Tail Glands of the Canidae." Journal of Mammalogy, 3 (August, 1923), 180-182.

______. The Ten Commandments in the Animal World. Garden City, N.Y.: Doubleday, Page and Co., 1923.

______. "Thanksgiving and the Yule-Log." Country Life in America, 19 (November, 1910), 37.

________. "They're Still Wild in Spots." Collier's, 76 (August 1, 1925), 12-13.

Thompson, Ernest Seton. "The Timmer-Doodle." Recreation, 7 (December, 1897), 445-446.

Seton-Thompson, Ernest. "Tito: The Story of a Coyote That Learned How: Part 1." Scribner's Magazine, 28 (August, 1900), 131-145.

________. "Tito: The Story of a Coyote That Learned How: Part 2." Scribner's Magazine, 28 (September, 1900), 316-325.

________. "To 'Jim'." Outing Magazine, 36 (April, 1900), 44.

Thompson, Ernest E. "Tracks in the Snow." St. Nicholas, 15 (March, 1888), 338-341.

Seton, Ernest Thompson. Trail and Camp Fire Stories. Edited by Julia M. Seton. New York: Appleton-Century, 1940.

Thompson, Ernest Seton. "The Trail of the Sandhill Stag." Scribner's Magazine, 26 (August, 1899), 191-204.

Seton-Thompson, Ernest. The Trail of the Sandhill Stag. New York: Scribner's, 1899.

Seton, Ernest Thompson. "The True Story of Daddy Binks." Recreation, 16 (January, 1902), 19-20.

Thompson, Ernest E. "The True Story of a Little Gray Rabbit." St. Nicholas, 17 (October, 1890), 953-55.

Seton, Ernest Thompson. "The Twelve Secrets of the Woods." Craftsman, 30 (June, 1916), 231-239, 329-331.

________. "Two Little Savages: Part 1." Ladies' Home Journal, 20 (January, 1903), 11-12.

________. "Two Little Savages: Part 2." Ladies' Home Journal, 20 (February, 1903), 11-12.

________. "Two Little Savages: Part 3." Ladies' Home Journal, 20 (March, 1903), 13-14.

________. "Two Little Savages: Part 4." Ladies' Home Journal, 20 (April, 1903), 11-12

________. "Two Little Savages: Part 5." Ladies' Home Journal, 20 (May, 1903), 11-12.

________. "Two Little Savages: Part 6." Ladies' Home Journal, 20 (June, 1903), 11-12.

________. "Two Little Savages: Part 7." Ladies' Home Journal, 20 (July, 1903), 11-12.

________. "Two Little Savages: Part 8." Ladies' Home Journal, 20 (August, 1903), 15-16, 32.

________. Two Little Savages: Being the Adventures of Two Boys Who Lived as Indians Do and What They Learned. New York: Doubleday, Page, 1903.

________. "Two Records for New Mexico." Journal of Mammalogy, 12 (May, 1931), 166.

________. "Two Successful Fur Farms." Country Life in America, 20 (September 1, 1911), 39-40.

________. "The Value of Moles." Journal of Mammalogy, 4 (February, 1923), 51.

________. "The Wapiti and His Antlers." Scribner's Magazine, 39 (January, 1906), 15-33.

________. The War Dance and the Fire-Fly Dance. New York: Doubleday, Page, 1910.

Thompson, Ernest E. S. "The Waterhen or Gallinule." Our Animal Friends, 27 (September, 1899), 14-15.

Seton-Thompson, Ernest. "A Welcome Superstition." Bird-Lore, 2 (October, 1900), 166.

Seton, Ernest E. T. "The Western Grebe in Manitoba." Auk, 2 (July, 1885), 314.

Thompson, Ernest E. "The Western Meadow-Lark." St. Nicholas, 16 (November, 1888), 63-64.

Seton, Ernest Thompson. "What Do Birds Signal With Their Tails?" Bird-Lore, 23 (November-December, 1921), 286-287.

________ [Sloat Hall]. "What is A True Sportsman?" Forest and Stream, 47 (September 26, 1896), 245.

Seton, Ernest Thompson. "What to do When Lost in the Woods." Country Life in America, 6 (August, 1904), 359.

_______. "The White Man's Last Opportunity." Canada West Magazine, 3 (April, 1908), 525-532.

_______. "The White-Tailed (Virginian) Deer and Its Kin." Scribner's Magazine, 40 (September, 1906), 321-341.

_______. "Why Do Birds Bathe? Part 1." Bird-Lore, 22 (November-December, 1920), 334-335.

_______. "Why Do Birds Bathe? Part 2." Bird-Lore, 23 (May-June, 1921), 124-137.

Thompson, Ernest E. "Why the Chickadee Goes Crazy Once a Year." Our Animal Friends, 21 (September, 1893), 17-18.

Seton, Ernest Thompson. "Why Wear Clothes?" Hearst's Magazine, August, 1920, pp. 21, 73.

_______. "A Wild Animal Bed Quilt." Ladies' Home Journal, 22 (January, 1905), 9.

Seton-Thompson, Ernest. "The Wild Animal Play." Ladies' Home Journal, 17 (July, 1900), 3-4, 28.

_______. The Wild Animal Play for Children. Philadelphia: Curtis Publishing Co., 1900. New York: Doubleday, Page, 1900.

Seton, Ernest Thompson. Wild Animal Ways. Garden City, N.Y.: Doubleday, Page, 1916.

_______. Wild Animals at Home. Garden City, N.Y.: Doubleday, Page, 1913.

Thompson, Ernest Seton. Wild Animals I Have Known. New York: Scribner's, 1898.

Seton, Ernest Thompson. Wild Animals I Have Known. Louisville, Kentucky: American Printing House for the Blind, 1900.

_______. "The Wild Geese of Wyndygoul." Country Life in America, 28 (April, 1916), 19-21.

_______. "Wild Life and the Motor Car." Journal of Mammalogy, 2 (November, 1921), 240.

Thompson, Ernest E. "Wild Rice." Forest and Stream, 39 (August 25, 1892), 157-158.

Seton, Ernest Thompson. "The Winnipeg Wolf." Ladies' Home Journal, 20 (December, 1902), 11-12.

Thompson, Ernest Seton. "The Wolf Question." Recreation, 8 (February, 1898), 126-127.

Seton, Ernest Thompson. "The Woman Bear: A Poem." Ladies' Home Journal, 34 (May, 1917), 11.

Thompson, Ernest E. S. "The Wood-Duck." Our Animal Friends, 24 (July, 1897), 256-257.

Thompson, Ernest E. "The Wood Rabbit or Hare." Our Animal Friends, 22 (November, 1894), 65.

________. "The Wood Thrush." Our Animal Friends, 25 (May, 1898), 207-208.

Seton, Ernest Thompson. Woodcraft Boys, Woodcraft Girls, How to Begin. New York: By the Author, 1915.

________. "Woodcraft Extolled as the Science that Makes Men: A Gospel for Out-of-Doors." Current Opinion, 73 (October, 1922), 461-464.

________. "The Woodcraft League or College of Indian Wisdom." Homiletic Review, 101 (June, 1931), 434-439.

________. The Woodcraft Manual for Boys. New York: Doubleday, Page, 1917.

________. The Woodcraft Manual for Girls. New York: Doubleday, Page, 1916.

________. "The Woodcrafter and the Stars." Country Life in America, 7 (November, 1904), 61.

________. Woodland Tales. Garden City, N.Y.: Doubleday, Page, 1921.

________. Woodmyth and Fable. New York: Century, 1905.

Seton-Thompson, Ernest. "The World's Most Wonderful Zoo." London Daily Express, August 16, 1901.

Seton, Ernest Thompson. "The Yak: A North American Opportunity." Country Life in America, 15 (February, 1909), 354-356.

Thompson, Ernest E. S. "The Yellowleg and the Hens." Our Animal Friends, 25 (November, 1897), 64-65.

(4) A Selected List of Secondary Works on Ernest Thompson Seton

"The Animal Story." Edinburgh Review, 214 (July, 1911), 94-118.

Atwood, Margaret. "Animal Victims." In Survival: A Thematic Guide to Canadian Literature. Toronto: Anansi, 1972. pp. 69-86.

Bell, Nancy. "The Work of Ernest Seton Thompson." The Humane Review, 4 (April, 1903), 11-20.

Bodsworth, Fred. "The Backwoods Genius with the Magic Pen." Maclean's Magazine, 72 (June 6, 1959), 22, 32, 34, 38, 39, 40.

Buckham, John Wright. "The Modern School of Nature Interpretation." Book Buyer, 20 (March, 1900), 108-113.

Burroughs, John. "Real and Sham Natural History." Atlantic Monthly, 91 (March, 1903), 298-309.

Chapman, Frank M. "A Champion of Ernest Thompson Seton." Saturday Review, 16 (October 2, 1937), 9.

________. "Naturalist, Artist, Author, Educator." Bird-Lore, 37 (July, 1935), 245-247.

Chesley, Albert M. "Laws of the Seton Indians." Social Activities for Men and Boys. New York: Association Press, 1919. pp. 261-269.

Devoe, Alan. "Artistic Wilderness." Saturday Review, 23 (December 7, 1940), 18.

________. "Mr. Mother Nature." Saturday Review, 37 (March 6, 1954), 20-21.

Emmons, Myra. "Ernest Seton Thompson--Artist, Naturalist, Writer." Recreation, 6 (May, 1897), 315-330.

________. "With Ernest Seton-Thompson in the Woods." Ladies' Home Journal, 18 (September, 1901), 3-4.

"Ernest Seton-Thompson." Bookman, 19 (January, 1901), 114-115.

"Ernest Seton-Thompson: Artist, Naturalist, Writer." Book News Monthly, 18 (May, 1900), 490-492.

"Ernest Thompson Seton." In Current Biography, 1943. New York: H. W. Wilson, 1944, pp. 685-688.

Garst, Shannon and Garst, Warren. Ernest Thompson Seton. New York: Messner, 1959.

Green, Roger Lancelyn. "Ernest Thompson Seton." Junior Bookshelf, 24 (December, 1960), 341-346.

Halsey, F. W. "Ernest Seton-Thompson: In Bryant Park, N.Y." American Authors and Their Homes. New York: James Pott, 1901, pp. 281-292.

Helgeland, Glenn B. "The National Wildlife Federation's Conservation Hall of Fame: Ernest Thompson Seton." National Wildlife, 6 (August, 1968), 24-25.

Hellman, George S. "Animals in Literature." Atlantic Monthly, 87 (March, 1901), 391-397.

Lucas, Alec W. "Nature Writers and the Animal Story." Literary History of Canada: Canadian Literature in English. Edited by Carl F. Klinck, et al. Toronto: University of Toronto Press, 1965.

MacArthur, James. "Wolf Thompson and His Wild Animals." Bookman, 9 (March, 1899), 71-74.

McCabe, Lida Rose. "At Windygoul with Ernest Thompson Seton." Book Buyer, 25 (1903), 21-28.

MacIntyre, F. Beatrice. "The Paleface Chief." New Mexico, 24 (April, 1948), 25, 35, 37.

Magee, William H. "The Animal Story: A Challenge in Technique." Dalhousie Review, 44 (Summer, 1964), 156-164.

Moore, Margaret. "Getting Acquainted with Ernest Thompson Seton." Normal Instructor and Primary Plans, 38 (October, 1929), 68-70.

Morley, Patricia. "Seton's Animals." Journal of Canadian Fiction, 2 (Summer, 1973), 195-198.

Morris, Brian. "Ernest Thompson Seton and the Origins of the Woodcraft Movement." Journal of Contemporary History, 5 (1970), 183-194.

"Note on Ernest Seton-Thompson." Bookman, 13 (March, 1904), 4-5.

Peattie, Donald Culross. "Nature and Nature Writers." Saturday Review, 16 (August 28, 1937), 10-11

"People Who Interest Us: Ernest Thompson Seton, Who Has Organized the Boy Scout Movement in the United States." Craftsman, 19 (October, 1910), 67.

Pigott, T. Digby. "A Canadian Naturalist." Living Age, 232 (January 25, 1902), 222-225.

Poirier, Michel. "The Animal Story in Canadian Literature: E. Thompson Seton and Charles G. D. Roberts." Queen's Quarterly, 34 (January-March, 1927), 298-312; 34 (April-June, 1927), 398-419.

Polk, James. "Lives of the Hunted." Canadian Literature, 53 (Summer, 1972), 51-59.

______. Wilderness Writers. Toronto: Clarke, Irwin, 1972.

Read, S. E. "Flight to the Primitive: Ernest Thompson Seton." Canadian Literature, 13 (Summer, 1962), 45-47.

Reid, Russell. "Ernest Thompson Seton." Museum Review, 2 (January, 1947), 1-5.

Rhodenizer, V. B. "Nature Writers 1: Writers of Animal Stories." A Handbook of Canadian Literature. Ottawa: Graphic Publishers Ltd., 1930. pp. 111-118.

Roberts, Charles G. D. "Ernest Thompson Seton." The Bookman, 45 (December, 1913), 147-149.

______. "The Home of a Naturalist." Country Life in America, 5 (December, 1903), 153-155.

"Salute to an Old Friend." The Saturday Evening Post, 211 (December 31, 1938), 22.

Samson, Jack. "Seton: America's Forgotten Naturalist." Field and Stream, 78 (April, 1974), 60-63, 191-194.

Sanford, Chester M., and Owen, Grace A. "Ernest Thompson Seton." In Modern Americans: A Biographical School Reader for the Upper Grades. Chicago, Philadelphia, Boston: Laurel Book Co., 1918. pp. 187-194.

Schmitt, Peter J. Back to Nature: The Arcadian Myth in Urban America. New York: Oxford University Press, 1969.

Seton, Julia Moss. By A Thousand Fires: Nature Notes and Extracts from the Life and Unpublished Journals of Ernest Thompson Seton. Garden City, N.Y.: Doubleday, 1967.

Stevenson, Lionel. "Nature in Canadian Prose." In Appraisals of Canadian Literature. Toronto: Macmillan, 1926

Storms, W. W. "The Woodcraft Indians." Holiday Magazine for Children, 3 (October, 1904), 76-79.

Sykes, M'Cready. "Let's Play Indian: Making a New American Boy Through Woodcraft." Everybody's Magazine, 23 (October, 1910), 473-483.

Tracy, Henry Chester. "Ernest Thompson Seton." In American Naturists. New York: E. P. Dutton, 1930. pp. 233-243.

Vogt, William. "Popularizing Nature." Saturday Review, 16 (September 18, 1937), 9.

Wallace, Ralph. "Wild Animals He Has Known." Reader's Digest, 49 (September, 1946), 59-63.

Whitelock, William W. "Ernest Seton-Thompson." The Critic, 39 (October, 1901), 320-325.

Zahniser, Howard. "Nature in Print." Nature Magazine, 34 (January, 1941), 7, 50.

________. "Nature in Print." Nature Magazine, 46 (November, 1953), 450-451.

(5) Wilderness, Ecology and History: Background

Atwood, Margaret. The Journals of Susanna Moodie. Toronto: Oxford University Press, 1970.

________. Surfacing. Toronto: McClelland and Stewart, 1972.

Bates, J. Leonard. "Fulfilling American Democracy: The Conservation Movement, 1907-1921." Mississippi Valley Historical Review, 44 (June, 1957), 29-57.

Berkhofer, Robert, Jr. "Space, Time, Culture and the New Frontier." Agricultural History, 38 (January, 1964), 21-30.

Brown, Robert Craig. "The Doctrine of Usefulness: Natural Resource and National Park Policy in Canada, 1887-1914." The Canadian National Parks: Today and Tomorrow. Edited by J. G. Nelson and R. C. Scace. Studies in Land Use History and Landscape Change, National Park Series, No. 3. 2 vols. Ottawa: National and Provincial Parks Association and the University of Calgary, 1968.

Cail, Robert E. Land, Man and the Law: The Disposal of Crown Lands in British Columbia, 1871-1913. Vancouver: Univeristy of British Columbia Press, 1974.

Careless, J. M. S. "Frontierism, Metropolitanism and Canadian History." Canadian Historical Review, 35 (March, 1954), 1-21.

________. "'Limited Identities' in Canada." Canadian Historical Review, 50 (March, 1969), 1-10.

Coleman, William. "Science and Symbol in the Turner Frontier Hypothesis." American Historical Review, 72 (October, 1966), 22-49.

Commager, Henry Steele. The American Mind: An Interpretation of American Thought and Character Since the 1880's. New Haven, Conn.: Yale University Press, 1950.

Crabbé, Philippe, and Spry, Irene, eds. Natural Resource Development in Canada. Social Science Studies, No. 8. Ottawa: University of Ottawa Press, 1973.

Cross, Whitney R. "Ideas in Politics: The Conservation Policies of the Two Roosevelts." Journal of the History of Ideas, 14 (1953), 421-438.

Darwin, Charles. The Descent of Man, and Selection in Relation to Sex. 2nd ed. London: John Murray, 1885.

Dickey, James. Deliverance. Boston: Houghton Mifflin, 1970.

Drew, Wayland. "Killing Wilderness." Ontario Naturalist, 12 (September, 1972), 19-23.

________. The Wabeno Feast. Toronto: Anansi, 1973.

________. "Wilderness and Culture." Alternatives, 3 (Summer, 1974), 8-11.

________. "Wilderness and Limitation." Canadian Forum, 52 (February, 1973), 16-19.

Dubos, René. "Man and His Environment: Adaptations and Interactions." The Quality of Man's Environment: Smithsonian Institution Symposium. Washington, D.C.: Smithsonian Institution Press, 1968.

Faulkner, William. "The Bear." The Portable Faulkner. Edited by Malcolm Cowley. Revised Edition; New York: Viking Press, 1967.

Fleming, Donald. "Roots of the New Conservation Movement." Perspectives in American History, 6 (1972), 7-91.

Frantz, Joe B. "Western Impact on the Nation." Western Historical Quarterly, 1 (July, 1970), 249-264.

Frye, Northrop. The Bush Garden: Essays on the Canadian Imagination. Toronto: Anansi, 1971.

Glacken, Clarence J. "Man Against Nature: An Outmoded Concept." The Environmental Crisis: Man's Struggle to Live with Himself. Edited by Harold W. Helfrich, Jr. New Haven, Conn.: Yale University Press, 1970.

Grant, George. Technology and Empire: Perspectives on North America. Toronto: Anansi, 1969.

Harris, Cole. "The Myth of the Land in Canadian Nationalism." Nationalism in Canada. Edited by Peter Russell. Toronto: McGraw Hill, 1966.

Hays, Samuel P. Conservation and the Gospel of Efficiency: The Progressive Conservation Movement, 1890-1920. New York: Atheneum, 1969.

Hazard, Lucy Lockwood. The Frontier in American Literature. New York: Frederick Ungar, 1961.

Higham, John. "The Reorientation of American Culture in the 1890's." The Origins of Modern Consciousness. Edited by John Weiss. Detroit, Mich.: Wayne State University Press, 1965.

Hofstadter, Richard. The Age of Reform, From Bryan to F.D.R. New York: Knopf, 1955.

________. Social Darwinism in American Thought. Boston: Beacon Press, 1955.

Holliday, J. S. "The Politics of John Muir." Sierra Club Bulletin, 57 (October-November, 1972), 10-13.

Jacobs, Wilbur. "Frontiersmen, Fur Traders and Other Varmints, An Ecological Appraisal of the Frontier in American History." American Historical Association Newsletter, 8 (November, 1970), 5-11.

Jones, D. G. Butterfly on Rock: A Study of Themes and Images in Canadian Literature. Toronto: University of Toronto Press, 1970.

Jennings, Francis. "Virgin Land and Savage People." American Quarterly, 23 (October, 1971), 519-541.

Kropotkin, Peter. Memoirs of a Revolutionist. Boston: Houghton, Mifflin, 1899.

________. "The Morality of Nature." Nineteenth Century, 57 (March, 1905), 407-426.

________. Mutual Aid: A Factor in Evolution. Edited by Paul Avrich. New York: New York University Press, 1972.

________. "Some of the Resources of Canada." Nineteenth Century, 43 (March, 1898), 494-514.

Krueger, Thomas A. "The Historians and the Edenic Myth: A Critique." Canadian Review of American Studies, 4 (Spring, 1973), 3-18.

Laing, R. D. The Politics of Experience. New York: Ballantine Books, 1967.

Leopold, Aldo. The Sand County Almanac, with Other Essays on Conservation from Round River. New York: Oxford University Press, 1966.

Lévi-Strauss, Claude. Mythologiques. 3 vols. Paris: Plon, 1964-68.

Livingston, John A. One Cosmic Instant: A Natural History of Human Arrogance. Toronto: McClelland and Stewart, 1973.

Mailer, Norman. Why Are We in Vietnam? New York: Putnam's, 1967.

Marx, Leo. "American Institutions and Ecological Ideals." Science, 170 (November 27, 1970), 945-952.

_______. The Machine in the Garden: Technology and the Pastoral Ideal in America. Galaxy Books. New York: Oxford University Press, 1967.

_______. "Pastoral Ideals and City Troubles." in The Quality of Man's Environment. Washington, D.C.: Smithsonian Institution Press, 1968.

McCloskey, Michael. "The Wilderness Act of 1964: Its Background and Meaning." Oregon Law Review, 45 (1966), 288-321.

Morton, W. L. "The 'North' in Canadian Historiography." Transactions of the Royal Society of Canada. Series IV, 8 (1970), 31-40.

Nash, Roderick. "Wilderness and Man in North America." The Canadian National Parks: Today and Tomorrow. Edited by J. G. Nelson and R. C. Scace. Studies in Land Use History and Landscape Change, National Park Series, No. 3. 2 vols. Ottawa: National and Provincial Parks Association and the University of Calgary, 1968.

_______. Wilderness and the American Mind. New Haven, Conn.: Yale University Press, 1967.

Nelles, H. V. The Politics of Development: Forests, Mines and Hydro-Electric Power in Ontario, 1849-1941. Toronto: Macmillan, 1974.

Neuhaus, Richard. In Defense of People: Ecology and the Seduction of Radicalism. New York: Macmillan, 1971.

New, W. H. Articulating West: Essays on Purpose and Form in Modern Canadian Literature. Toronto: New Press, 1972.

Nye, Russel B. "The American View of Nature." This Almost Chosen People: Essays in the History of American Ideas. East Lansing, Mich.: Michigan State University Press, 1966.

Pearce, David. "Is Ecology Elitist?" Ecologist, 3 (February, 1973), 61-63.

O'Riordan, Timothy. "The Third American Conservation Movement: New Implications for Public Policy." Journal of American Studies, 5 (August, 1971).

Ricou, Laurence. Vertical Man, Horizontal World: Man and Landscape in Canadian Prarie Fiction. Vancouver: University of British Columbia Press, 1973.

Rubinoff, Lionel. "Nationalism and Celebration: Reflections on the Sources of Canadian Identity." Queen's Quarterly, 82 (Spring, 1975), 1-13.

Schmitt, Peter J. Back to Nature: The Arcadian Myth in Urban America. New York: Oxford University Press, 1969.

Scholte, Bob. "Lévi-Strauss' Penelopean Effort: The Analysis of Myths." Semiotica, 1 (1969), 99-124.

Sills, David L. "The Environmental Movement and Its Critics." Human Ecology, 3 (January, 1975), 1-41

Smith, Henry Nash. Virgin Land: The American West as Symbol and Myth. Cambridge: Harvard University Press, 1950.

Thorpe, F. J. "Historical Perspective on the Resources for Tomorrow Conference." Resources for Tomorrow Conference, Background Papers. 3 vols. Ottawa: Queen's Printer, 1961-62.

Turner, Frederick Jackson. The Frontier in American History. New York: Holt, Rinehart, Winston, 1962.

Turrentine, Jackson W. "The Creation of Yellowstone National Park." Mississippi Valley Historical Review, 29 (1942), 187-206.

Ward, Barbara and Dubos, René. Only One Earth: The Care and Maintenance of a Small Planet. New York: Norton, 1972.

White, G. Edward. The Eastern Establishment and the Western Experience: The West of Frederic Remington, Theodore Roosevelt and Owen Wister. New Haven, Conn.: Yale University Press, 1968.

White, Hayden. "The Forms of Wildness: Archaeology of an Idea." The Wild Man Within: An Image in Western Thought from the Renaissance to Romanticism. Edited by Edward Dudley and Maximillian E. Novak. Pittsburgh, Pa.: University of Pittsburgh Press, 1972.

White, Lynn, Jr. "The Historical Roots of Our Ecologic Crisis." Science, 155 (March 10, 1967), 1203-1207.

Woodcock, George. Anarchism: A History of Libertarian Ideas and Movements. Meridian Books. Cleveland, Ohio: World Publishing Co., 1962.

________. "A Plea for the Anti-Nation." Canadian Forum, 52 (April, 1972), 16-19.

Woodcock, George, and Avakumovic, Ivan. The Anarchist Prince: A Biographical Study of Peter Kropotkin. London: J. V. Boardman, 1950.

Zaslow, Morris. The Opening of the Canadian North, 1870-1914. Toronto: McClelland and Stewart, 1971.

(6) Art

Ackerman, Gerald M. "Gérôme: Reassessing a Scorned Painter." Art News, 72 (January, 1973), 32-35.

Anderson, A. A. Experiences and Impressions: The Autobiography of A. A. Anderson. Freeport, N.Y.: Books for Libraries Press, 1970.

"The Art Anatomy of Animals." Quartier Latin, 2 (1897), 177.

Art Gallery of Ontario. The Canadian Collection. Toronto: McGraw-Hill, 1970.

Boime, Albert. The Academy and French Painting in the Nineteenth Century. London: Phaidon, 1971.

Bouret, Jean. The Barbizon School and Nineteenth Century French Landscape Painting. Greenwich, Conn.: New York Graphic Society, 1973.

Bourinot, John G. Our Intellectual Strength and Weakness. Montreal: Foster Brown, 1893.

Bryan's Dictionary of Painters and Engravers, 1964 ed. S.v. "Stubbs, George," by E. R. Dibdin.

Bryan's Dictionary of Painters and Engravers, 1964 ed. S.v. "Wertheimer, Gustave."

Bryan's Dictionary of Painters and Engravers, 1964 ed. S.v. "Wolf, Joseph," by George C. Williamson.

Clement, Clara Erskine and Hutton, Laurence. Artists of the Nineteenth Century and Their Works. 2 vols. St. Louis, Mo.: North Point, 1969.

Coke, Van Deren. The Painter and the Photograph from Delacroix to Warhol. Albuquerque, N.M.: University of New Mexico Press, 1972.

Colgate, William. Canadian Art: Its Origin and Development. Toronto: Ryerson, 1943.

Davis, Ann. "The Wembley Controversy in Canadian Art." Canadian Historical Review, 54 (March, 1973), 48-74.

Dictionnaire critique et documentaire des peintres, sculpteurs, dessinateurs et gravures, Nouvelle édition, 1966. S.v. "Frémiet (Emmanuel)."

Dictionnaire critique et documentaire des peintres, sculpteurs, dessinateurs et gravures, Nouvelle édition, 1966. S.v. "Morot (Aimé Nicolas)."

Dictionnaire critique et documentaire des peintres, sculpteurs, dessinateurs et gravures, Nouvelle édition, 1966. S.v. "Mosler (Henry)."

Fairbairn, Margaret L. "A Decade of Canadian Art." Canadian Magazine, 17 (June, 1901), 159-163.

Ford, Harriet. "The Royal Canadian Academy of Arts." Canadian Magazine, 3 (May, 1894), 45-50.

Forster, J. W. L. Under the Studio Light: Leaves from a Portrait Painter's Sketch Book. Toronto: Macmillan, 1928.

Galton, Francis. "Conventional Representation of the Horse in Motion." Nature, 26 (July 6, 1882), 228-29.

Gilbey, Sir Walter. Animal Painters of England. 3 vols. London: Vinton, 1900-1911.

________. The Life of George Stubbs, R.A. London: Vinton, 1898.

Goodrich, Lloyd. Thomas Eakins: His Life and Work. New York: Whitney Museum of American Art, 1933.

Greenhill, Ralph. Early Photography in Canada. Oxford University Press: Toronto, 1965.

Harper, J. Russell. Early Painters and Engravers in Canada. Toronto: University of Toronto Press, 1970.

________. Homer Watson, RCA, 1855-1936. Ottawa: National Gallery of Canada, 1963.

________. Painting in Canada: A History. Toronto: University of Toronto Press, 1966.

Harris, Lawren. "Creative Art and Canada." In The Yearbook of the Arts in Canada, 1928-1929. Edited by Bertram Brooker. Toronto: Macmillan, 1929, pp. 179-186.

Henri, Robert. The Art Spirit. Philadelphia: Lippincott, 1930.

Homer, William Innes. Robert Henri and His Circle. Ithaca, N.Y.: Cornell University Press, 1969.

Jussim, Estelle. Visual Communication and the Graphic Arts: Photographic Technologies in the Nineteenth Century. New York: R. R. Bowker, 1974.

Lambert, Richard S. The Adventure of Canadian Painting. Toronto: McClelland and Stewart, 1947.

Lankester, E. Ray. "The Muybridge Photographs." Nature, 40 (May 23, 1889), 78-80.

Larkin, Oliver W. Art and Life in America. New York: Holt, Rinehart, Winston, 1960.

Maas, Jeremy. Victorian Painters. London: Barrie and Rockliff, 1969.

Macdonald, Colin S., comp. A Dictionary of Canadian Artists. 4 vols. Ottawa: Canadian Paperbacks, 1971-74.

McInnes, Graham. Canadian Art. Toronto: Macmillan, 1950.

MacTavish, Newton. The Fine Arts in Canada. Toronto: Macmillan, 1925.

________. "Maurice Cullen: A Painter of the Snow." Canadian Magazine, 38 (April, 1912), 537-544.

Miner, Muriel Miller. G. A. Reid. Canadian Artist. Toronto: Ryerson, 1946.

Mortimer-Lamb, H. "The Art of Curtis Williamson, RCA." Canadian Magazine, 32 (November, 1908), 65-71.

Murray, Joan. Ontario Society of Artists: 100 Years. Toronto: Art Gallery of Ontario, 1972.

Murray, Joan. "Victorian Canada: Part 1." Canadian Antiques Collector, 5 (February, 1970), 10-14.

_______. "Victorian Canada: Part 2." Canadian Antiques Collector, 5 (March, 1970), 16-20.

_______. "Victorian Canada: Part 3." Canadian Antiques Collector, 5 (April, 1970), 16-20.

Palmer, A. H. Life of Joseph Wolf. London: Longmans, 1895.

Pevsner, Nikolaus. Academies of Art Past and Present. New York: Da Capo, 1973.

Reid, Dennis. A Concise History of Canadian Painting. Toronto: Oxford University Press, 1973.

"Review of Studies in the Art Anatomy of Animals." Athenaeum, 110 (July 31, 1897), 165-166.

Rewald, John. Post Impressionism from Van Gogh to Gauguin. New York: Museum of Modern Art, 1962.

Reynolds, Graham. Victorian Painting. London: Studio Vista, 1966.

Richardson, E. P. Painting in America from 1502 to the Present. New York: Thomas Y. Crowell, 1969.

Rothenstein, William. Men and Memories: Recollections, 1872-1900. London: Faber and Faber, 1931.

Sherwood, W. A. "The Influence of the French School on Recent Art." Canadian Magazine, 1 (October, 1893), 638-641.

_______. "A National Spirit in Art." Canadian Magazine, 3 (October, 1894), 498-501.

Smith, James. "The Royal Canadian Academy." Canadian Magazine, 9 (1897), 300-312.

Stephens, Frederic George. Sir Edwin Landseer. London: S. Low, Marston, Searle and Rivington, 1880.

Thomas, J. Sharkey. "Animals in Art." Nature Canada, 4 (July-September, 1975), 37-46.

Trapp, Frank Anderson. The Attainment of Delacroix. Baltimore: Johns Hopkins Press, 1970.

White, Harrison C. and White, Cynthia A. Canvases and Careers: Institutional Change in the French Painting World. New York: John Wiley, 1965.

Williamson, Moncrieff. Robert Harris, 1849-1919: An Unconventional Biography. Toronto: McClelland and Stewart, 1970.

Yriarte, Charles. "E. Meissonier: Personal Recollections and Anecdotes." Nineteenth Century, 43 (May, 1898), 822-840.

(7) Youth Work and the Cultural Context

"Are You a Seton Indian?" New York Herald, 11 October 1903.

Ariès, Philippe. Centuries of Childhood: A Social History of Family Life. Translated by Robert Baldick. London: Jonathon Cape, 1962.

Baden-Powell, Robert S. S. Aids to Scouting for NCO's and Men. London: Gale and Polden, 1909.

________. "Educational Possibilities of the Boy Scouts' Training." Nineteenth Century, 70 (August, 1911), 293-305.

________. Lessons from the 'Varsity' of Life. London: C. Arthur Pearson, 1933.

________. Scouting and Youth Movements. New York: Jonathon Cape, 1931.

________. Scouting for Boys. London: C. Arthur Pearson, 1909.

________. Scouting for Boys: A Handbook for Instruction in Good Citizenship. London: Horace Cox, 1908.

Bailey, Liberty Hyde. The Nature-Study Idea. New York: Doubleday, Page, 1903.

Beach, Thomas W. "A Wonderful Home: How a Great Naturalist Lives." London Daily Mail, 29 November 1909.

Beard, Daniel Carter. "The Boy Scouts." Outlook, 95 (July 23, 1910), 696-697

________. "The Boy Scouts of America." American Review of Reviews, 44 (October, 1911), 429-438.

________. The Buckskin Book of the Boy Pioneers of America. n.p., 1911.

________. Hardly a Man is Now Alive: The Autobiography of Dan Beard. New York: Doubleday, Doran, 1939.

Blumenfeld, Ralph D. "The Boy Scouts." Outlook, 95 (July 23, 1910), 617-629.

Bok, Edward. The Americanization of Edward Bok: The Autobiography of a Dutch Boy Fifty Years After. New York: Scribner's, 1920.

Bowers, C. A. "The Ideologies of Progressive Education." History of Education Quarterly, 7 (Winter, 1967), 452-73.

Boy Scouts of America. Boy Scouts of America: The Official Handbook for Boys (Garden City, N.Y.: Doubleday, Page, 1911.

________. The Official Handbook for Boys. Revised Edition; New York: Grosset and Dunlap, 1914.

Bryer, Jackson R. and Harding, Eugene. Hamlin Garland and the Critics: An Annotated Bibliography. Troy, N.Y.: Whitston, 1973.

Burton, F. R. American Primitive Music; with Especial Attention to the Songs of the Ojibways. New York: Moffat, Yard, 1909.

Clark, Ella Elizabeth. Indian Legends of Canada. Toronto: McClelland and Stewart, 1960.

Clark, William Philo. The Indian Sign Language. Philadelphia, Pa.: L. R. Hamersly, 1885.

Clemens, Cyril and Carroll Sibley. Uncle Dan: The Life Story of Dan Beard. New York: Thomas Y. Crowell.

Cohen, Brownwen J. "Nativism and Western Myth: The Influence of Nativist Ideas on the American Self-Image." Journal of American Studies, 8 (April, 1974), 23-39.

Cohn-Bendit, Daniel and Cohn-Bendit, Gabriel. Obsolete Communism: The Left Wing Alternative. Translated by Arnold Pomerans. London: Penguin, 1969.

Cremin, Lawrence A. The Transformation of the School. New York: Knopf, 1961.

Curti, Merle. Social Ideas of American Educators. Totowa, N.J.: Littlefield, Adams, 1966.

Curtis, Edward S. The North American Indian. Edited by F. W. Hodge. 20 vols. Cambridge: Cambridge University Press, 1907-1930.

Curtis, Natalie, ed. The Indian's Book: An Offering by the American Indians of Indian Lore, Musical and Narrative, to Form a Record of the Songs and Legends of Their Race. New York: Harper, 1907.

Demos, John. "Developmental Perspectives on the History of Childhood." Journal of Interdisciplinary History, 2 (Autumn, 1971), 315-327.

________. A Little Commonwealth: Family Life in Plymouth Colony. New York: Oxford University Press, 1970.

Dewey, John. Democracy and Education. New York: Macmillan, 1964.

________. "Interpretation of Savage Mind." Psychological Review, 9 (May, 1902), 217-230.

________. "My Pedagogic Creed." in Dewey on Education. Edited by Martin S. Dworkin. New York: Teachers' College Press, 1959, pp. 19-32.

Dickson, Lovat. Wilderness Man: The Strange Story of Grey Owl. Toronto: Macmillan, 1973.

Dorgan, Ethel Josephine. Luther Halsey Gulick, 1865-1918. New York: Teachers' College, Columbia University, 1934.

Dorsey, James O. "A Study of Siouan Cults." Eleventh Annual Report of the Bureau of Ethnology to the Secretary of the Smithsonian Institution, 1889-90. Washington: Government Printing Office, 1894, pp. 351-553.

Dorson, Richard M. American Folklore and the Historian. Chicago: University of Chicago Press, 1971.

Doubleday, F. N. The Memoirs of a Publisher. Garden City, N.Y.: Doubleday, 1972.

Dunkel, Harold B. Herbart and Herbartianism: An Educational Ghost Story. Chicago: University of Chicago Press, 1970.

________. "Herbartianism Comes to America." History of Education Quarterly, 9 (Summer, 1969), 202-233.

Eastman, Charles A. From the Deep Woods to Civilization: Chapters in the Autobiography of an Indian. Boston: Little Brown, 1916.

________. Indian Boyhood. New York: McClure, Phillips, 1902.

________. The Soul of the Indian: An Interpretation. Boston: Houghton Mifflin, 1911.

Ellis, Havelock. My Life: Autobiography of Havelock Ellis. Boston: Houghton, Mifflin, 1939.

Emerson, Ellen Russell. Indian Myths. Boston: J. R. Osgood, 1884.

Engels, Friedrich. Origin of the Family, Private Property and the State, in Light of the Researches of Lewis Henry Morgan. Translated by Ernest Untermann. Chicago: C. H. Kerr, 1902.

Erikson, Erik. Childhood and Society. 2nd ed. New York: Norton, 1963.

________. Ghandi's Truth: On the Origins of Militant Non-violence. New York: Norton, 1969.

________. Identity, Youth and Crisis. New York: Norton, 1968.

________. Young Man Luther: A Study in Psychoanalysis and History. New York: Norton, 1958.

"Ernest Thompson Seton's Statement Answered by Executive Board of the National Council." New York Times, 11 December 1915.

Fletcher, Alice. Indian Story and Song from North America. Boston: Small, Maynard, 1900.

Forbush, William Byron. The Boy Problem. Boston: Pilgrim Press, 1907.

Fortes, Meyer. Kinship and the Social Order: The Legacy of Lewis Henry Morgan. Chicago: Aldine, 1969.

Gardner, Brian. Mafeking: A Victorian Legend. London: Cassell, 1966.

Garland, Hamlin. Book of the American Indian. New York: Harper, 1923.

________. The Captain of the Gray Horse Troop. New York: Harper, 1902.

________. Companions on the Trail. New York: Macmillan, 1931.

________. Crumbling Idols: Twelve Essays on Art Dealing Chiefly with Literature, Painting and the Drama. Edited by Jane Johnson. Cambridge, Mass.: Harvard University Press, 1960.

________. Hamlin Garland's Diaries. Edited by Donald Pizer. San Marino, California: Huntingdon Library, 1968.

________. "The Red Man's Present Needs." North American Review, 174 (April, 1902), 476-488.

________. Roadside Meetings. New York: Macmillan, 1930.

George, Henry. Progress and Poverty. New York: D. Appleton, 1880.

Gibson, H. W. "The History of Organized Camping." Outdoor Education. Edited by Donald R. Hammerman and William M. Hammerman. Minneapolis, Minn.: Burgess, 1968, pp. 89-103.

Gillis, John R. "Conformity and Rebellion: Contrasting Styles of English and German Youth, 1900-1933." History of Education Quarterly, 13 (Fall, 1973), 249-260.

Hall, G. Stanley. Adolescence: Its Psychology and Its Relations to Physiology, Anthropology, Sociology, Sex, Crime, Religion and Education. 2 vols. New York: D. Appleton, 1904.

Hillcourt, William. Baden-Powell: The Two Lives of a Hero. London: Heinemann, 1964.

Hodge, C. F. Nature Study and Life. Boston: Ginn and Co., 1902.

Hodge, F. W. et al. Handbook of American Indians North of Mexico. 2 vols. Washington: Government Printing Office, 1907-1910.

Hofman, Charles, ed. Frances Densmore and American Indian Music; a Memorial Volume. New York: Museum of the American Indian, Heye Foundation, 1968.

Hopkins, C. Howard. A History of the YMCA in North America. New York: Association Press, 1951.

Howard, O. O. Famous Indian Chiefs I Have Known. New York: Century, 1908.

Huizinga, Johan. Homo Ludens: A Study of the Play Element in Culture. Boston: Beacon Press, 1955.

"Indians at Wyndygoul." Greenwich News, 23 October 1903.

Johnson, E. Pauline. Canadian Born. Toronto: Morang, 1903.

________. Flint and Feather. Toronto: Musson, 1912.

________. The Moccasin Maker. Toronto: William Briggs, 1913.

________. Ojistoh. New York: Werner, 1911.

________. The White Wampum. London: Lane, 1895.

Jones, J. Walter. Fur Farming in Canada. 2nd ed. Ottawa: Canada, Commission of Conservation, 1914.

Keniston, Kenneth. "Psychological Development and Historical Change." Journal of Interdisciplinary History, 2 (Autumn, 1971), 329-345.

________. "Youth: A 'New' Stage of Life." American Scholar, 39 (Autumn, 1970), 631-654.

Knebel, A. G. Four Decades With Men and Boys. New York: Association Press, 1936.

La Barre, Weston. The Ghost Dance: Origins of Religion. New York: Delta, 1972.

Leupp, Francis E. In Red Man's Land. New York: Fleming H. Revell, 1914.

________. The Indian and His Problem. New York: Scribner's, 1910.

Levy, Harold P. Building a Popular Movement: A Case Study of the Public Relations of the Boy Scouts of America. New York: Russell Sage Foundation, 1944.

Lewin, Herbert S. "Hitler Youth and the Boy Scouts of America: A Comparison of Aims." Human Relations, 1 (1947), 206-227.

Lifton, Robert Jay. Boundaries. Toronto: Canadian Broadcasting Corporation, 1969.

________. History and Human Survival. New York: Random House; Vintage Books, 1971.

McCraye, Walter. Pauline Johnson and Her Friends. Toronto: Ryerson, 1947.

McLaughlin, James. My Friend the Indian. Boston: Houghton, Mifflin, 1910.

Macleod, David Irving. "Good Boys Made Better: The Boy Scouts of America, Boys' Brigades and YMCA Boys' Work, 1880-1920." Unpublished Ph.D. dissertation, University of Wisconsin, 1973.

Mair, Charles. Tecumseh, A Drama, and Canadian Poems. Toronto: Radisson Society, 1926.

Mallery, Garrick. "Picture Writing of the American Indians." Tenth Annual Report of the Bureau of Ethnology to the Secretary of the Smithsonian Institution, 1888-89. Washington: Government Printing Office, 1893, pp. 25-822.

________. "Sign Language Among North American Indians, Compared with that Among Other Peoples and Deaf Mutes." First Annual Report of the Bureau of Ethnology to the Secretary of the Smithsonian Institution, 1879-80. Washington: Government Printing Office, 1881, pp. 269-552.

Marriott, Alice and Rachlin, Carol K. American Indian Mythology. New York: Thomas Y. Crowell, 1968.

Matthews, Washington. "Navajo Weavers." Third Annual Report of the Bureau of Ethnology to the Secretary of the Smithsonian Institution, 1881-82. Washington: Government Printing Office, 1884, 375-391.

Meyers, Ira Benton. "The Evolution of Aim and Method in the Teaching of Nature-Study." Elementary School Teacher, 11 (Dec., 1910-Jan., 1911), 205-213, 237-248.

Mooney, James. "The Ghost Dance Religion and the Sioux Outbreak of 1890." Fourteenth Annual Report of the Bureau of Ethnology to the Secretary of the Smithsonian Institution, 1892-93. Washington: Government Printing Office, 1894, 641-1136.

Morgan, Lewis Henry. Ancient Society, or Researches in the Lines of Human Progress from Savagery through Barbarism to Civilization. New York: Henry Holt, 1877.

________. League of the Ho-de-no-sau-nee, or Iroquois. 2 vols. Rochester, N.Y.: Sage, 1851.

Morse, Richard C. History of the North American Young Men's Christian Associations. New York, 1913.

Murray, William D. As He Journeyed: The Autobiography of William D. Murray. New York: Association Press, 1929.

________. The History of the Boy Scouts of America. New York: Boy Scouts of America, 1937.

Neblett, Thomas F. "Youth Movements in the United States." Annals of the American Academy of Political and Social Science, 194 (November, 1937), 141-151.

Oursler, Will. The Boy Scout Story. Garden City, N.Y.: Doubleday, 1955.

Pilkington, John. "Fuller, Garland, Taft and the Art of the West." Papers on Language and Literature, 8 (Fall Supplement, 1972), 39-56.

Presbrey, Frank. "The Associate Editors of Boy's Life," Boy's Life, 4 (May, 1914), 6.

Resek, Carl. Lewis Henry Morgan, American Scholar. Chicago: University of Chicago Press, 1960.

Reynolds, E. E. B-P. The Story of His Life. London: Oxford University Press, 1943.

________. The Scout Movement. London: Oxford University Press, 1950.

Riis, Jacob. "The Boy Scouts." Outlook, 105 (October 25, 1913), 412-421.

Robinson, Edgar M. The Early Years: The Beginnings of Work with Boys in the Young Men's Christian Association, New York: Association Press, 1950.

Ross, Dorothy. G. Stanley Hall: The Psychologist as Prophet. Chicago: University of Chicago Press, 1972.

Roszak, Theodore. The Making of a Counter Culture: Reflections on the Technocratic Society and Its Youthful Opposition. Garden City, N.Y.: Doubleday, 1969.

"Seton Still Insists on Quitting Scouts." New York Times, 6 December 1915.

Shrive, Norman. Charles Mair: Literary Nationalist. Toronto: University of Toronto Press, 1965.

________. "What Happened to Pauline?" Canadian Literature, 13 (Summer, 1962), 25-38

Springhall, J. O. "The Boy Scouts Class and Militarism in Relation to British Youth Movements, 1908-1930." International Review of Social History, 16 (1971), 125-158.

Steffens, Lincoln. The Autobiography of Lincoln Steffens. 2 vols. New York: Harcourt, Brace and World, 1931.

Stokoe, William C. Semiotics and Human Sign Languages. The Hague: Mouton, 1972.

Van Steen, Marcus. Pauline Johnson: Her Life and Work. Toronto: Musson, 1965.

Wade, E. K. Twenty-One Years of Scouting: The Official History of the Boy Scout Movement from Its Inception. London: C. Arthur Pearson, 1929.

Walker, James R. The Sun Dance and Other Ceremonies of the Oglala Division of the Teton Dakota. New York: Trustees of the American Museum of Natural History, 1917.

Welter, Rush. Popular Education and Democratic Thought in America. New York: Columbia University Press, 1962.

"West Says Seton is Not a Patriot." New York Times, 7 December 1915.

White, G. Edward. "The Social Values of the Progressives: Some New Perspectives." South Atlantic Quarterly, 70 (Winter, 1971), 62-76.

White, Morton. Pragmatism and the American Mind. New York: Oxford University Press, 1973.

Whitmore, Allan Richard. "Beard, Boys and Buckskins: Daniel Carter Beard and the Preservation of the American Pioneer Tradition." Unpublished Ph.D. dissertation, Northwestern University, 1970.

"Why Mr. Seton is Not Chief Scout." Boy's Life, 6 (January, 1916), 28.

Wilkinson, Paul. "English Youth Movements, 1908-1930." Journal of Contemporary History, 4 (April, 1969), 3-23.

World Book, 1918 ed. S.v. "The Boy Scouts," by Daniel Carter Beard.

Zachariah, Mathew. "The Impact of Darwin's Theory of Evolution on Theories of Society." Social Studies for Teachers and Administration, 62 (February, 1971), 69-77.

(8) Ecology, Ethology and the Animal Story

Allen, J. A. The American Bisons Living and Extinct. Cambridge, Mass.: Welch, Bigelow, 1876.

________. "Biographical Memoir of Elliott Coues, 1842-1899." National Academy of Sciences, Biographical Memoirs, 6 (1909), 397-446.

________. "The Geographic Distribution of the Mammalia Considered in Relation to the Principal Ontological Regions of the Earth, and the Laws that Govern the Distribution of Animal Life." Bulletin of the United States Geological and Geographical Society of the Territories, 4 (May 3, 1878), 313-377.

________. "Geographic Distribution of North American Mammals." Bulletin of the American Museum of Natural History, 4 (December 29, 1892), 199-243.

Anderson, R. M. Catalogue of Canadian Recent Mammals. Ottawa: National Museum of Canada, Bulletin No. 102, 1946.

Armstrong, Edward A. A Study of Bird Song. London: Oxford University Press, 1963.

Audubon, John James. Ornithological Biography, or an Account of the Habits of the Birds of the United States. 5 vols. Edinburgh: Adam Black, 1831-1839.

Audubon, John James and Bachman, John. The Quadrupeds of North America. 3 vols. New York: V. G. Audubon, 1849-1854.

Bailey, Vernon. "Revision of the American Voles of the Genus Evotomys." Proceedings of the Biological Society of Washington, 11 (May 13, 1897), 113-138.

Banfield, A. W. F. The Mammals of Canada. Toronto: University of Toronto Press, 1974.

________. A Revision of the Reindeer and Caribou, Genus Rangifer. Ottawa: National Museum of Canada, Bulletin No. 177, 1961.

Barrus, Clara. The Life and Letters of John Burroughs. 2 vols. Boston: Houghton, Mifflin, 1925.

Beer, C. G. "Ethology--The Zoologist's Approach to Behavior." Tuatara, 11 (1963), 170-177; 12 (1964), 16-39.

Bird, Charles D. "Stuart Criddle, 1877-1971." Canadian Field Naturalist, 87 (1973), 71-76.

Broadhead, Michael. "Elliott Coues and the Sparrow War." New England Quarterly, 44 (1971), 420-432.

Burroughs, John. Leaf and Tendril. Boston: Houghton, Mifflin, 1908.

________. "The Literary Treatment of Nature." Atlantic Monthly, 94 (July, 1904), 38-43.

________. "Roosevelt as a Nature Lover and Observer." Outlook, 86 (July 13, 1907), 547-553.

________. The Summit of the Years. Boston: Houghton, Mifflin, 1913.

________. Ways of Nature. Boston: Houghton, Mifflin, 1905.

Cappon, James. Charles G. D. Roberts. Toronto: Ryerson, n.d.

Carrigher, Sally. Wild Heritage. Boston: Houghton, Mifflin, 1965.

Chapman, Frank M. Autobiography of a Bird-Lover. New York: Appleton-Century, 1933.

________. Camps and Cruises of an Ornithologist. New York: Appleton, 1908.

________. "The Case of William J. Long." Science, 19 (March 4, 1904), 387-389.

________. Handbook of Birds of Eastern North America. New York: Dover Books, 1966.

________. "In Memoriam: Joel Asaph Allen." Auk, 39 (January, 1922), 1-14.

________. "In Memoriam: Louis Agassiz Fuertes." Auk, 47 (January, 1928), 1-26.

Clark, Edward B. "Roosevelt on the Nature Fakirs." Everybody's Magazine, 16 (June, 1907), 770-774.

________. "Real Naturalists on Nature Faking." Everybody's Magazine, 17 (September, 1907), 423-427.

Criddle, Stuart. "The Prairie Pocket Gopher." Journal of Mammalogy, 11 (1930), 275-280.

________. "A Study of the Snowshoe Rabbit." Canadian Field-Naturalist, 52 (March, 1938), 31-40.

Cutright, Paul Russell. Theodore Roosevelt: The Naturalist. New York: Harper, 1956.

Dall, W. H. Spencer Fullerton Baird, A Biography. Philadelphia: Lippincott, 1915.

Darlington, P. J. Zoogeography: The Geographic Distribution of Animals. New York: Wiley, 1957.

Darwin, Charles. The Expression of the Emotions in Man and Animals. Chicago: University of Chicago Press, 1965.

Davis, William Harper. "Natural and Unnatural History." Science, 19 (April 22, 1904), 667-675.

de la Mettrie, J. O. Man a Machine. Translated by Gertrude C. Bussey and Mary W. Calkins. Chicago: Open Court, 1912.

Dymond, J. R. "Zoology in Canada." In A History of Science in Canada. Edited by H. M. Tory. Toronto: Ryerson, 1939, pp. 41-57.

Eaton, R. L. "An Historical Look at Ethology: A Shot in the Arm for Comparative Psychology." Journal of the History of the Behavioral Sciences, 6 (1970), 176-187.

Elliott, D. G. "In Memoriam: Elliott Coues." _Auk_, 18 (January, 19091), 1-11.

Evans, Mary Alice and Evans, Howard Ensign Evans. _William Morton Wheeler, Biologist_. Cambridge, Mass.: Harvard University Press, 1970.

Fisher, A. K. "In Memoriam: George Bird Grinnell." _Auk_, 56 (January, 1939), 1-12.

Flader, Susan L. "Thinking Like a Mountain: A Biographical Study of Aldo Leopold." _Forest History_, 17 (April, 1973), 14-28.

Fowler, Kenneth. "The Story of the Camp Fire Club." _Field and Stream_, 15 (November, 1910), 601-615.

Gasman, Daniel. _The Scientific Origins of National Socialism: Social Darwinism in Ernst Haeckel and the German Monist League_. London: Macdonald, 1971.

Gibson, A. and Crawford, H. G. "Norman Criddle." _Canadian Entomologist_, 65 (1933), 193-200.

Gold, Joseph, "The Precious Speck of Life." _Canadian Literature_, 26 (Aug., 1965), 22-32.

Graham, Frank. _Man's Dominion: The Story of Conservation in America_. New York: Evans, 1971.

Gray, Peter, ed. _The Encyclopedia of the Biological Sciences_. New York: Van Nostrand, Reinhold, 1970.

Gray, Philip Howard. "The Early Animal Behaviorists: Prolegomenon to Ethology." _Isis_, 59 (1968), 372-383.

________. "The Nature of Man or Man's Place in Nature? Bruno Spinoza and la Mettrie." _Journal of General Psychology_, 76 (1967), 183-192.

________. "The Problem of Free Will in a Scientific Universe: René Descartes to John Tyndall." _Journal of General Psychology_, 80 (1969), 57-72.

________. "Spalding and His Influence on Research in Developmental Behavior." _Journal of the History of the Behavioral Sciences_, 3 (1967), 168-179.

Gregory, W. K. "Biographical Memoir of Frank M. Chapman." _National Academy of Sciences, Biographical Memoirs_, 25 (1948), 111-145.

Grinnell, George Bird and Roosevelt, Theodore, eds. Trail and Camp-Fire. New York: Forest and Stream Publishing Co., 1897.

Grinnell, George Bird and Sheldon, Charles, eds. Hunting and Conservation. New Haven: Yale University Press, 1925.

Grinnell, Hilda W. "Bibliography of Clinton Hart Merriam." Journal of Mammalogy, 24 (November, 1943), 436-457.

Haeckel, Ernst. The Evolution of Man. 2 vols. New York: H. L. Fowle, 1897.

________. The History of Creation. London: H. S. King, 1876.

________. The Riddle of the Universe at the Close of the Nineteenth Century. New York: Harper, 1900.

________. The Wonders of Life. London: Watts, 1905.

Hagmeier, Edwin M. "A Numerical Analysis of the Distributional Patterns of North American Mammals, 2: Re-evaluation of the Provinces." Systematic Zoology, 15 (1966), 279-299.

Hagmeier, Edwin M. and Stults, C. Dexter. "A Numerical Analysis of the Distributional Patterns of North American Mammals." Systematic Zoology, 13 (1964), 125-155.

Hall, E. Raymond and Kelson, Keith R. The Mammals of North America. 2 vols. New York: Ronald Press, 1959.

Hallowell, Irving A. "Ojibwa Ontology, Behavior, and World View." In Culture in History: Essays in Honor of Paul Radin. Edited by S. Diamond. New York: Columbia University Press, 1960, pp. 19-52.

Harris, Harry. "Robert Ridgway, with a Bibliography of His Published Writings." Condor, 30 (January, 1928), 1-118.

Hays, H. R. Birds, Beasts and Men: A Humanist History of Zoology. Baltimore: Penguin, 1973.

Henshaw, H. W. "In Memoriam: William Brewster." Auk, 37 (January, 1922), 1-23.

Hewitt, C. Gordon. The Conservation of the Wild Life of Canada. New York: Scribner's, 1921.

Hinde, R. A. "Appetitive Behaviour and Consummatory Act." Behaviour, 5 (1953), 189-224.

_______, ed. Bird Vocalizations: Their Relations to Current Problems in Biology and Psychology. Cam-Bridge: Cambridge University Press, 1969.

Hornaday, William T. The American Natural History: A Foundation of Useful Knowledge of the Higher Animals of North America. 4 vols. New York: Scribner's, 1914.

_______. The Mind and Manners of Wild Animals. New York: Scribner's, 1922.

_______. Our Vanishing Wild Life: Its Extermination and Preservation. New York: Scribner's, 1913.

Howard, H. E. Territory in Bird Life. London: Murray, 1920.

Hume, David. A Treatise of Human Nature. Garden City, N.Y.: Doubleday, 1961.

Hutt, Corinne. "Exploration and Play in Children." In Play, Exploration and Territory in Mammals. Edited by P. A. Jewell and Caroline Loizos. Symposia of the Zoological Society of London, No. 18. London: Academic Press, 1966, pp. 61-81.

Huxley, J. S. "The Courtship Habits of the Great Crested Grebe (Podiceps cristatus); with an Addition to the Theory of Sexual Selection." Proceedings of the Zoological Society of London, 2 (1914), 491-562.

_______, ed. "A Discussion of Ritualization of Behaviour in Animals and Man." Philosophical Transactions of the Royal Society of London, Series B772, 251 (1966), 247-526.

_______, ed. The New Systematics. New York: Oxford University Press, 1940.

Jenness, Diamond. "The Indian Interpretation of Man and Nature." Transactions of the Royal Society of Canada, 24 (1930), 57-62.

Keith, W. J. Charles G. D. Roberts. Toronto: Copp Clark, 1969.

Lack, David. The Natural Regulation of Animal Numbers. Oxford: Clarendon Press, 1954.

Langley, S. P. "Report of the Secretary of the Smithsonian Institution." Annual Report of the Board of Regents of the Smithsonian Institution, 1890. Washington: Government Printing Office, 1891, pp. 1-92.

Lehrman, Daniel. "A Critique of Konrad Lorenz's Theory of Instinctive Behavior." Quarterly Review of Biology, 28 (December, 1953), 337-363.

Leroy, C. G. The Intelligence and Perfectibility of Animals from a Philosophic Point of View with a Few Letters on Man. Translated by Mrs. Richard Congreve. London: Chapman and Hall, 1870.

Lloyds, H. "Norman Criddle, 1875-1933." Canadian Field Naturalist, 57 (1933), 145-147.

Loeb, Jacques. The Mechanistic Conception of Life. Edited by Donald Fleming. Cambridge, Mass.: Harvard University Press, 1964.

Long, William J. "The Modern School of Nature Study and Its Critics." North American Review, 176 (May, 1903), 688-698.

________. School of the Woods: Some Life Studies of Animal Instincts and Animal Training. Boston: Ginn, 1902.

________. "Science, Nature and Criticism." Science, 19 (May 13, 1904), 760-767.

Lorenz, Konrad. "Imprinting." Auk, 54 (1937), 245-273.

________. King Solomon's Ring: New Light on Animal Ways. New York: Signet, 1972.

________. "McDougall's Instinct Theory." In Instinctive Behavior: The Development of a Modern Concept. Edited by Claire H. Schiller. New York: International Universities Press, 1957, pp. 156-163.

________. On Aggression. London: Methuen, 1966.

McDougall, William. Body and Mind: A History and a Defense of Animism. London: Methuen, 1911.

________. An Introduction to Social Psychology. London: Methuen, 1908.

________. "The Use and Abuse of Instinct in Social Psychology." Journal of Abnormal Social Psychology, 16 (1922), 285-333.

McIlwraith, Thomas. The Birds of Ontario. 2nd ed. Toronto: William Briggs, 1894.

Macoun, John. Manitoba and the Great North-West. Guelph, Ontario: World Publishing Co., 1882.

Macoun, John and Macoun, James. Catalogue of Canadian Birds. Ottawa: Government Printing Bureau, 1909.

Mayr, Ernst. Animal Species and Evolution. Cambridge, Mass.: Harvard University Press, 1966.

Merriam C. Hart. "The Geographic Distribution of Animals and Plants in North America." Yearbook of the Department of Agriculture, 1894. Washington: Government Printing Office, 1895, pp. 207-214.

________. "The Geographic Distribution of Life in North America, with Special Reference to the Mammalia." Annual Report of the Board of Regents of the Smithsonian Institution, 1891. Washington: Government Printing Office, 1893, pp. 365-415.

________. "Laws of Temperature Control of the Geographic Distribution of Terrestrial Animals and Plants." National Geographic Magazine, 6 (December, 1894), 229-238.

________. Life Zones and Crop Zones of the United States. Washington: United States Department of Agriculture, Division of Biological Survey, Bulletin No. 10, 1898.

________. Results of a Biological Survey of the San Francisco Mountain Region and Desert of the Little Colorado, Arizona. Washington: United States Department of Agriculture, North American Fauna, No. 3, 1890.

________. Review of the Grizzly and Big Brown Bears of North America (Genus Ursus), with a Description of a New Genus, Vetularctos. Washington: United States Department of Agriculture, North American Fauna, No. 41, 1918.

________. "Species and Subspecies: A Reply to Mr. Conn." Science, 12 (December 7, 1888), 279.

Misiak, H. and Sexton, Virginia. History of Psychology: An Overview. New York: Grune and Stratton, 1966.

Morgan, Lewis Henry. The American Beaver and His Works. Philadelphia: Lippincott, 1868.

Murphy, R. C. "Frank Michler Chapman, 1864-1945." Auk, 67 (July, 1950), 307-315.

Oberholser, Harry C. "Robert Ridgway: A Memorial Appreciation." Auk, 50 (April, 1933), 159-169.

Oehser, Paul H. "In Memoriam: Florence Merriam Bailey." Auk, 69 (January, 1952), 19-26.

Osgood, W. H. A Biological Reconnaissance of the Base of the Alaska Peninsula. Washington: United States Department of Agriculture, North American Fauna, No. 24, 1904.

Palmer, T. S. "In Memoriam: Clinton Hart Merriam." Auk, 71 (April, 1954), 130-136.

Pomeroy, E. M. Sir Charles G. D. Roberts: A Biography. Toronto: Ryerson, 1943.

Preble, E. A. A Biological Investigation of the Athabasca-Mackenzie Region. Washington: United States Department of Agriculture, North American Fauna, No. 27, 1908.

________. A Biological Investigation of the Hudson Bay Region. Washington: United States Department of Agriculture, North American Fauna, No. 22, 1902.

Raven, C. E. John Ray, Naturalist: His Life and Works. Cambridge: Cambridge University Press, 1942.

Rhoads, S. N. "Bison bison athabascae." Proceedings of the Academy of Natural Science, 49 (1898), 498.

Ridgway, Robert. "Spencer Fullerton Baird." Auk, 5 (January, 1888), 1-14.

Roberts, Charles G. D. Eyes of the Wilderness. Toronto: Macmillan, 1933.

________. The Kindred of the Wild. Boston: Page, 1902.

________. The Last Barrier and Other Stories. Toronto: McClelland and Stewart, 1958.

________. The Watchers of the Trails. Toronto: Copp Clark, 1904.

Roe, F. G. The North American Buffalo: A Critical Study of the Species in Its Wild State. Toronto: University of Toronto Press, 1970.

Roosevelt, Theodore. "Nature Fakers." Everybody's Magazine, 17 (September, 1907), 427-430.

Ross, Alexander Milton. The Birds of Canada: With Descriptions of Their Plumage, Habits, Food, Song, Nests, Eggs, Times of Arrival and Departure. Toronto: Rowsell and Hutchinson, 1872.

Scott, John Paul. Animal Behavior. 2nd ed. Chicago: University of Chicago Press, 1972.

Sebeok, Thomas A. and Ramsay, Alexandra, eds. Approaches to Animal Communication. The Hague: Mouton, 1969.

Sibley, C. Lintern. "The Vogue of the Nature Story." Canadian Magazine, 38 (January, 1912), 287-292.

Snyder, L. L. "In Memoriam: James Henry Fleming." Auk, 58 (January, 1941), 1-12.

Soper, J. Dewey. "The Mammals of Manitoba." Canadian Field-Naturalist, 75 (October-December, 1961), 171-219.

Spalding, D. A. "Instinct, with Original Observations on Young Animals." Macmillan's Magazine, 27 (1873), 282-293.

Sterling, Keir B. Last of the Naturalists: The Career of C. Hart Merriam. New York: Arno Press, 1974.

Streseman, E. "Baron von Perneau, Pioneer Student of Bird Behavior." Auk, 64 (1947), 35-52.

Taverner, P. A. "Memories of William Edwin Saunders, 1861-1943." Auk, 61 (July, 1944), 345-51.

Thorpe, W. H. Biology and the Nature of Man. London: Oxford University Press, 1962.

________. Bird-Song: The Biology of Vocal Communication and Expression in Birds. Cambridge: Cambridge University Press, 1961.

________. Learning and Instinct in Animals. Cambridge, Mass.: Harvard University Press, 1963.

________. Science, Man and Morals. London: Methuen, 1965.

Tinbergen, N. "'Derived' Activities; Their Causation, Biological Significance, Origin and Emancipation During Evolution." Quarterly Review of Biology, 27 (March, 1952), 1-32.

________. The Herring Gull's World: A Study of the Social Behavior of Birds. New York: Basic Books, 1960.

________. Social Behaviour in Animals: With Special Reference to Vertebrates. London: Methuen, 1965.

________. The Study of Instinct. New York: Oxford University Press, 1951.

Tobey, Ronald C. The American Ideology of National Science, 1919-1930. Pittsburgh: University of Pittsburgh Press, 1971.

Trefethen, James B. Crusade for Wildlife: Highlights in Conservation Progress. Philadelphia: Stackpole, 1961.

True, Webster Prentiss. The Smithsonian Institution. Washington: Smithsonian Institution, 1934.

Wallace, W. Stewart, ed. The Royal Canadian Institute Centennial Volume, 1849-1949. Toronto: Royal Canadian Institute, 1949.

Wetmore, Alexander. "Robert Ridgway, 1850-1929." National Academy of Sciences, Biographical Memoirs, 15 (1931), 57-101.

Wheeler, William Morton. Ants: Their Structure, Development and Behavior. New York: Columbia University Press, 1910.

________. "'Natural History,' 'Oecology' or 'Ethology'?" Science, 15 (June 20, 1902), 971-976.

________. "The Obligations of the Student of Animal Behavior." Auk, 21 (1904), 251-255.

________. "On Instincts." Journal of Abnormal Psychology, 15 (1921), 295-318.

________. "Vestigial Instincts in Insects and Other Animals." American Journal of Psychology, 19 (1908), 1-13.

________. "What is Natural History?" Bulletin of the Boston Society of Natural History, 59 (1931), 3-12.

________. "Woodcock Surgery." Science, 19 (February 26, 1904), 347-350.

Whitman, C. O. Posthumous Works of Charles Otis Whitman. 3 vols. Washington: Carnegie Institution of Washington, 1919.

Willughby, Francis. The Ornithology of Francis Willughby, Wherein All the Birds Hitherto Known are Accurately Described. Edited by John Ray. London: J. Martyn, 1678.

Wrigley, Robert E. "Mammals of the Sandhills of Southwestern Manitoba." Canadian Field-Naturalist, 88 (January-March, 1974), 21-39

Wynne-Edwards, V. C. Animal Dispersion in Relation to Social Behavior. New York: Hafner, 1962.

(9) Miscellaneous Materials

Barnett, J. M. "Ashbridge's Bay." Ontario Naturalist, 11 (December, 1971), 24-26.

Bowsfield, Hartwell, ed. The James Wickes Taylor Correspondence, 1859-1870. Winnipeg: Manitoba Record Society, 1968.

Canada. Department of Agriculture. What Farmers Say of Their Personal Experience in the Canadian North-West. Ottawa: Queen's Printer, 1882.

Carnegie, Andrew. The Autobiography of Andrew Carnegie. Boston: Houghton, Mifflin, 1920.

Carr, Violet M., ed. Ops: Land of Plenty. Lindsay, Ont.: Ops Township Council, 1968.

Charlesworth, Hector. Candid Chronicles: Leaves from the Notebook of a Canadian Journalist. Toronto: Macmillan, 1925.

Christy, Robert Miller. Manitoba Described. London: Wyman and Sons, 1885.

Dickie, Francis. "World Famous Manitoban." Winnipeg Free Press, 16 November 1963.

Dictionary of National Biography, S.v. "Cameron, Archibald (1707-1753)," by Thomas Finlayson Henderson.

_______, S.v. "Cameron, Donald (1695-1748)," by John Westby-Gibson.

_______, S.v. "Seton, George, Fifth Earl of Winton (d.1749)," by Thomas Finlayson Henderson.

Gundy, H. Pearson. "The Development of Trade Book Publishing in Canada." In Ontario, Royal Commission on Book Publishing. Background Papers. Toronto: Queen's Printer, 1972, pp. 1-37.

Kirckonnel, Watson. County of Victoria Centennial History. 2nd ed. Lindsay, Ont.: Victoria County Council, 1967.

[Legal Notice of Name Change.] New York Evening Journal, 29 October 1901.

Martin, Chester. "Dominion Lands" Policy. Edited by Lewis H. Thomas. Toronto: McClelland and Stewart, 1973.

Middleton, Jesse Edgar. The Municipality of Toronto: A History. 3 vols. Toronto: Dominion Publishing Co., 1923.

Morgan, H. J. Men and Women of the Time: A Handbook of Canadian Biography. Toronto: William Briggs, 1898.

Mott, Frank Luther. A History of American Magazines, 1885-1905. Cambridge, Mass.: Harvard University Press, 1957.

Mulvany, C. Pelham. Toronto: Past and Present. Toronto: W. E. Caiger, 1884.

Nute, Grace Lee. "Down North in 1892." Beaver, 279 (June, 1948), 42-46.

_______. "Paris to Peel's River in 1892." Beaver, 278 (March, 1948), 19-23.

_______. "To Edmonton in 1892." Beaver, 281 (June, 1950), 3-5.

Roosevelt, Theodore. The Letters of Theodore Roosevelt. 8 vols. Edited by Elting Morison. Cambridge, Mass.: Harvard University Press, 1951.

Seton, Robert. An Old Family; or, the Setons in Scotland and America. New York: Brentanos, 1899.

Warkentin, John. "Steppe, Desert and Empire." In Prairie Perspectives 2. Edited by Anthony W. Rasporich and Henry C. Klassen. Toronto: Holt, Rinehart, Winston, 1973.

________, ed. The Western Interior of Canada. Toronto: McClelland and Stewart, 1964.

Weir, T. R. "Pioneer Settlement of Southwest Manitoba, 1879-1901." Canadian Geographer, 8 (1964), 64-71.

BIOLOGISTS AND THEIR WORLD

An Arno Press Collection

Adler, Kraig, editor. **Early Herpetological Studies and Surveys in the Eastern United States.** 1978

Adler, Kraig, editor. **Herpetological Explorations of the Great American West.** Two vols. 1978

Agassiz, [Jean] Louis [Rodolphe]. **Contributions to the Natural History of the United States of America.** Four vols. in two. 1857/1860/1862

Allard, Dean Conrad, Jr. **Spencer Fullerton Baird and the U.S. Fish Commission.** 1978

Altum, Bernard. **Der Vogel Und Sein Leben.** 1868

Azara, Don Felix de. **Apuntamientos Para La Historia Natural De Los Quadrúpedos Del Paragüay Y Rio De La Plata.** Two vols. in one. 1802

Baer, Karl Ernst v[on]. **Reden Gehalten in Wissenschaftlichen Versammlungen und Kleinere Aufsaetze Vermischten Inhalts.** Three vols. in two. 1864/1876/1873

Barrett-Hamilton, Gerald E. H. and Martin A. C. Hinton. **A History of British Mammals.** Two vols. 1910-1921

Boulenger, G[eorge] A[lbert]. **The Tailless Batrachians of Europe.** Two parts in one. 1897/1898

Brocchi, [Paul]. **Mission Scientifique Au Mexique Et Dans L'Amérique Centrale,...** Étude Des Batraciens De L'Amérique Centrale. 1882

Buffon, [Georges L. L.]. **The History of Singing Birds.** Translated from the French. 1791

Buffon, [Georges L. L.]. **The Natural History of Oviparous Quadrupeds and Serpents.** Translated by Robert Kerr. Four vols. in two. 1802

The Cabinet of Natural History and American Rural Sports. Three vols. in one. 1830/1832/1833

Candolle, A[ugustin] P[yramus] de and K[urt] Sprengel. **Elements of the Philosophy of Plants.** Translated from the German. 1821

Cassin, John. **United States Exploring Expedition During the Years 1838, 1839, 1840, 1841, 1842 Under the Command of Charles Wilkes, U. S. N.:** Mammalogy and Ornithology. Two vols. 1858

Chapman, Frank M. **Essays in South American Ornithogeography.** Edited by Keir B. Sterling. 1978

Cope, Edward D[rinker]. **The Vertebrata of the Tertiary Formations of the West.** One vol. in two. 1883

Cuvier, [Georges]. **The Class Mammalia.** Five vols. 1827

Donovan, E[dward]. **The Natural History of British Fishes.** Five vols. in two. 1802/1803/1804/1806/1808

Duméril, Auguste [Henri André], [Marie-Firmin] Bocourt and [François] Mocquard. **Mission Scientifique Au Mexique Et Dans L'Amérique Centrale,...** Étude Sur Les Reptiles. Two vols. 1870-1909

Flower, William Henry and Richard Lydekker. **An Introduction to the Study of Mammals, Living and Extinct.** 1891

Forbush, Edward Howe. **Birds of Massachusetts and Other New England States.** Three vols. 1925/1927/1929

Girard, Charles. **United States Exploring Expedition During the Years 1838, 1839, 1840, 1841, 1842 Under the Command of Charles Wilkes, U. S. N.:** Herpetology. Two vols. 1858

Grinnell, Joseph. An Account of the Mammals and Birds of the Lower Colorado Valley. 1914

Howard, H[enry] Eliot. **Territory in Bird Life.** 1920

Hume, Edgar Erskine. **Ornithologists of the United States Army Medical Corps.** 1942

Huxley, Leonard. **Life and Letters of Sir Joseph Dalton Hooker.** Two vols. 1918

LeConte, John L. and George H. Horn. **Classification of the Coleoptera of North America.** 1883

Linsley, E. Gorton, editor. **Beetles From the Early Russian Explorations of the West Coast of North America, 1815-1857.** 1978

Linsley, E. Gorton, editor. **The Principal Contributions of Henry Walter Bates to a Knowledge of the Butterflies and Longicorn Beetles of the Amazon Valley.** 1978

Murray, Andrew. **The Geographical Distribution of Mammals.** 1866

Osborn, Henry Fairfield. **Cope:** Master Naturalist. 1931

Peale, Titian R. **United States Exploring Expedition During the Years 1838, 1839, 1840, 1841, 1842 Under the Command of Charles Wilkes, U. S. N.:** Mammalogy and Ornithology. 1848

Phillips, John C. **American Game Mammals and Birds.** 1930

Rafinesque: Autobiography and Lives. 1978

Ray, John. **Synopsis Methodica Animalium Quadrupedum et Serpentini Generis.** 1693

Ray, John. **Synopsis Methodica Avium & Piscium.** [Edited by William Derham]. 1713

Richardson, John. **Fauna Boreali-Americana;** Or the Zoology of the Northern Parts of British America, Part Third: The Fish. 1836

Richardson, John, William Swainson and William Kirby. **Fauna Boreali-Americana;** Or the Zoology of the Northern Parts of British America, Part Four: Insecta, 1837

Riley, Charles V[alentine]. **[Nine] Annual Report[s] on the Noxious, Beneficial and Other Insects of the State of Missouri, 1869-1877** *and* **General Index and Supplement.** Ten vols. in three. 1869-1877/1881

Say, Thomas. **The Complete Writings of Thomas Say on the Entomology of North America.** Edited by John L. LeConte. Two vols. 1859

Schuchert, Charles and Clara Mae LeVene. **O. C. Marsh:** Pioneer in Paleontology. 1940

Sclater, William Lutley and Philip Lutley Sclater. **The Geography of Mammals.** 1899

Seton, Ernest Thompson. **Trail of an Artist-Naturalist:** The Autobiography of Ernest Thompson Seton. 1940

Smith, James Edward. **A Selection of the Correspondence of Linnaeus and Other Naturalists, From the Original Manuscripts.** Two vols. 1821

Stuckey, Ronald L., editor. **Development of Botany in Selected Regions of North America Before 1900.** 1978

Stuckey, Ronald L., editor. **Essays on North American Plant Geography From the Nineteenth Century.** 1978

Stuckey, Ronald L., editor. **Scientific Publications of Charles Wilkins Short.** 1978

Swammerdam, John. **The Book of Nature.** 1758

Wadland, John Henry. **Ernest Thompson Seton:** Man in Nature and the Progressive Era, 1880-1915. 1978

Waterhouse, G[eorge] R[obert]. **A Natural History of the Mammalia.** Two vols. 1846/1848

Weiss, Harry B. and Grace M. Ziegler. **Thomas Say:** Early American Naturalist. 1931

Wheeler, Geo[rge] M., [editor]. **[Reports Upon Insects Collected During Geographical and Geological Explorations and Surveys West of the One Hundredth Meridian During the Years 1872, 1873, and 1874].** 1875

Willughby, Francis. **De Historia Piscium** *and* **Icthyographia Ad Amplisimum Virum Dnum., Samuelem Pepys, Presidem Soc. Reg.** Two vols. in one. 1686/1685

Youmans, William Jay, editor. **Pioneers of Science in America.** Revised edition. 1896